Anke Schwarz
Demanding Water

MEGACITIES AND GLOBAL CHANGE

MEGASTÄDTE UND GLOBALER WANDEL

herausgegeben von

Frauke Kraas, Martin Coy, Peter Herrle und Volker Kreibich

Band 22

Anke Schwarz

Demanding Water

A Sociospatial Approach to Domestic Water Use in Mexico City

Umschlagabbildung:
Drinking water for sale at a local purification plant in the Ermita Zaragoza housing complex, Iztapalapa, Mexico City, 2014. © Anke Schwarz

Bibliografische Information der Deutschen Nationalbibliothek:
Die Deutsche Nationalbibliothek verzeichnet diese Publikation in der Deutschen Nationalbibliografie; detaillierte bibliografische Daten sind im Internet über <http://dnb.d-nb.de> abrufbar.

Druck: Hubert & Co, Göttingen
Gedruckt auf säurefreiem, alterungsbeständigem Papier.
Printed in Germany.
ISBN 978-3-515-11686-2 (Print)
ISBN 978-3-515-11690-9 (E-Book)

In memory of Guadalupe Hernández Oliver,
Jorge Legorreta, and Berta Cáceres

ABSTRACT

In the essentially water-rich basin of Mexico City, water taps are now installed in most homes. Yet in many of the city's poorer neighborhoods in particular, water is supplied intermittently and taps often remain dry. How does such a socially constructed water scarcity affect water-related everyday practices in the home? And, more in general, what is the relation between urban space and domestic practices of water use? In accordance with these questions, the present study aims to explore how people's everyday practices are linked to urban space, understood as a social product always in the making. A sociospatial approach is employed, setting Pierre Bourdieu's Theory of Practice in conversation with a relational understanding of space. Rather than identifying a 'habitus of water use', this involves a reflection on the implications of Bourdieu's famous statement that the habitus makes the habitat.

The research design is based on empirical fieldwork involving a set of qualitative methods. As it is assumed that both past and current water supply conditions in the home are key parameters for domestic practices of water use, the sphere of the dwelling lies in the focus. Taking subjective experiences and everyday practices as a starting point, it primarily draws upon in-depth interviews, conducted at the interviewee's home to allow for a simultaneous participatory observation. A total of 53 residents from several neighborhoods in two different boroughs within the jurisdiction of the Federal District, Mexico City, were interviewed to provide a range of different urban living conditions. Through the newly developed tool of habitat biographies capturing people's past dwelling experiences, the interviews also adopt a historical perspective. Focus group discussions and other empirical methods complete the picture, tackling practices of water use with respect to questions of materiality, knowledge and meaning.

This book demonstrates how using water can turn into a demanding everyday task even in cities where virtually all dwellings do have water taps. It sheds a light on everyday practices of water use in Mexico City in the realms of drinking, personal hygiene, and domestic storage, and their relation to past and current supply conditions. Across all these sets of practices, influential urban imaginations with respect to the logic of urban water supply and a widespread mistrust in the potability of tap water are at play, and marked differences arise between waters used for body-related and more technical purposes. Other than expected, current water supply conditions seem to influence much stronger on domestic practices of water use than people's past experiences. Keeping water in domestic storage tanks, waiting for water provision to resume, reusing domestic grey water, and the ubiquitous consumption of bottled water have all become essential everyday practices in Mexico City amidst unequal and often unreliable patterns of water supply reinforced by processes of neoliberal urbanization and infrastructural unbundling. Women in poorer parts of the metropolis are the ones bearing the brunt of such exclusionary configurations of urban space. Serving as a reminder that a water tap in the home alone does not guarantee proper access to water, the present book develops a sociospatial, subject-based approach to explore everyday practices and experiences in the realm of water. Such an approach is relevant beyond the example of Mexico City presented here, as a permanent water supply of a potable quality remains the exception rather than the rule in many cities around the world.

KURZFASSUNG DER ARBEIT

Heute verfügen die meisten Haushalte im (an sich wasserreichen) Einzugsgebiet von Mexiko-Stadt über einen Wasseranschluss. Dennoch bleiben die Hähne oft trocken, da die Wasserversorgung besonders in vielen ärmeren Stadtvierteln unstetig bis erratisch ist. Welche Auswirkungen hat diese – keineswegs natürliche, sondern sozial konstruierte – Wasserknappheit auf wasserbezogene Alltagspraktiken im Haushalt? Allgemeiner gesprochen: Wie ist das Verhältnis zwischen urbanem Raum und Praktiken der Wassernutzung im Haushalt? Die vorliegende Arbeit geht diesen Fragen mit dem Ziel nach, zu untersuchen, in welcher Weise Alltagspraktiken mit dem städtischen Raum zusammenhängen; letzterer wird dabei als soziales Produkt andauernder Prozesse der Raumproduktion verstanden. Sie entwirft einen explorativen sozialräumlichen Zugang, der Pierre Bourdieus Theorie der Praxis mit einem relationalen Raumverständnis verbindet. Dies beinhaltet weniger die Suche nach einem ‚Habitus der Wassernutzung' als vielmehr eine Reflektion über die Implikationen des berühmten Bourdieuschen Diktums, dass der Habitus das Habitat mache.

Das Forschungsdesign basiert auf empirischer Feldforschung unter Einsatz unterschiedlicher qualitativer Methoden. Die Arbeit ist auf der Ebene des Wohnraums angesiedelt, wobei davon ausgegangen wird, dass sowohl vergangene wie auch gegenwärtige Wasserversorgungsbedingungen wichtige Parameter für die Haushaltswassernutzung sind. Da subjektive Erfahrungen und Alltagspraktiken als Ausgangspunkt dienen, stützt sie sich in erster Linie auf individuelle Tiefeninterviews, die direkt in den Wohnungen der Interviewten durchgeführt wurden, was zugleich eine teilnehmende Beobachtung ermöglichte. Um eine Bandbreite an unterschiedlichen urbanen Lebensbedingungen abzubilden, wurden insgesamt 53 Personen aus mehreren Stadtteilen in zwei Bezirken des *Distrito Federal,* Mexiko-Stadt, interviewt. Mittels des neuentwickelten Instruments der Wohnbiographie, das vergangene Wohnerfahrungen graphisch dokumentiert, wurde dabei auch eine historische Perspektive eingenommen. Fokusgruppendiskussionen und andere empirische Methoden vervollständigen das Bild, indem sie Fragen der Materialität sowie der Wissens- und Bedeutungsproduktion im Kontext der Wassernutzungspraktiken angehen.

Dieses Buch macht deutlich wie Wassernutzung sogar in solchen Städten zu einer Herausforderung werden kann, in denen so gut wie alle Gebäude über einen Wasseranschluss verfügen. Es beleuchtet alltägliche Praktiken der Wassernutzung den Bereichen des Trinkens, der Körperpflege und der Vorratshaltung und setzt diese ins Verhältnis zu gegenwärtigen und früheren Versorgungsbedingungen in Mexiko-Stadt. Quer zu diesen Praktiken liegen einflussreiche *imaginarios urbanos* zur infrastrukturellen Versorgungslogik und ein generelles Misstrauen in die Trinkbarkeit des Leitungswassers. Dabei werden deutliche Unterschiede gemacht zwischen Wasser, welches für körperbezogene Zwecke verwendet wird, und solchem, das im Haushalt als Nutzwasser dient. Anders als erwartet scheinen aktuelle Versorgungsbedingungen diese Wassernutzungspraktiken weitaus stärker zu beeinflussen als frühere Erfahrungen. Praktiken wie das Speichern von Wasser in privaten Zisternen, das Warten auf die Wiederaufnahme der Wasserversorgung, die Weiterverwendung von Grauwasser im Haushalt und der allgegenwärtige Flaschenwasserkonsum sind vor dem Hintergrund ungleicher und oft unzuverlässiger Versorgungsmuster, die durch neoliberale Stadtentwicklungsprozesse noch verstärkt werden, für viele Bewohnerinnen und Bewohner Mexiko-Stadts unentbehrlich geworden. Die Hauptlast solcher ausschließender Konfigurationen des urbanen Raumes tragen Frauen in den ärmeren Stadtteilen. Die vorliegende Arbeit erinnert somit daran, dass ein Wasseranschluss an sich noch keine Garantie für einen angemessenen Zugang zu Wasser ist. Darüber hinaus entwickelt sie einen sozialräumlichen, subjekt-orientierten methodischen Zugang zu Alltagspraktiken und Alltagserfahrungen im Bereich der Haushaltswassernutzung. Da eine permanente Wasserversorgung in Trinkwasserqualität in vielen Städten weltweit eher die Ausnahme als die Regel ist, erscheint ein solcher wissenschaftlicher Zugang auch jenseits des hier untersuchten Beispiels Mexiko-Stadt als überaus relevant.

RESUMEN

En el Valle de México, esencialmente rico en agua, hoy casi todas las viviendas cuentan con una llave de agua. Sin embargo, el agua está suministrada de manera intermitente y las llaves frecuentemente quedan secas, sobre todo en muchas de las colonias pobres. ¿De que manera afecta tal escasez de agua – siempre entendida como un producto social – a las prácticas cotidianas de uso del agua en el hogar? Y, más en general ¿cuál es la relación entre el espacio urbano y tales prácticas domésticas? Acorde a estas preguntas, el presente libro tiene como objetivo explorar la manera en que las prácticas cotidianas se vinculan con el espacio urbano, entendiéndolo como un producto social en constante proceso de producción. Se emplea un enfoque socio-espacial basado en la introducción de un concepto relacional del espacio a la teoría de la práctica, de Pierre Bourdieu. En lugar de identificar algún 'hábitus del uso de agua' esto implica una reflexión sobre las repercusiones de su famosa frase que fuera el hábitus que haga el hábitat.

El diseño de la investigación se basa en trabajo de campo, involucrando una serie de métodos empíricos cualitativos. Se enfoca al ámbito de la vivienda, asumiendo que tanto las actuales como las pasadas condiciones de abastecimiento en el hogar son parámetros principales para las prácticas domésticas de uso del agua. Tomando las experiencias subjetivas y las prácticas cotidianas como puntos de partida, el estudio recurre principalmente a entrevistas profundas realizadas directamente en el hogar de la persona entrevistada para permitir, al mismo tiempo, una observación participativa. Un total de 53 habitantes de diversas colonias en dos delegaciones del Distrito Federal, Ciudad de México, han sido entrevistados para incluir una gran variedad de condiciones de vida en la urbe. Además, las entrevistas toman una perspectiva histórica a través del nuevo método de las biografías del hábitat él que documenta las experiencias del habitar. Para completar el panorama se realizaron grupos focales y otros métodos empíricos dedicados a las prácticas de uso del agua con respecto a las materialidades, conocimientos y significados involucrados.

El presente libro muestra como el uso del agua se puede convertir en una demandante tarea cotidiana hasta en las ciudades donde casi todas las viviendas están conectadas a la red de suministro de agua. Arroja luz sobre las prácticas cotidianas de uso del agua en la Ciudad de México en los ámbitos de su ingesta, de la higiene personal y del almacenamiento doméstico, y a su relación con las condiciones de abastecimiento tanto actuales como pasadas. También existen poderosos imaginarios urbanos con respecto a la lógica de suministro de agua y una desconfianza generalizada sobre la potabilidad del agua de la llave, los que atraviesan todas estas prácticas. Como resultado surgen diferencias marcadas entre las aguas consideradas aptas para los usos relacionados al cuerpo humano y las empleadas en usos más técnicos en el hogar. Inesperadamente parece que en el uso del agua, las actuales condiciones de suministro de agua en el hogar influyen mucho más sobre las prácticas domésticas de la gente que sus experiencias históricas. El almacenamiento en cisternas domésticas, la espera al restablecimiento del suministro, el reuso de aguas grises en el hogar, y el omnipresente consumo de agua embotellada, todas se han convertido en prácticas cotidianas esenciales en la Ciudad de México bajo patrones de suministro desiguales y a menudo erráticos reforzados por los procesos de urbanización neoliberal y la privatización gradual de la infraestructura urbana. Son las mujeres en las colonias pobres las que cargan el mayor peso de estas configuraciones excluyentes del espacio urbano. El presente libro sirve para recordar que la mera presencia de una llave de agua en el hogar no garantiza un acceso adecuado al agua. Desarrolla un enfoque socio-espacial basado en la percepción subjetiva para explorar las prácticas y experiencias cotidianas en el ámbito del agua. Dado que un suministro permanente de agua de calidad potable, sigue siendo la excepción más que la norma para la una parte significativa de la población urbana alrededor del mundo, tal enfoque parece de suma importancia mucho más allá del ejemplo de la Ciudad de México.

ACKNOWLEDGEMENTS

The present book is mainly based on intense research in Mexico, and many people have contributed to its success along the way. First of all, I owe credit to my supervisors Christof Parnreiter and Kathrin Wildner for all their encouragement, advice and patience during this project. Moreover, I want to express my gratitude to Timothy Moss for his comments on earlier versions of the manuscript. Invaluable feedback was also granted by Patricia de Guadalupe Mar Velasco, Roberto Eibenschutz Hartman, and Flor López Guerrero.

I am deeply indebted to Carmen Hernández Oliver for everything and her bright heart and mind, to Monika Streule for co-producing urban trajectories in multiple ways, and to Friedrich Tietjen for his refreshing thoughts and comments. Without the unrestrained support of Margarita García Arteaga and Ana Laura Landa Chávez, I hardly would have been able to engage in this project as fully as I did. I also wish to thank Alejandro Cerda García, Anna Kunath, Catalina Rojas Ugarte, Chloe Begg, Christian Hartz, Clara Salazar Cruz, Cristina Guerra Mancilla, Daniel Salazar Nuñez, Fredi Maier, Gertrud Neumann-Denzau & Helmut Denzau, Giovanni Civardi, Jana Ringer, Katja Fritsche, Kerstin Krellenberg, Maricruz Carmona, Melanie Kryst, Nathalie Jean-Baptiste, Oscar Pérez Jímenez, Ricardo Landa, Rocío García Arteaga, Ruth Schaldach, and Xanath Sánchez Serrano for their comments, support and encouragement at different stages of this project. Luise Bartels and Aurelia Markwalder of Ernst und Mund, Leipzig, did a great book design and always had the freshest ideas for graphics and else. Sylvia James fine-tuned my English and edited the manuscript with great care. Claudia Rojo Hernández did an excellent job transcribing endless hours of interview material and providing support during focus group discussions, together with Óscar Torres Arroyo and Luis Alvarado Jauriguiberri. My colleagues from the Department Urban and Environmental Sociology at the Helmholtz Centre for Environmental Research encouraged my work, and Gabriela Torres fine-tuned some of the habitat biography figures. As always, I am fully responsible for the content of the book and any remaining errors are entirely my own. In addition, I would like to thank the people from FARO del Oriente in Iztapalapa, Luis de la Cruz Fierro and Gerardo Palacios Sánchez from Agua, Trabajo, Servicio y Vida A.C., Jaime Rello of the Unión Popular Revolucionaria Emiliano Zapata, Brenda Rodríguez from Mujer y Medio Ambiente A.C., and the International Network for Urban Research and Action (INURA). The Heinrich Böll Foundation, Berlin, provided funding for my research in Mexico City and a three-year scholarship, which were essential for the realization of this project. But most of all, I wish to thank my interview partners in Mexico City, for welcoming me to their homes and sharing their experiences and thoughts about water – this is what made the present book possible.

CONTENTS

ABBREVIATIONS

CEPAL	Comisión Económica para América Latina y el Caribe
CONAGUA	Comisión Nacional del Agua
CONAPO	Consejo Nacional de la Población
CEDS	Consejo de Evaluación del Desarrollo Social del Distrito Federal
FAO	Food and Agriculture Organization
FG	Focus group discussion
GDF	Gobierno del Distrito Federal
IADB	Inter-American Development Bank
IMSS	Instituto Mexicano del Seguro Social
INEGI	Instituto Nacional de Estadística y Geografía
INFONAVIT	Instituto del Fondo Nacional de la Vivienda para los Trabajadores
ISSSTE	Instituto de Seguridad y Servicios Sociales de los Trabajadores del Estado
OECD	Organisation for Economic Co-Operation and Development
PNUMA	Programa de las Naciones Unidas para el Medio Ambiente
SACM	Sistema de Aguas de la Ciudad de México
WHO	World Health Organization
ZMCM	Zona Metropolitana de la Ciudad de México

1. INTRODUCTION

Tap water may be scarce in some parts of Mexico City but there is definitely no lack of it in literature on the subject. Whether adverting to water stress, water crisis, water conflicts, water wars, or even water torture – a steady stream of academic reports, books, newspaper articles and all sorts of publications referring to Mexico City's water supply situation in similarly dramatic terms is constantly being fed by new releases (see Peña Ramírez 2004, Perló Cohen/González Reynoso 2005, de la Luz González 20.01.2009, Oswald Spring 2011a, Castano 22.02.2012, Hollander 05.02.2014, Watts 12.11.2015, Kimmelman 17.02.2017, to name only a few[1]). Many (though not all) of these writings seem either oddly detached from the urban, or are somehow fascinated by the extreme, the scandalous, by a 'megacity' or 'megalopolis' painted in vivid colors as an exotic monstrosity and imagined by some as the outcome of a largely uncontrollable urban growth beyond all 'reasonable' scale (see Davis 2006, Burdett/Sudjic 2007). Without any doubt, water supply in Mexico City is a formidable, controversial and, above all, immensely complex issue. The urban water system and the enormous hydraulic engineering efforts on which it depends to provide tap water for a city of roughly 22 million inhabitants (and dispose of its waste water) resemble a "paradox" (Connolly 1999: 61), local groundwater sources are overexploited, and social struggles about an ever-increasing water extraction in the region are ongoing. Within the metropolis, water distribution is all but equal, and a dry water tap is a daily reality for many of its residents. Still, it can be argued that a 'scandalizing' approach – illustrated by terms such as 'water crisis' – tends to obstruct the view on Mexico City's water questions in a dual sense.

First, both the depletion of natural resources and the production of highly differential urban living conditions are not the result of some disastrous yet natural process of urban growth. A host of research has already been undertaken in this area – in what follows, I will only mention some of the most relevant literature for the present venture. The urbanization of nature, and with it the construction and operation of large technical infrastructures such as water supply networks, is product of a certain path of capitalist development and are essential for its survival (see Smith 1984, Harvey 1985 and 1996, Santos 1996). In other words, the production of modern cities was enabled by a simultaneous production and regulation of nature particularly through urban infrastructure networks. For urban water supply in particular, the power relations engrained in modernist supply regimes and their role as enablers of capitalist urbanization is analyzed in the writings of Erik Swyngedouw

1 This is a only a small selection of academic and non-academic publications on Mexico City's water supply situation bearing these terms in their title. A profound analysis of the involved metaphors and discourses would be interesting; along with the concepts of water scarcity and water stress (as coined by Falkenmark/Lindh 1976), many of these gloomy terms were apparently borrowed from earlier publications such as Gleick (1993) and Shiva (2002).

et al. (2002), Matthew Gandy (2004), and Maria Kaika (2005), amongst others. In consequence, it seems appropriate to "move away from thinking of water as a resource that is external to social relations, towards one in which social relations are *embedded*" (Budds 2009: 420). Upon this backdrop, a lack of water at the domestic tap can be read as the result of a socially constructed water scarcity (see Mehta 2010) rather than of natural limits to the resource. As we will see later, this is probably even more evident in the essentially water-rich basin of Mexico City.

Effects of the current logic of supply are, however, directly experienced by Mexico City's residents, leading to a second aspect potentially hidden behind the narrative of a 'water crisis': the way such infrastructures influence everyday life. The urbanization of nature through infrastructures and social practices, or, in other words, the production of urban space and everyday life are clearly linked – something Henri Lefebvre set out thoroughly in his writings (see Lefebvre 1974 and 1991). When I first learned about often haphazard and erratic water supply patterns as part of people's everyday experience during previous research in one of Mexico City's *colonias populares*[2] in the borough of Iztapalapa (see Schwarz 2009), I recall being impressed by how my interview partners tended to treat them essentially as something to be taken for granted. After an initial period without any proper access to utility grids, they had eventually achieved a connection to the electricity and later the water supply network for their neighborhood. When I visited in 2008, water taps seemed to be installed in all dwellings. During our interviews, however, people mentioned rather casually that water provision was restricted to one or two days per week, and even then, water would be available for no more than a couple of hours. The way they put it, this seemed to be the natural course of things – yet it left me puzzling which role water actually played in their daily life. While reports on water fetching and other strategies imposed by a lack of piped water abound (see Sorenson *et al.* 2011), these were households that the World Health Organization (WHO) would, according to their own definition, file as having access to 'improved drinking water sources' as they dispose of a water tap in their dwelling. In addition, some of my interview partners back in 2008 emphasized that they actually experienced a deterioration of living conditions in comparison to former dwellings when moving into that Iztapalapan neighborhood. Along with reports on socially differentiated water supply and consumption patterns in Mexico City, which were usually quite general and not scaled down so much as to consider individual neighborhoods or dwellings (see Libreros Muñoz 2004, Tortajada 2006), this took me to consider the sociospatial nature of everyday water use in the domestic realm. There is, I would argue, a need to gain a better understanding of everyday experiences with water supply and of domestic practices of water use under these conditions.

2 As a product of popular urbanization (see Schteingart 1989, Azuela de la Cueva 1989, Gilbert/Varley 1991), this kind of self-built habitat serves as the main source of housing for those excluded from the formal housing market. It was the dominant mode of urbanization in Mexico City (and elsewhere) during the second half of the 20th century; in the year 2000, *colonias populares* were home to more than half of all inhabitants of the Metropolitan Region of Mexico City (ZMCM), housing some 9.2 million people in total (see Duhau/Giglia 2008: 177).

Research Question and Rationale

The present work explores links between the production of space and domestic practices of water use in Mexico City, employing the following research question: *What is the relation between urban space – as a social product always in the making – and domestic everyday practices of water use in Mexico City?* First of all, this approach draws on a relational concept of space. Urban space is, in other words, not a simple container within which social relations are spun, but it is itself a social process always in the making (see Massey 1999: 6 ff.). Space is, therefore, essentially about the social practices which constantly form and constitute it, and an ever-unfinished product of everyday practices (though it is, of course, also produced on other scales). As such everyday practices form the centerpiece of the present work, the term water use is employed deliberately, following Pierre Bourdieu's conceptualization of practices as routinized, non-intentional strategies (see Bourdieu 1977: 177 f., and chapter 2.1). In contrast to a purely (market)economic perspective embodied by the term *water consumption* – which is widely employed for instance in research involved in modelling residential water demand (see Worthington/Hoffman 2008) and, unsurprisingly, in studies on consumers and consumption (see Warde 2005, Halkier *et al.* 2011) – the present work is mainly interested in people's everyday practices involving water. Reducing water use to quantifiable values (such as metered and consumed volumes of water) would miss the point, particularly as such approaches tend to be based on the idea of rational choice. Rather than looking at questions of water consumption, the present work hence opts for the term water use. The latter is conceptualized as a set of everyday practices involving (tap) water which are always of a social as well as spatial nature. Water use forms part of everyday life, which under capitalist conditions indeed essentially constitutes a space of consumption and social reproduction – but also one of lived experience and potential social transformation (see Lefebvre 1991: 352 ff.). The domestic sphere is thus not only of relevance as cities are usually considered one of the key water consumers along with the industrial and agricultural sectors, but also as a crucial product and site of these everyday routines. A research perspective following this line of thought is concerned with subjective experiences, perceptions and narratives involving water. Rather than striving to generate the picture of an average water consumer, it asks how water is used on the micro scale, in the home on a daily basis in order to identify social regularities and patterns of practices (see Fam *et al.* 2015: 642). For this purpose, the present work takes subjective experiences of Mexico City's inhabitants as a starting point, and employs a sociospatial approach combining Bourdieu's Theory of Practice – which provides a social contextualization of everyday practices (see Bourdieu 1977) – with a relational understanding of space.

Why locate such a study on domestic water use in Mexico City? Given the city's location in an essentially water-rich basin, it displays a surprising array of socially produced water problems, many of which are the result of a regime of large-scale technical water regulation running back centuries. Local fresh and salt water lakes have been drained ever since the early colonial period, and the very same logic of water governance continues into the 21st century (for a historical overview, see Legorreta 2006). Today, the region is well-known for its technically highly complex trans-basin water transfers, particularly its large-scale exports of untreated waste water to neighboring basins, and a severely diminished groundwater recharge due to the massive size of urban built-up area. Amidst all this, local aquifers suffer an overexploitation, as the city's water resources are put under a pressure of 173% (see Oswald Spring 2011b: 500, Peña Ramírez 2012). In consequence, Mexico City suffers continuous soil subsidence, groundwater contamination, and periodic flooding in some areas; it provides an example of environmental problems which are also experienced in other urban areas in Mexico and elsewhere (see Peña Ramírez/López López 2004: 160). But perhaps more importantly for the present study, there are strong indications that access to Mexico City's tap water is both socially stratified and spatially differentiated (see Constantino Toto *et al.* 2010: 250, Consejo de Evaluación del Desarrollo Social del Distrito Federal (CEDS) 2010: 94). Water supply problems in the largest Mexican city are in fact a recurrent issue in public discourse, and (local) press reports on the topic invariably show pictures of water tankers dispensing the liquid to plastic barrels lined up in the streets. Yet the image of the water tanker is somehow misleading – at least when it comes to the Federal District[3], which was chosen as the focus of the present work as it is provided by one sole water utility, the *Sistema de Aguas de la Ciudad de México* (SACM), and represents a more or less homogenous entity in terms of water policies[4]. In the Federal District, domestic water connections are by now an almost universal feature: less than 3% of all inhabited dwellings within its jurisdiction had no water tap installed in 2010 (see Instituto Nacional de Estadística y Geografía (INEGI) 2010). In consequence, supply by water tankers is by now mainly a substitute during temporary supply disruptions locally known as *cortes de agua*, which

3 Rather than covering the entire metropolis, the present study is focused on the situation in the Federal District, which with its 8.8 million inhabitants is home to roughly half of Mexico City's population; the other half lives in the surrounding Estado de México.

4 Narrowing down the scope of the present work to the Federal District is mainly motivated by the strong political division between the capital district and the surrounding Estado de México – each with its own regulatory framework regarding water supply (along with the federal policies implemented by the *Comisión Nacional del Agua* (CONAGUA), the federal water agency). The water supply networks serving the Federal District and those serving the adjoining urban municipalities, for instance, are not physically interconnected in any way. Most likely, this is an indirect result of longstanding political tensions between these two entities dating back to at least the 1950s, and thus predating the strongest period of industrialization and rapid urban expansion of Mexico City. With the *Sistema de Aguas de la Ciudad de México* (SACM), there is, moreover, a single water utility serving the Federal District. SACM is in public ownership as a decentralized institution under the head of the Federal District's environmental ministry. However, a partial privatization began in 1994 with an outsourcing of some of SACM's tasks via the granting of concessions to four private joint ventures. In a first step, concessions were granted for ten years, and then renewed subsequently in 2004 and 2014. These 'unbundled' tasks initially included metering, billing and the repair of leakages, and were extended to the operation and maintenance of the secondary water network (and hence part of water distribution) in the latest concession round, which indicates an even more increased opening to private capital under the slogan of 'decentralization' in the future (see Pradilla Cobos 1994, CEDS 2010: 125 f., Romero Lankao 2011).

are scheduled by the federal water agency CONAGUA and affect large parts of the metropolis several times per year. On a day-to-day basis, however, the main issue in the Federal District is not the absence of a domestic tap but questions of water pressure and steadiness of supply, along with reservations over tap water quality. According to official census data from 2010, as much as 18% of all dwellings connected to the Federal District's public water network were not supplied permanently (see INEGI 2010). Non-permanent supply is mostly a matter of rationing schemes imposed by the water utility (see Sistema de Aguas de la Ciudad de México (SACM) 2013) and shows a clear spatial bias as it spares the centrally located boroughs[5]. In this sense, the modern infrastructural ideal, which sought to advance 'social progress' and homogenize space through universal, hierarchically organized infrastructural networks (see Graham/Marvin 2001: 40 ff.) has never been fulfilled for each and every one living in Mexico City, particularly when it comes to water. This observation undoubtedly applies to a majority of cities in the so-called Global South and highlights long-standing theoretical shortcomings of this paradigm, which is already subject to a much-needed reconsideration (see Furlong 2010 and 2014). With domestic taps failing to fulfill their promise, or doing so only erratically, it can be assumed that water plays a particular role in the everyday life of the Federal District's residents, in particular when it comes to securing water availability for daily practices in the home. More details on the Federal District's landscape of water supply, including the characteristics of supply in terms of water quantity and quality are provided in chapter 4.1; later chapters illustrate how everyday practices of water use form part of this landscape.

State of Research

Large technical infrastructures facilitating 'modern' urban life and improving living conditions in today's cities have become a common research topic in several of the academic fields bearing the 'urban' in name – be it geography, planning or urban studies. This is hardly a surprise, given the changes all kinds of utility grids, from telecommunications and electric power to sewage and water, have undergone in the context of neoliberal reforms over the last decades (see for instance Bakker 2003). In what follows, I will provide a brief overview on the existing literature concerned with urban water supply and water use, discussing in how far these approaches treat questions of space and spatiality. The literature roughly falls into three domains: a socio-geographical perspective focusing on water supply regimes and urbanization processes on the meso scale, an economic perspective concerned with quantifying and modelling water demand, and a qualitative, practice-centered perspective.

First, relations between water infrastructures and processes of urban development are tackled from a socio-geographical perspective specifically for the Mexico City context by a number of authors (see Peña Ramírez 2004 and 2012, Legorreta

5 The Federal District is subdivided in 16 administrative boroughs *(delegaciones)*, which are partially responsible for duties held by municipalities elsewhere in the country (water supply not being amongst them in the Federal District).

2006, Barkin 2006, Fuerte Celis 2013, as well as Ward 1998). The works of José Castro in particular link social contestations around Mexico City's water to questions of citizenship and power (see Castro 2004, 2006 and 2007). Others have analyzed the neoliberalization of the Mexican water sector (see Pradilla Cobos 1994, Wilder/Romero Lankao 2006, de Alba *et al.* 2006, Montero Contreras 2009), though not necessarily from a spatial perspective. More in general, an entire section of urban geography is dedicated to the links between spatial development and technical infrastructures as well as the power relations engrained in these supply regimes, with literature on the urbanization of nature cited above as a point of departure. With respect to water networks in particular, the works of Timothy Moss (2000), Karen Bakker (2003), Erik Swyngedouw (2004), Maria Kaika (2005) and Kathryn Furlong (2006) should be mentioned as examples for this approach. And Stephen Graham and Simon Marvin, with their seminal 2001 book *Splintering urbanism*, drew attention to the inherent social logic of networked urban infrastructures (including water), and their interdependency with processes of urban development, particularly under conditions of neoliberal restructuration and global transformation (see Graham/Marvin 2001). Infrastructural unbundling and urban fragmentation should be studied as parallel, interconnected processes – this claim has been highly influential and inspired a critical debate which makes reference to different case studies around the world[6]. Urban water networks and their relevance for processes of urban differentiation and segregation are a centerpiece of this critical debate (see *Geoforum* 39 (2008)[7], Moss 2008, Naumann 2009). The analysis in Keller Easterling's 2014 *Extrastatecraft* extends this path with respect to other emergent infrastructural elements of globalization and their politics (see Easterling 2014). The entire body of literature in this field is characterized by a clear geographical approach drawing strongly on a relational understanding of space. Its main contribution to geographical thought lies in its ability to analyze cities and their (water) infrastructures as dynamic, integrated sociotechnical processes rather than treating each of them as separate entities. Primarily concerned with the spatiality of infrastructural unbundling from a supply perspective, empirical research in this realm is usually located on the meso scale, choosing the infrastructural networks of entire cities or regions as their objects of inquiry. The present work seeks to complement this body of work by shifting attention to practices of water use on the micro scale. Infrastructural conditions form an integral part of this sociospatial approach, keeping processes of potential 'unbundling'

6 Yet to my knowledge, so far there are no studies involving a Splintering Urbanism perspective for the Mexico City context. As regards water, one can only speculate whether this is related to the relatively low-profile character of water infrastructure unbundling in the Federal District, which the local government has been careful not to tag as privatization, – or simply to an absence of the modern infrastructural ideal of universal water supply to begin with.

7 The 2008 special issue of Geoforum (Volume 39, Issue 6) contains empirical explorations of the Splintering Urbanism approach (see Botton/Gouvello 2008, Jaglin 2008, Kooy/Bakker 2008, MacKillop/Boudreau 2008, Pflieger/Matthieussent 2008). Product of an international workshop in 2005, these contributions coincide in voicing a strong critique on Graham and Marvin's assumed universality of the modern infrastructural ideal (see Coutard 2008).

in Mexico City in mind without assuming a general applicability of the modern infrastructural ideal to begin with.

Second, turning to publications concerned with water use on the micro scale (rather than supply) it can be stated that domestic water use was so far largely an object of economic inquiry with respect to levels of consumption, willingness to pay, and the modelling of future demand. Numerous studies are dedicated to quantifying household water consumption through indicator-based modelling feeding on surveys or statistical data (see Morales Novelo/Rodríguez Tapia 2007 and Adler 2011 for the Mexico City context, and for an overview: Nieswiadomy 1992, Worthington/Hoffman 2008). In this context, social status is often treated as one of many indicators influencing on domestic demand. As a result, urban water consumption is usually thought to be socially differentiated in some way or the other (see Anand 2007, Kenney *et al.* 2008). Dwelling-specific indicators and water-consuming household devices are also considered at times, as well as certain types of water use assumed to have a strong influence on consumption levels, for instance car washing (see Mylopoulos *et al.* 2004). However, these studies tend to treat some selected elements of urban space as static parameters (if not ignoring spatiality altogether), and are usually not interested in gaining an understanding of water-using practices itself but only in quantifying consumption.

Third, there is a body of literature dedicated to water use as a social practice. The work of Elizabeth Shove and colleagues was paramount in shifting attention to the consumption side, giving rise to the field of social practice theory (see Shove/Pantzar 2005; Shove *et al.* 2012). This focus on sustainable consumption and the role of infrastructures (see Southerton *et al.* 2004) broadened the perspective in water research, complementing the body of works related to the Splintering Urbanism paradigm and others centered on socio-technical systems and water governance from a perspective of supply. Hereby, social practice theory contributed a great deal to an understanding of the emergence, evolution and transformation of social practices, including those related to domestic water consumption. Certain sets of water-using practices such as showering (see Hand *et al.* 2005, Berker 2013), or the irrigation of private gardens under drought conditions (see Chappells *et al.* 2011) were analyzed from this perspective. Yet in these studies, urban space often seems to feature in passing. Apparently, it may alternatively (and at times simultaneously) represent a resource for practices, their stage, and/or their product (see Shove *et al.* 2012: 130).

The micro level of the home – as a site of gendered carework – is a central element of research dedicated to questions of water and gender (often in combination with questions of poverty or class), which are concerned with the reproduction of social inequalities through a limited access to water (see Bennett 1995, Cleaver 1998, Crow/Sultana 2002, Bapat/Agarwal 2003, Dugard/Mohlakoana 2009) – though space is usually not a core issue of these studies. Contributions from cultural studies on water use are also numerous, for instance in the anthropological and ethnological field (see Stoffer 1966, Böhme 1988, Bergua Amores 2008; and for an overview on more recent water-related anthropological research, Orlove/Caton 2010), or with respect to water use and health (see White *et al.* 1972) – but again, these works are usually not directly concerned with spatial questions.

The focus on practices eventually leads us to Bourdieu-inspired research on water questions, which are of particular relevance to the present work which rests in part on his praxeological approach. Most do not address the household level but study the reproduction and transformation of social practices on a meso scale in processes of societal transition. Agricultural practices and their impact on or relation to land use and water resources are a common topic, analyzed for instance in the context of post-socialist societies (see Orderud/Polickova-Dobiasova 2010: 205, Eichholz *et al.* 2012), with respect to local responses to the neoliberalization of the agrarian sector and its impact on Mexican *ejidos*[8] (see Wilshusen 2010), or under conditions of climate change (see Beilin *et al.* 2012). The praxeological approach was also employed to tackle questions of urban water governance; more precisely decentralized water supply solutions in La Paz (see Eichholz 2012). These works coincide with other practice-centered approaches introduced earlier as they typically treat space as either a static, non-relational matter framing social developments, or as their expression – which is unsurprising given the rather abstract and two-dimensional conceptualization of space in Bourdieu's writings (see 2.4.1). In this linear interpretation, these practices are not understood as something that itself shapes or interacts with urban space. The same accounts for another study on showering as a 'symbolic' practice (see Jensen 2008), one of the few publications dedicated explicitly to domestic water use that directly draws upon Bourdieu's Theory of Practice. The work of Louise Askew and Pauline McGuirk comes closest to a relational understanding of space in that it explores the irrigation of suburban gardens in Australia as a practice seeking to accumulate cultural capital and thus as a tool of social distinction (see Askew/McGuirk 2004). Though not explicitly Bourdieu-based, there is also the inspiring work by Jeff Wiltse on how social distinction along the categories of race and class was negotiated in the realms of public and private swimming pools in post-WWII cities in the United States (see Wiltse 2007). As for literature on domestic water use in Mexico City in particular, again there are, to my knowledge, only few academic publications approaching the issue from a spatial perspective on the micro level – Enrique Ayala Alonso for example discusses the strive for social distinction and modernization represented in early 19th century bourgeois dwellings in Mexico City and the installation of the first private bathrooms (see Ayala Alonso 2010: 53 ff.). From an equally historical

8 In close resemblance to other contemporary forms of collective land use such as the Soviet *kolkhoz*, the agricultural collectives of the Mexican *ejido* were essentially a territorialization of demands for land reform by parts of the revolutionary movement. Expropriating land from huge haciendas, around 28,000 ejidos – under public ownership, but with partly private, partly collective usufruct – were established during and after the government of Lazaro Cardenas in 1930s, re-inventing pre-Columbian forms of collective land use. Afterwards, this ejidal land played an ambiguous role in Mexico City's urbanization as it turned into a prime source for (irregular) industrial and urban construction during the era of import-substituting industrialization, from the 1950s onwards. Whereas most of these land use changes contradicted formal plans, a 1992 reform of ejido laws led to a drastic change of land use regulations, allowing for a further commodification of communal land (see Cymet 1992, Salazar Cruz 2014a).

perspective, Sharon Bailey Glasco analyzes water use in public bathhouses during the colonial period (see Bailey Glasco 2010: 91 ff.). But as already mentioned, a vast majority of the literature is concerned with supply questions and the interaction between processes of infrastructural and urban development on a city-wide scale or beyond.

Aims of Research

Generally speaking, little research in the field of urban geography has so far been dedicated to everyday practices of domestic water use from a sociospatial perspective. The present work develops a sociospatial approach that aims to take us beyond earlier works by shifting attention from the meso scale to everyday practices of water use on the micro scale, employing a qualitative, subject-centered perspective. For this purpose, it drafts a research design based on existing and new empirical methods able to grasp everyday practices of water use in their past and present spatiality. Following calls for a down-scaling of research on domestic water use (see Fam *et al.* 2015), it aims to complement the existing geographical literature on urban water infrastructures, supply regimes and power relations by re-introducing everyday practices. It thus strives to contribute to a more empirically grounded water research, which nevertheless keeps the social fault lines along which differential urban water supply is organized in mind.

There have long been calls to link Bourdieu's approach to social practice with a concept of relational space (see Painter 2000: 258), or more in general, to employ time and space as core elements in a conceptualization of social practice and social structuration (see Thrift 1996: 71). According to both geographical (see Haferburg 2007: 342) and sociological literature (see Schroer 2006: 176), there is a need to further clarify in particular the relation between habitat and habitus drawn from Bourdieu's praxeological approach. As already mentioned, the here proposed sociospatial approach to everyday practices draws upon a combination of the Bourdieuian Theory of Practice with the concept of relational space. I would argue that it is precisely such a perspective that will allow us to conceptualize tap water as something which "captures and embodies processes that are simultaneously material, discursive and symbolic" (Swyngedouw 2004: 28).

For the present purpose, this theoretical approach is to be operationalized through a research design able to capture everyday practices of water use in both their past and current spatiality. It draws upon fieldwork-based qualitative empirical methods which seem apt for such an explorative venture. These are, first and foremost, individual in-depth interviews, based on a semi-structured interview guideline and conducted at the interviewee's home to allow for a simultaneous participatory observation, and focus group discussions. Through the newly developed tool of habitat biographies capturing people's past dwelling experiences, the interviews also adopt a historical perspective. An explorative though theory-informed approach, putting an emphasis on understanding relations between everyday practices and spatiality, seems appropriate in that there is, at least to my knowledge, no prior research on domestic water use in Mexico City which would explicitly study these links. Exploring the demanding character of water and water use in Mexico City therefore is a matter of studying the differences made by limitations in domestic water supply and hereby "exposing the arbitrariness of the taken for granted" (Bourdieu 1977: 169). The present study strives to do so from a sociospatial perspective, with a focus on the everyday practices and experiences of those living in Mexico's largest metropolis.

Structure of this Book

The present study is organized as follows: chapter 2 is dedicated to the theoretical framework, linking the Theory of Production of Space with the habitus approach, and defining some key concepts. The following chapter introduces the research design and the employed empirical methods, before chapter 4 provides an overview on the water supply situation in the Federal District in general and in the boroughs of Iztapalapa and Cuauhtémoc in particular, and introduces the twelve studied neighborhoods. Turning to the empirical findings, chapter 5 covers the domestic practices of drinking (5.1), hygiene and cleaning (5.2) and storing water (5.3) as well as the imagined landscapes of supply (5.4), with a focus on the current socio-spatial setting. With reference to people's past experiences with water supply limitations, different types of habitat biographies are developed in chapter 6, and the following chapter sets them in relation to selected water-using practices to explore the influence of past experiences on current practices. Chapter 8 offers a reflection on the findings and research strategy of the present work as well as an outlook to future studies, before chapter 9 proceeds to reflect the conceptual approach linking habitus and habitat, and the potential of the empirical instrument of habitat biographies, followed by some concluding remarks in chapter 10.

2. THEORETICAL FRAMEWORK

How is water used by people in their homes? The present chapter will link this apparently profane question to the urban, and emphasize that studying water use from a sociospatial perspective means approaching it simultaneously as an everyday practice *and* a spatial practice, and adopting a perspective which takes each individual subject and their actions as a starting point. This has two consequences as regards the theoretical approach to explore the relation between practices of domestic water use and urban conditions in Mexico City. On the one hand, it calls for a conceptualization able to grasp the social function and contextualization of everyday practices; on the other hand, for an understanding of urban space as a product always in the making rather than a given entity. In fact, both approaches are intrinsically entwined, as this chapter will expose. To meet these requirements, the present work rests on the idea of enriching Pierre Bourdieu's theoretical approach with a relational understanding of space – something already proposed in earlier writings (see Harvey 1989: 214 ff., Schwarz 2009: 38 ff.). A discussion of key concepts for the present work, in 2.1, is followed by an introduction to Bourdieu's praxeological approach with a particular emphasis on the concept of habitus (chapter 2.2). This theoretical strand is merged with a relational understanding of space in chapter 2.3, which provides an operationalization of the sociospatial approach to domestic practices of water use.

2.1 DISCUSSION OF CONCEPTS

Some key concepts need to be clarified before introducing the theoretical framework proper: the links between everyday practices, strategies and routines on the one hand, and whether it is adequate to use the term 'class' discussing contemporary social inequalities in Mexico, on the other. In the end, it is Bourdieu's praxeological approach which informs both.

2.1.1 Of Practices, Strategies and Routines

What is a social practice? As a starting point, it could be said that (everyday) practices are social in as far as they are always embedded in a social context. Yet to overcome tautologies of this type, Bourdieu's praxeological approach seems most adequate for the present purpose as it is based on a conceptualization of the relation between social status and praxis through the notions of capitals, habitus and field. Applied to the present study, this adheres to the idea that practices of water use do not happen in a void but are related to patterns of action we have learned and experienced before in our specific social context (or field, as Bourdieu would call it)

– and incorporated as a set of social dispositions or rules, which form a specific habitus (these key concepts are introduced in 2.2). Actors hold a certain social position (which can be identified through the amount and composition of capitals, see 2.1.2), and they act from this very position. In this sense, everyday practices can be understood as products of habitus, while also playing a part in its reproduction or transformation. More precisely, everyday practices are at the very core of this process, as it is human agency that reproduces social structures – and, given the duality of both social structure and praxis as proposed by Roy Bhaskar, the latter also has the potential to transform the former:

> "Society is both the ever-present *condition* (material cause) and the continually reproduced *outcome* of human agency. And praxis is both work, that is, conscious *production*, and (normally unconscious) *reproduction* of the conditions of production, that is society." (Bhaskar 1998: 34–35)

When it comes to a theory of practice, Andreas Reckwitz offers a widely cited definition, which is of interest in that it links the dimensions of material, knowledge and lived experience – an aspect which will be revisited in the context of spatial practices (see 2.3):

> "A 'practice' (Praktik) is a routinized type of behavior which consists of several elements, interconnected to one other: forms of bodily activities, forms of mental activities, 'things' and their use, a background knowledge in the form of understanding, know-how, states of emotion and motivational knowledge. (…) The single individual (…) is not only a carrier of patterns of bodily behavior, but also of certain routinized ways of understanding, knowing how and desiring. (…) A practice is thus a routinized way in which bodies are moved, objects are handled, subjects are treated, things are described and the world is understood." (Reckwitz 2002: 249–250)

As is highlighted here, routines are an essential feature of practices rather than an activity to be treated separately. In practice theory, social relations are reproduced via a routinization of practices:

> "For practice theory, the nature of social structure consists in routinization. Social practices are routines: routines of moving the body, of understanding and wanting, of using things, interconnected in a practice. Structure […can be found] in the routine nature of action. Social fields and institutionalized complexes (…) are 'structured' by the routines of social practices. Yet the idea of routines necessarily implies the idea of a temporality of structure: *routinized social practices occur in the sequence of time, in repetition; social order is thus basically social reproduction.* For practice theory, then, the 'breaking' and 'shifting' of structures must take place in everyday crises of routines" (Reckwitz 2002: 255; emphasis added).

We will return to the question of routines and the role of crises or ruptures as potential sources of social transformation in the context of Bourdieu's concept of habitus, discussed in the following subchapter. Reckwitz relates to the works of Bourdieu, Anthony Giddens (1984) and Theodore Schatzki (1996), which brings us back to the initial question: how can practices and strategies be understood in a Bourdieuian sense? Rather than being mere sets of rules, strategies involve temporality, rhythm and direction, he states:

> "To substitute strategy for the rule is to reintroduce time, with its rhythm, its orientation, its irreversibility. (...) practices [are] defined by the fact that their temporal structure, direction, and rhythm are constitutive of their meaning." (Bourdieu 1977: 9)

Therefore, practices may be understood as both routines and strategies. Whereas routines are the vehicles of socialization and incorporation of social rules, as will be discussed below, the Bourdieuian idea of a strategy differs from the definition as provided for instance by the Concise Oxford English Dictionary: "a plan designed to achieve a particular long-term aim" (Soanes/Stevenson 2008: 1425). In contrast to this common notion, strategies in the Bourdieuian sense are essentially non-intentional – they are, in other words, usually not planned in a conscious manner. However, this apparent non-intentionality is not to be confused with an absence of interest, according to Bourdieu. Strategies are not based on a 'calculation' in the strict economic sense yet still driven by interests in a symbolic sense:

> "in a universe characterized by the more or less perfect interconvertibility of economic capital (in the narrow sense) and symbolic capital, the economic calculation directing the agents' strategies takes indissociably into account profits and losses which the narrow definition of economy unconsciously rejects as unthinkable and unnamable, i.e. as economically irrational. In short, (…) *practice never ceases to conform to economic calculation* even when it gives every appearance of dis-interestedness by departing from the logic of interested calculation (…) and playing for stakes that are non-material and not easily quantified." (Bourdieu 1977: 177)

The interests inherent to practices are hence based on the convertibility of economic and symbolic capital, and concealed through routinization and the incorporation of social rules, so that social practice is not to be conceived as a result of a conscious compliance with rules or of strategic planning by any individual subject:

> "social praxis can be understood neither as an obedient actualization of a given regulating system, which does presume an exact knowledge of this system and a more or less conscious willingness to obey (…), nor as an aim-oriented practice which implies a conscious planning (…), nor as a strategically drafted interaction." (Fuchs-Heinritz/König 2011: 116)[1]

Therefore, routines of the everyday play a crucial role in Bourdieu's Theory of Practice as a moment of socialization in the sense of training and incorporation of social rules, wherein habitus is conceptualized as a mechanism enabling the functioning and orderliness of social practice (see ibid. and 2.2). Notably, this leads us directly to Henri Lefebvre's seminal analysis of everyday life not only as a site of the reproduction of capitalism but also as a potential site of social resistance and change (see Lefebvre 1974). When everyday practices are understood as the very momentum where social structure is reproduced and potentially also transformed, the domestic realm becomes a prime site for an exploration of these practices (see Fam *et al.* 2015: 642). Taking one example in particular, practices of water use form part of domestic labor as a crucial everyday practice (reproductive and care work) and as strategy of survival which (still) is predominantly a women's task (see Adler Lomnitz 1993, González de la Rocha 1994; for a literature review on female domestic labor as a strategy of survival, see Salazar Cruz 1999: 24 ff.). Upon this backdrop, studying the domestic realm remains valid even if not sharing the kind of household composition or family model implied as a norm by Bourdieu's writings, wherein the home is the realm of a nuclear heterosexual family with female home-

1 Unless otherwise indicated, all translations from texts other than English were performed by the author.

maker, male breadwinner, and children (see Silva 2005: 97). Regardless of the kind of household structure concerned, the home is and remains to be one of the crucial realms of everyday life in terms of social reproduction and care work. Moreover, it is in domestic routines that structural discrimination becomes tangible (though not necessarily in a conscious way) as, for example, "the home is the principal medium through which poverty is experienced and reproduced" (Ward 1998: 241). Situating the inquiry of practices in the home, as the present study does, thus recognizes the role of domestic practices such as water use for social reproduction and, potentially, social transformation. In short, everyday practices are routinized and strategical in an (apparently) non-intentional way, and while tending to reproduce social relations, they also contain a moment of potential transformation and change.

2.1.2 Popular Class and Middle Class

To what extent is it adequate to speak of social classes when studying contemporary Mexico City – and how could such a class be defined? First and foremost, the Bourdieuian concept of class[2] is of interest for the present work because it provides a processual, subject-based perspective. In a Bourdieuian sense, social classes do not exist *ex ante*, but are continuously evolving from social practice, as individual and society mutually produce and reproduce themselves (see Bourdieu 2006: 365). As an ongoing process of social construction, these classes are 'probable' rather than 'real' (see Bourdieu 1985: 12). As Mike Savage – having conducted a range of Bourdieu-based empirical sociological studies along with his colleagues in the British context[3] – has put it,

> "class does not stand like a puppet-master above the stage, pulling the strings of the dolls from on high: rather it works through the medium of individualized processes." (Savage 2000: 95)

Social formation is hence wrought over in the realm of individual, socially framed practices. Moreover, Bourdieu has stressed the relational character of social classification, working particularly through processes of distinction:

> "Each and every social position is hence defined by the entirety of what it is not (...) social identity gains shape and is confirmed by difference" (Bourdieu 1982: 279).

Following Bourdieu, it can be argued that on the micro level of the everyday, social practices are related to social status in a dual manner: first, via the 'objective' living

2 The author is aware that the concept of class has been a highly controversial issue; here is not the place to resume this debate whose cornerstones were set by the Marxist and the Weberian perspectives respectively (see White 2000 and Brown 2006 for an overview, along with Saar 2008 and Bescherer *et al*. 2014 for an ongoing debate in the German-spoken context).

3 As recent works by Mike Savage, Fiona Devine, Niall Cunningham and colleagues – including the 'Great British Class Survey' – demonstrate, employing a class perspective and conducting empirical class analysis is by no means outdated, apparently all the more so in the British context (see Savage *et al*. 2013, Savage *et al*. 2015). Notably, these scholars draw directly on Bourdieu in what they define as a strategically inductive, culturally informed analysis of class formation.

conditions (including the possession of devices involved in these practices, and other forms of capitals – see below), and second, as these practices are framed by a set of incorporated social rules. Thus, referring to this processual conceptualization of class again brings us back to the concept of habitus, which can be deemed crucial for the dialectic between social status and (everyday) practices. Along with fields and capitals, habitus is a key element of Bourdieu's praxeological approach as a way of conceptualizing the reproduction of social classifications (see 2.2). As will be discussed later, its process-orientation in particular is what allows this approach to enter in a dialog with the concept of relational space (see 2.3). But to get back to the point: hereafter, the term class is employed in the present manuscript in a descriptive manner[4] to delineate a certain social group as defined by a similar social position. According to Bourdieu, social position is defined by different forms of capitals:

> "The social position of an actor is to be defined by (…) the distributional structure of (…) the instruments of power: Primarily, economic capital (in its various forms), then cultural and social capital; finally also symbolic capital, as a form of the three aforementioned capitals perceived and recognized as legitimate (commonly referred to as prestige…)." (Bourdieu 1985: 10)

Though economic capital forms the base and starting point of Bourdieu's logic, symbolic capital in particular is thus central to his approach to social formation, which integrates both social status and questions of identity (see Savage *et al.* 2005: 42 f.). Symbolic capital, in the sense of prestige or social appreciation, is not a dimension of social inequality in its own right, but emerges from economic, social and cultural capital as well as corresponding to a position in social hierarchy (see Kreckel 2009: 154). Translating this into an empirical perspective, the present study refers to social status as roughly pre-defined through occupation, type of neighborhood and material characteristics of the dwelling, including potential overcrowding (see 3.2.3 and 3.2.4). Together with health insurance, this covers questions of economic capital, whereas cultural capital is represented by educational background[5]. With respect to Bourdieu's concept of capitals, social status is hence based on multiple dimensions, notably a broad economic (or material) base, and translates into questions of meaning and social identity (though not necessarily in terms of a class 'for itself' or class awareness to begin with).

Upon this backdrop, the term class is employed in the present work to delineate two of the social groups which are relevant for the Mexican context: popular class and middle class. Though the term is apparently not common in Mexican literature on social inequality today – which typically refers to social strata or social groups (see for instance Pichardo Hernández/Hurtado Martín 2010, Rubalcava/Schteingart 2012) – it seems adequate to speak of class when analyzing urban Mexico for two 'empirically grounded' reasons. On the one hand, a profound social differentiation

4 That said, it remains for others to judge whether it is possible to speak of any class 'for itself', i.e. in the sense of class awareness in the case of Mexico.

5 A specific indicator for social capital was not explicitly integrated into the short-survey, as it goes beyond the scope of the present work to analyze social contacts, networks and networking capabilities of the interview partners.

can be observed in the Mexican society, and on the other, there is a combination of material aspects and meaning in the description of social groups which becomes evident in the commonly used term 'popular'.

An articulate social stratification of the Mexican society is manifest in spheres as diverse as access to housing and urban infrastructures, social security, health and education (see Schteingart *et al.* 1997, Ward 1998: 187 ff., Braig 2004, Mejía Montes de Oca 2010, Rubalcava/Schteingart 2012). This stratification has been analyzed and described through a number of different methodological-conceptual approaches; it is expressed, amongst others, in considerable levels of poverty from a multidimensional perspective (see Boltvinik 2012), or, in other terms, of marginalization – particularly in the peripheral parts of the metropolis[6] (see Consejo Nacional de Población (CONAPO) 2013: 57). In terms of inequality, Mexico displays the most pronounced income inequality – as measured by the Gini coefficient[7] – amongst all OECD members, ahead of Turkey and Russia, with a value[8] of 0.474 in 2004 and 0.466 in 2010 (see OECD 2015). In fact, inequality has not decreased but remained widely stable[9] over the last decades in the context of neoliberal restructuration of public policies since the late 1980s (see Farfán Mendoza 2010). In addition to income and socioeconomic status, social inequality in Mexico also continues to be strongly organized along questions of gender, ethnicity and color (see Frias 2008, Cerda García 2011), along with a considerable rural-urban divide in terms of living conditions (see Rubalcava/Schteingart 2012).

The term 'popular' is commonly employed in Mexico to describe a low socioeconomic status. It contains both stigma and (positive) identity – as in the colonias populares, the (predominantly) irregular settlements of Mexico City: "In a social or political context, (…) popular is laden with connotations, ranging from the folkloric and traditional to low-cost (…) and, in general, to denote lower social classes" (Connolly 2004: 83). The terms *colonias populares* and *popular urbanization* are widely used to grasp and characterize processes of irregular urbanization (see Duhau 1997, Salazar Cruz 1999, Ziccardi 2000, Horbarth Corredor 2003, Duhau/Giglia 2008: 329 ff.). Moreover, both everyday discourse and official publications (such as Consejo de Evaluación del Desarrollo Social del Distrito Federal (CEDS) 2010) frequently use the term. It is also employed as a label in the Federal District's socially differentiated water tariffs[10], and in the basic health insurance scheme *Seguro Popular* introduced by the federal government in 2004, both of which serve as (approximate) markers of popular class in the context of the present study[11]. The latter aims to

6 According to CONAPO's calculations based on the 2010 census, around a quarter of the population of the *Zona Metropolitana del Valle de México* was living in conditions of high or very high marginalization (see CONAPO 2013: 57).

7 The Gini coefficient is a widely used measure of national income distribution, with a value of 0 expressing total equality and a ratio of 1 total inequality.

8 Based on the pre-2012 income definition of the Organization for Economic Co-Operation and Development (OECD).

9 According to Cortés (as quoted in Duhau/Giglia 2008: 106), the Gini index for Mexico ranged widely unchanged around 0.5 between the early 1960s and 2000.

10 Based on the *Índice de Desarrollo e Infraestructura de la Ciudad* (IDS) index (see 4.2).

11 Either that or not having any access at all to health insurance.

cover a limited array of health services for the share of population not part of the formal labor market and hence without any access to the respective insurances. Social security affiliation is a strong indicator of social status given Mexico's stratified health system, which results in "unequal access, financing, and health outcomes as a result of segmentation in the delivery of services", as the WHO put in a report that already includes the *Seguro Popular* scheme (see WHO April 2006). The Mexican housing market displays an equally strong social stratification (and has done so since decades); with those without formal work contracts having only very limited access to the formal housing market – one of the main drivers of irregular 'popular' urbanization, which has been quietly tolerated by the state as a form of passive housing policies for the urban poor (see Azuela de la Cueva 1989, Gilbert/Varley 1991, Adler Lomnitz 1993, Duhau 1997, Ward 1998, Connolly 2004). Upon this backdrop, the term popular class bears a clear social meaning, referring to those with a lower socio-economic status and grasping their often precarious situation as regards employment, housing and general living conditions as well as (a lack of) social security. In a sense, this is the urban section of all those left behind in the cracks of Mexico's fragmented modernization (see Braig 2004) – though they could as well be read as their product. Mexico City's middle class, in contrast, represents mainly those formally employed in the public and private sector, and it is not least their income and access to social security which has often allowed them to obtain and maintain better living conditions[12] when compared to popular class residents. Two social insurance schemes cover mainly this social group since as long as the 1950s: the *Instituto de Seguridad y Servicios Sociales de los Trabajadores del Estado* (ISSSTE) for employees of the public sector, and the *Instituto Mexicano del Seguro Social* (IMSS) covering (formal) employees of the private sector. The Mexican elites and parts of the better-off middle class traditionally rely on private insurers and clinics, and do even more so since the privatization of these institutions in the mid-1990s (see Braig 2004: 294 ff.). Whereas ISSSTE and IMSS offer a broad range of medical and social services to the middle class, this stratification is paralleled by similar structures in the housing sector, where the *Instituto del Fondo Nacional de la Vivienda para los Trabajadores* (INFONAVIT) provides housing mortgages for employees of the formal sector since the 1970s. Consequently, members of the Mexican middle class can typically be identified by their access to certain housing markets and their type of health insurance, along with educational levels.

A short introduction such as the one presented here can obviously not replace a proper analysis of social structuration – it is merely meant to provide a rough approximation to the two social groups whose practices of water use will be studied in what follows. They were chosen as objects of inquiry for the present study not so much with the aim of tackling their difference in a comparative stance but because

12 Though the opening of the state-sponsored sections of the Mexican housing market (as represented by INFONAVIT) to private capital since the mid-1990s, with its industrialized mass-production of low-quality and low-density dwellings in the urban peripheries brought some dramatic changes (see Coulomb/Schteingart 2006, Hastings García 2007, Salazar Cruz 2014b, and Rodriguez/Sugranyes 2005 for a similar situation in Chile, where neoliberal housing policies of this type were implemented first).

they represent a majority of Mexico City's population. This is of interest not so much due to their sheer numbers, but as it can be assumed that these people's perspectives provide an insight as to how ordinary residents of the Federal District[13] experience water supply and handle water at home. In other words, harsh differences in water use between Mexico City's elite residing in gated communities and apartment towers adorned with private gardens and pools, and the inhabitants of the most peripheral and precarious colonias populares, recently built and still without any urban infrastructure, were to be expected. By exploring the water-using practices of residents from Mexico City's middle and popular class, the present work opts for a more subtle approach, which is valid precisely as it represents the common everyday experience of millions. Finally, it should again be stressed that a use of the term class in the present work does not intent to create the illusion of the existence of a homogenized popular or middle class *per se*, let alone some kind of homogenous class consciousness. Yet a Bourdieuian approach seems promising precisely as it drafts social status as something continuously (re)generated by social practice.

2.2 HABITUS: THE INCORPORATION OF SOCIETY

In the present work, practices are understood in a Bourdieuian sense as routinized and strategical in an (apparently) non-intentional manner (see 2.1.1). However, this non-intentionality is neither to be confused with an absence of regularity nor a mere product of rational choice. It is the notion of habitus as a key concept of Bourdieu's Theory of Practice which is able to explain how practices may be socially pre-structured without being directly determined. An introduction to the notion of habitus is also of interest here with respect to its integration to a sociospatial approach at a later point (chapter 2.3).

A Sense of Reality

Resorting to Bourdieu's notion of habitus as an incorporation of society provides a subject-based approach to the relation between everyday practices and social status. Habitus can be understood as a 'sense of limits' which is able to exert an influence on practices and is typically related to the dominant social order:

> "Every established order tends to produce (...) the naturalization of its own arbitrariness. Of all the mechanisms tending to produce this effect, the most important and the best concealed is undoubtedly the *dialectic of the objective chances and the agents' aspirations*, out of which arises the sense of limits, commonly called the *sense of reality*. (...) when there is a quasi-perfect correspondence between the objective order and the subjective principles of organization (...) the natural and social world appears as self-evident." (Bourdieu 1977: 164; emphasis added)

13 In contrast to those inhabitants of Mexico City residing outside of the Federal District's jurisdiction, it can be assumed that those within are subject to the same water policies, and supplied by the same water utility (albeit in a sociospatially differentiated manner) (see also 4).

As a kind of 'second nature', habitus is based on experienced social rules turned into a sort of incorporated knowledge, which is, as such, subject to a "genesis amnesia" (Bourdieu 1977: 79). Habitus is hence a concealed or 'unconscious' principle, wherein

> "the 'unconscious' is never anything other than the forgetting of history which history itself produces by incorporating the objective structures it produces in the second natures of habitus" (Bourdieu 1977: 78–79).

It is this unconsciousness – which is not to be confused with the Freudian concept – that makes habitus a particularly strong mechanism of the reproduction of social relations. It does so as it serves as a kind of implicit guideline for practices, working through a range of schemata pre-structuring the realms of practice, perception and thought.

A Generative Principle

Due to its unconsciousness, the literature argues, habitus is not accessible to self-reflection. That is, within the concept in the strict Bourdieuian sense, it is widely impossible for any individual to grasp his or her own habitus. This directly follows from its conceptualization as a set of incorporated rules which frames practices – including perception:

> "habitus provides us with an internalized system of *classificatory schemes through which we interpret new situations* by relating them to similar situations we have experienced in the past." (Weiss 2008: 230; emphasis added)

Perceptions, including those of own practices, are therefore never quite independent from the workings of habitus. This generally non-reflective nature also has methodological implications for any research seeking to grasp habitus as an object of inquiry[14] (as the present study does not). Moreover, everyday practices in their

14 As habitus is conceptualized to be operating widely in the realm of unconsciousness, any direct access through empirical methods such as interviews is limited (see Fuchs-Heinritz/König 2011: 117). Given the inaccessibility of habitus via self-reflection, employing direct questions about habitus itself, for instance in an interview guideline, would evade a direct 'naming' of the rules of which habitus consists, as Bourdieu noted: "Invited by the anthropologist's questioning to effect a reflexive and quasi-theoretical return on to his own practice, the best-informed informant produces a discourse which compounds two opposing systems of lacunae. Insofar as it is a discourse of familiarity, it leaves unsaid all that goes without saying: the informant's remarks (...) are inevitably subject to the censorship inherent in their habitus, a system of schemes of perception and thought (…) producing an unthinkable and an unnamable" (Bourdieu 1977: 18). In consequence, two different methodological approaches can be observed in the growing number of truly empirical works on habitus in the field of geography and urban studies. First, a vast majority chooses to reconstruct habitus through a hermeneutic interpretation, i.e. a research design based on semi-structured individual interviews about or observing practices and beliefs, thereafter assembling it to one or several habitus specific for the studied field (see Askew/McGuirk 2004, Waterson 2005, Orderud/Polickova-Dobiasova 2010). Second, a few specific methods to cope explicitly with the unconsciousness of habitus and gain access to the 'incorporated society' starting from a subject-based perspective have been developed in the last years,

routinized and usually unreflected, often unplanned manner, seem particularly prone to operate within the limits indicated by the habitus. As it functions as their generative principle (see Bourdieu 1982: 277), actual practices are wherein habitus is expressed most clearly:

> "If agents are possessed by their habitus more than they possess it, this is because it acts within them as the *organizing principle of their actions*, and because this *modus operandi* informing all thought and action (including thought of action) reveals itself only in the *opus operatum* [the practice]." (Bourdieu 1977: 18; emphasis added)

Habitus provides a general framework guiding and potentially influencing people's perceptions, thoughts, feelings and actions, yet it does not determine them (see Steinrücke 2006: 68 ff.). Instead, myriad variations of practices are possible within each habitus, which, understood as a guideline, is shaping more their form than their content (see Schwingel 2005: 71). Habitus is constituted by a system of dispositions – that is, incorporated tendencies or inclinations (see Bourdieu 1977: 214) – to act in a certain manner in certain situations. In other words, habitus is "a psychosomatic memory of sorts" (Rehbein 2006: 90), located in the body itself. Such an incorporated system of dispositions tends to pre-structure people's practices in a way specific for a certain era, class, gender or profession, that is, a certain field (see Steinrücke 2006: 65). Habitus thus allows for an "orchestrated improvisation" (Bourdieu 1977: 17) of practices within the limits set by habitus itself, and by field and capitals. Through the interplay of these three key concepts which form the base of Bourdieu's Theory of Practice[15], practices are always framed by an actor's social position. The habitus as a *structured structure* relates to the social position of a person, which in turn is measurable in the quantity and composition of his or her capital (see 2.1.2). Simultaneously, such a context specific habitus serves as a *structuring structure*, retranslating into practices specific to a certain social group. In Bourdieu's own words:

> "The structures constitutive of a particular type of environment (...) produce habitus, systems of durable, transposable dispositions, *structured structures predisposed to function as structuring structures*, that is, as principles of the generation and structuring of practices and representations which can be objectively 'regulated' and 'regular' without in any way being the product of obedience to rules" (Bourdieu 1977: 72; emphasis added).

In short, habitus has a dual character: it evolves from social structure and also tends to reproduce it. Practices generated through habitus are always strategical (in the non-intentional sense) and correspond to a certain field – that is, a certain part of objective structures roughly delineated against other fields by a common logic (see

for instance reflexive photography (see Dirksmeier 2007: 77, Dirksmeier 2009) or the integration of visual and creative methods in focus group discussions (see for instance Abrahams/Ingram 2013). Sociological studies on habitus, in the UK context, frequently employ focus group discussions with socially more or less homogeneous groups, which are meant to reproduce collective narratives related to social position (see for instance Reay 2005, Merryweather 2010, Wills *et al*. 2011).

15 Along with *doxa*, that is, the experience of a "quasi-perfect correspondence between the objective order and the subjective principles of [social] organization" (Bourdieu 1977: 164) without any antagonistic principles at work, i.e. an undisputed reproduction of the social *status quo*.

Schwingel 2005: 82, Rehbein 2006: 106). Hence there is a dialectic between field and habitus, in which

> "people's competence to participate in fields is critically related to (…) their socially and historically acquired dispositions. Capitals can only be mobilized in particular fields and by people with appropriate habitus." (Devine/Savage 2005: 13)

Following the example of Wendy Wills and colleagues (2011), the present study is set in the specific field of the habitat, or domestic sphere of each household, as will be explained below.

A Durable Mechanism?

This apparently smooth interplay between field, capitals and habitus in particular appears to have given rise to the question of determinism: to what extent does this process represent a mechanistic logic of social reproduction? Has the concept of habitus, in consequence, "more to say about social reproduction than social change", as Silva (2005: 90) has put it? From a praxeological perspective, it can be assumed that processes of social reproduction are open to change as they are related to practices in the very same dual sense represented by the habitus concept:

> "social structures (…) are constituted by human practices, and yet at the same time they are the very medium of this constitution. […individuals] therefore have the possibility, as, in some sense, capable and knowing agents, of reconstituting or even transforming this structure." (Thrift 1996: 69)

Generally spoken, this approach is based on the idea that society is shaped through practices, which are always also socially framed – it is "an ensemble of structures, practices and conventions that individuals reproduce or transform" (Bhaskar 1978: 12). Therefore, the reproduction of social positions and relations through practices in which habitus is at play is not to be conceptualized as being of a purely deterministic nature; it leaves some space for progressive action and change precisely due to its historical character:

> "[social] status, as well as the habitus developed therein, is a product of history – and thus (as difficult as it may be) historically transformable" (Bourdieu 1985: 37).

Nonetheless, Bourdieu apparently felt a need to stress this point again rather explicitly in one of his very last texts:

> "The habitus is not a fate, not a destiny. I must insist on this (...) against the interpretation which was proposed and imposed by some of the first reviews of my work (...). The model of (…) the vicious cycle of structure producing habitus which reproduces structure *ad infinitum* is a product of commentators." (Bourdieu 2005: 45)

Notably, he made these comments in response to his critics at a time[16] when the habitus concept was already being debated and applied widely, increasingly also in

16 This text was first published in 2002, the year of Pierre Bourdieu's death, in the first edition of Hillier/Rooksby (2005).

an English-spoken context. This debate is of interest for the present purpose only in so far as it reflects and highlights the habitus concept's dialectical position between structure and agency. Orthodox structuralists might seize the prominent role of practices as an indicator for voluntarism, while at the same time, and more importantly, Bourdieu's praxeological approach has been accused of being overly structuralistic, deterministic and mechanistic (see for instance Jenkins 1992, Calhoun 1993, Turner 1994, Mouzelis 2007; or from within theories of practice: Schatzki 1996: 137 f.). Others question whether the Bourdieuian habitus concept is rooted in a biologistic understanding of culture, taking (sexed) bodies for granted and ignoring the notion of 'doing' gender (see Silva 2005: 92); or whether it merely lifts the dichotomy between subject and structure to a different level by conceptualizing the body as open to social determination and the mind as closed (see Farías 2010: 28). This is certainly not the place to deepen these debates. The present work employs the term habitus in the sense of a set of incorporated social rules pre-structuring (rather than determining) social practices. As a concept striving to transcend the contradictions between objectivism and subjectivism (see Painter 2000: 241, Schwingel 2005: 57, Fuchs-Heinritz/König 2011: 125), it seems apt for the present purpose: to provide a conceptual approach which allows exploring the relation between everyday practices and sociospatial settings from a subject-based perspective (see 2.3).

Yet there remains the question of the durability of habitus: whether and to what extent habitus is transformable when confronted with a different field, i.e. under changing circumstances. In short, how stable is habitus as a framing mechanism of practice? It is not just those who criticize the Bourdieuian concept as an alleged mechanistic conceptualization of social reproduction who share the idea of habitus as a relatively permanent and unalterable principle (see for instance Gale 2005). There are other voices, most notably the one of Gail Weiss, claiming that habitus is subject to change to a certain degree as it expands continuously with experience (see Weiss 2008: 229 ff.). She goes on to specify that it is particularly during sudden, unexpected events[17] that habitus "as a dynamic rather than static structure" (Weiss 2008: 231) may be radically transformed:

> "unexpected events (...) have the power to radically affect and even transform the habitus itself. (...) Sudden upheavals in the habitus undoubtedly challenge our ability to maintain habitual routines." (Weiss 2008: 231)

This reminds of the above cited "everyday crises of routines" (Reckwitz 2002: 255) as ruptures which may constitute moments of transformation of social structure – also and particularly in the form of a sense of social limits. But apart from these extraordinary events, it also appears as if habitus has – though always within its own limits – a certain dynamic quality. As a set of acquired rather than natural dispositions, it is long-lasting but may be transformed by gaining awareness and specific training, as Bourdieu has stated:

17 She refers to the example of the 9/11 attacks in Manhattan, and their impact on the American society.

> "being a product of history (...), [habitus] may be *changed by history*, that is by new experiences, education or training (which implies that aspects of what remains unconscious in habitus be made at least partially conscious and explicit). Dispositions are long-lasting, they tend to (...) reproduce themselves, but they are not eternal. *They may be changed by historical action oriented by intention and consciousness and using pedagogic devices*. (One has an example in the correction of an accent of pronunciation)." (Bourdieu 2005: 45; emphasis added)

Without entering into the realm of arbitrariness, it thus seems as if the concept of habitus is open to social transformation to some degree, especially in as far as its implicit nature and 'genesis amnesia' may be overcome by generating awareness. This directly leads us to consider more in detail how this system of dispositions is incorporated in the first place.

The Symbolic Gymnastics of Habitus Acquisition

Some argue that through the interplay of capitals, habitus and field, Bourdieu's Theory of Practice is able to replace the need for any explicit theory of socialization (see Krais/Gebauer 2002: 61). Yet it seems surprising that – despite the broad stream of publications on habitus-related topics that has developed over the last three decades – one question still remains widely unanswered: how exactly habitus is acquired. Peter Dirksmeier, for instance, emphasizes such a lack of proper conceptualization of habitus generation in Bourdieu (see Dirksmeier 2007: 86; an overview on this critique in the literature is provided in Fuchs-Heinritz/König 2011: 137 f.). The impression that his praxeological approach represents a rather incomplete theory of socialization is deepened when Bourdieu himself calls the relation between objective structures and habitus a "black box" (Bourdieu 1997: 93)[18]. The vast literature on Bourdieu contains several attempts to conceptualize this process of incorporation. Some of them remain vague, as when stating that the incorporated system of dispositions is achieved and appropriated during the trajectory of an individual biography (see Steinrücke 2006: 65). The most advanced conceptualization so far appears in Gerhard Fröhlich's readings of Bourdieu, in which he drafts habitus as being incorporated in three different manners: through subtle learning, through a direct transmission of norms and rules of behavior, and through orderly training, typically in the form of games (see Fröhlich 1994: 39). Children, to whom Fröhlich mainly applies this theory, would thus acquire practical principles through which they are afterwards able to generate a large amount of different practices 'informed' by these habitus. Others cite direct experiences as well as institutionalized education and education in the family as mechanisms through which habitus is generated (see Fuchs-Heinritz/König 2011: 135 ff.). If habitus is a biographical achievement, this would also apply to continuous experience and learning beyond the primary socialization during childhood. It seems as if the literature does agree that habitus operates within each subject and is learned through a process of incorporation. Incorporation, as the term implies, is an action which directly involves the human body. How, then, is this 'second nature' acquired? Elsewhere, Bourdieu has

18 As cited in Fuchs-Heinritz/König (2011: 137).

hinted to routinization as a form of (bodily) learning, which primarily draws on a Aristotelian and Platonian concept of *mimesis*: a "sort of symbolic gymnastics" (Bourdieu 1977: 2). It is, in other words, a form of learning through a repetition of the practices of others:

> "In societies[19] which lack any other recording and objectifying instrument, inherited knowledge can survive only in its embodied state. Among other consequences, it follows that it is never detached from the body which bears it and which (...) can deliver it only at the price of a sort of gymnastics intended to evoke it: *mimesis*. The body is thus continuously mingled with all the knowledge it reproduces" (Bourdieu 1977: 218).

Following this logic, it appears as if habitus is acquired to some extent through repetition in everyday routines, and stabilized through this sort of repeated training. Bourdieu reminds us that this is a form of the law of learning-by-doing, which

> "every made product (...) exerts by its very functioning, particularly by the use made of it, an educative effect which helps to make it easier to acquire the dispositions necessary for its adequate use." (Bourdieu 1977: 217)

Tracing the idea of repetitive learning hence brings us back to questions of routinized practices. Social practices, understood as non-intentional strategies evolving from a certain type of habitus, are always interested – even though they may appear disinterested through routinization:

> "The importance of routine as a (...) 'officialising' strategy [Bourdieu, 1977: 22] cannot be overestimated, since routine can quickly make pieces of behavior appear not only natural but also disinterested." (Thrift 1996: 81)

This provides the opportunity to ask not only how but also where such routinized 'concealing' through the incorporation of these principles of practice happens. With reference to the domestic realm as a Bourdieuian field with a specific logic, the present work is a proposal to take urban space explicitly into consideration as a part and context of this process of incorporation. Indeed, there are some hints to the role of space[20] in Bourdieu's own writings, for example when stating that "the structures constitutive of a particular type of environment (e.g. the material conditions of existence characteristic of a class condition) produce habitus" (Bourdieu 1977: 72). By describing habitus as a "collective product of living conditions" (Steinrücke 2006: 68), others strike a similar tone. Upon this backdrop, it could be argued that habitus is constituted through processes of learning in a certain social and thus always also spatial context. We will return to this aspect in chapter 2.3, which explains how the present work – rather than searching for a certain habitus of water use (with all the methodological problems this would carry) – is aimed at unearthing the conditions of its production. The concept of habitus is used here as a vehicle to explore the relation between sociospatial conditions and everyday practices from a subject-based perspective, as the following section illustrates.

19 Here, Bourdieu makes reference to the Kabyle society he studied in *Outline of a Theory of Practice*.

20 On the concept of space in Bourdieu, see 2.3.1.

2.3 A SOCIOSPATIAL APPROACH TO EVERYDAY PRACTICES

The main aim of the present study is to analyze practices of water use as sociospatial practices, that is, for their relation to the sociospatial setting. Conceptually, this exploration is based on a relational concept of space, and on Bourdieu's notions of habitus and habitat. This theory-derived research approach pursues a spatialization of Bourdieu's concept of habitus by encouraging a conversation between theories of both practice and space, in what will be called a sociospatial approach to everyday practices.

2.3.1 Making Habitat? The Question of Space in Bourdieu

As introduced above, Bourdieu's Theory of Practice is of great avail to grasp the social pre-structuration of practices. However, the underlying concept of space in Bourdieu seems to be widely a metaphorical one (see also Painter 2000: 254 f.). It is based on a purely abstract definition of space which in consequence appears somehow two-dimensional (see also Fröhlich 1994: 41). Bourdieu's *abstract social space* is conceptualized as a representation of society basically limited to two dimensions: it is defined by the volume of capitals each subject disposes of, and their composition. As we can observe in the famous illustration of an abstract social space in *Distinction: A Social Critique of the Judgment of Taste* which he derived from his empirical research on the French society (see Bourdieu 1982: 212 f.), it resembles a map more than a space, as Boike Rehbein observes (see Rehbein 2006: 167). In other words, the disparities of urban space symptomatic for a highly differentiated capitalist society are understood as a direct expression of unequal social relations (see Bourdieu 1991: 26 f., Haferburg 2003: 83). This closely follows a Durkheimian concept of space as an expression of social structure (see Schroer 2006: 49 ff.). Consequently, geographical space in such a society is always socially differentiated in that it both materializes and symbolizes abstract social space:

> "inhabited (respectively appropriated) space [acts] as a kind of spontaneous symbolization of social space (…). In a hierarchized society, there is no space that would not be hierarchized and not express hierarchies and social distances." (Bourdieu 2010a: 118)

As such spatial disparities form part of the dialectic between social position and incorporated dispositions, they essentially appear as reflections (albeit often indirect or concealed ones) of social hierarchies in most of his writings. It fits within the same logic that Bourdieu continuously writes about appropriated physical – rather than (socially) produced – space (see Dirksmeier 2009: 93), a matter culminating in a statement in which appropriated physical space and "reified social space" (Bourdieu 1991: 29) are used as synonyms. I thus agree with Joe Painter in that Bourdieu's concept of space seems to be a rather limited one, as it "tends to be seen exclusively in terms of distributions, distances and arrangements" (Painter 2000: 255).

However, there are some openings to a potential production of space, most notably with reference to the notion of habitat. The latter gained a certain prominence through one of Bourdieu's most cited expressions according to which the habitus makes the habitat (see Bourdieu 1991: 32). On the one hand, this can be read as a statement that the production of habitat, and therefore urban space, is also guided by the incorporated dispositions forming a habitus. If making habitat thus appears as a strategical practice in the sense of Bourdieu, it could be argued that this also applies to the production of space (understood as a spatial practice). On the other hand, it seems as if there is not much else Bourdieu has written to clarify what exactly he means by 'habitat' – and as far as the author is aware, the secondary literature has remained largely silent on this topic. Given the rather metaphorical reference to spatiality in Bourdieu's writings, this seems unsurprising. However, there are some indications in his work, again related to the production or 'making' of this habitat: it seems to involve an act of arrangement or placing, of choosing certain elements, and thus an appropriation of space and things (see Bourdieu 2010a: 118). This reveals a close similarity to the habitus-guided logic of practice wherein, according to Bourdieu, lifestyles are made up of an "ensemble of persons, goods and practices chosen for himself by the actor" (Bourdieu 2006: 360). It should have become clear by now that in the logic of the Bourdieuian approach, this 'choice' is not literally a matter of rational choice but a non-intentional strategical practice guided by the working of incorporated dispositions. The habitat then appears as something which results from and is related to people's practices. In this sense, the Bourdieuian 'making habitat' may be interpreted as a production of space located first and foremost at the micro scale of the home[21]. This is where, according to Bourdieu, the habitus operates through the practices of choosing, arranging and appropriating spatial features. Such a reading is also inspired by Andrea Giglia's interpretation of the production of habitat as a production of domesticity, including both the material construction of dwellings and the production of meaning through their symbolic appropriation and use (see Giglia 2013: 182). In the present work, the term habitat – or, as a synonym, the sociospatial setting – is therefore used to grasp what Bourdieu refers to as 'living conditions' (see Bourdieu 1982: 281) in a wider sense. For the present purpose, the concept of habitat is widened to explicitly include different moments of production of space. This encloses not only the housing conditions *per se* (materiality, residential density and the like), but also infrastructural conditions of water supply, type of neighborhood and available household goods, as well as questions of knowledge and meaning. Only the entirety of these aspects forms a habitat as an ensemble of sociospatial practices, as the author understands it. This again brings about the question how the relation between habitus and habitat is to be conceptualized in detail, and whether it is of a dialectical nature.

21 It seems as if to Bourdieu, the production of space on other scales – one example would be a city's infrastructural policies – might only be tackled in a more vague sense through the concept of fields.

2.3.2 Towards a Sociospatial Approach

From an epistemological perspective, the origin of practice theories, including Bourdieu's praxeological approach, and of theories of production of space can be traced back, in part, to historical materialism. As Karl Marx established in his *Theses on Feuerbach*, the material world is a product of social practice (see Marx 1978: 5). This is precisely why the author is of the opinion that reflecting these theories conjointly benefits both of them. One potential link lies in enlightening the Bourdieuian understanding of space by the way of theories of the social production of space[22], thus providing a theoretical link between spatial and social development. These theories rest on a relational and processual understanding of urban space: it is a social product and as such, always in the making (see Massey 1999: 6 f.). A relational concept of space involves the idea that space and time are to be thought together, or in the words of Milton Santos,

> "time, space and world are realities which must be mutually convertible if our epistemological concern is all-encompassing. *Human society in progress, that is, in its materialization, is always the starting point*. This realization happens on a material base: space and its use, time and its use, materiality and its various shapes, actions and their various aspects." (Santos 1996: 47; emphasis added)

Rather than a mere materialization of social relations, space itself is a social process, as it "implies, contains and dissimulates social relationships – and this despite the fact that a space is not a thing but rather a set of relationships between things" (Lefebvre 1991: 82). From such a relational perspective, social inequalities and uneven development are not merely expressed mirror-like in spatial differences; as Doreen Massey has put it, "those unequal relationships (...) are organized spatially" (Massey 1994: 87). Urban space and the ongoing process of urbanization are crucial for the process of capitalist development[23] (see Smith 1984, Harvey 1985, Soja 1989, Merrifield/Swyngedouw 1997). With spatial and social development inseparably entangled, the production of space can be understood as a social practice precisely in this sense. In the present work, it is from such a perspective that

22 There is a marked difference between this concept of a 'production of space' and earlier spatial theory in the disciplines of so-called spatial sciences *(Raumwissenschaften)*, which employ an objectivist (or naturalizing) rather than relational understanding of space (see Harvey 1989: 203, Lefebvre 1991: 368).

23 Under capitalist conditions, the production of space bears a dual and contradictory character: it accelerates the accumulation of capital only to slow it down once built, consequently calling for a 'creative destruction' of these physical constraints if growth is to be continued within that same logic. Urbanization processes spur economic growth through, for example, the construction of infrastructure – yet once this infrastructure is built, it serves as an asset slowing down further capital circulation. This contradictory nature of urban space is what David Harvey captured in the term 'spatial fix' (see Harvey 1996: 412). Reading space as a form of 'fixed' capital is of particular interest with regard to urban infrastructures – such as water and drainage networks – with their strong path-dependence in terms of materiality and regulation. Processes of social differentiation may be perpetuated through such large technical infrastructures, and thus prove rather stable once they are inscribed in urban space (see Graham/Marvin 2001: 190 ff.).

domestic practices of water use are analyzed not simply as social but as sociospatial practices (see Gottdiener 1994). When space is defined as an ensemble of interrelated social practices in the realms of materiality, knowledge and meaning (see Lefebvre 1991: 33 f.), making space appears simultaneously as a matter of physical construction and one of knowledge production through academic discourses, urban design, mapping and other regimes of regulation such as urban and infrastructural planning (see Crampton *et al.* 2007). At the same time, lived space is created through the assignation of symbolic meaning to physical space, for instance through its daily use. It is in particular the ethnographic concept of *imaginarios urbanos* (urban imaginations) as developed by Armando Silva Téllez, Néstor García Canclini, and colleagues (see Silva Téllez 1992, García Canclini 1997, Silva Téllez 2003, Lindón 2007, Hiernaux 2007) which provides an instrument to explicitly tackle this 'lived' dimension of the production of space. This socio-symbolic approach has broadened the horizon of urban research by including subjective experiences whilst keeping social context in mind (see Lindón 2007: 7). *Imaginarios urbanos*[24] refer to a collectively produced and shared meaning of space, thus expressing a "collective subjectivity" (García Canclini 2013: 40). Yet these imaginarios are more than mere symbolic representations of the city and its streets, squares and neighborhoods – they may even serve as operational guidelines for (future) social practice (see Hiernaux 2007: 20; García Canclini 2013: 39). Both aspects are crucial – as Alicia Lindón puts it, imaginarios

> "are always a product of social interaction (…). They are constructed through discourses, rhetoric and social practices. Once made, they possess the ability to influence and give orientation to practices and discourses" (Lindón 2007:10).

Imaginarios serve, in other words, as a kind of collectively shared mental maps which are used to navigate the city (see Márquez 2007), thus forming an intrinsic part of processes of spatial production. However, they are not to be understood as mere representations or 'authentic' reflections of urban phenomena but may be quite distorted or even entirely decoupled from any concrete reference (see Lindón 2007: 10). The present work tackles such imaginarios through a reconstruction of people's imagined landscapes of water supply in Mexico City, which are assumed to form an integral part of domestic practices of water use (see 5.4).

Integrating such a relational understanding of space into an analysis of social practices transcends Bourdieu in as much as it fills his abstract metaphor of space with content (see Harvey 1989: 214 ff., Schwarz 2009: 38 ff.). This approach seems promising to fully grasp the spatiality of social processes. When space is understood as an ongoing social process rather than a container of social practices it allows the assumption that as with any other social practice, habitus is also at work in spatial practices. In spite of his apparently two-dimensional notion of space, Bourdieu's famous notion that "it is the habitus which makes the habitat" (Bourdieu 1991: 32) already points in this direction. As a key element of Bourdieu's theoretical thought, habitus is a crucial element for studying the dialectic between social

24 I prefer the Spanish term used for this concept over the slightly awkward translation into 'urban imaginations'.

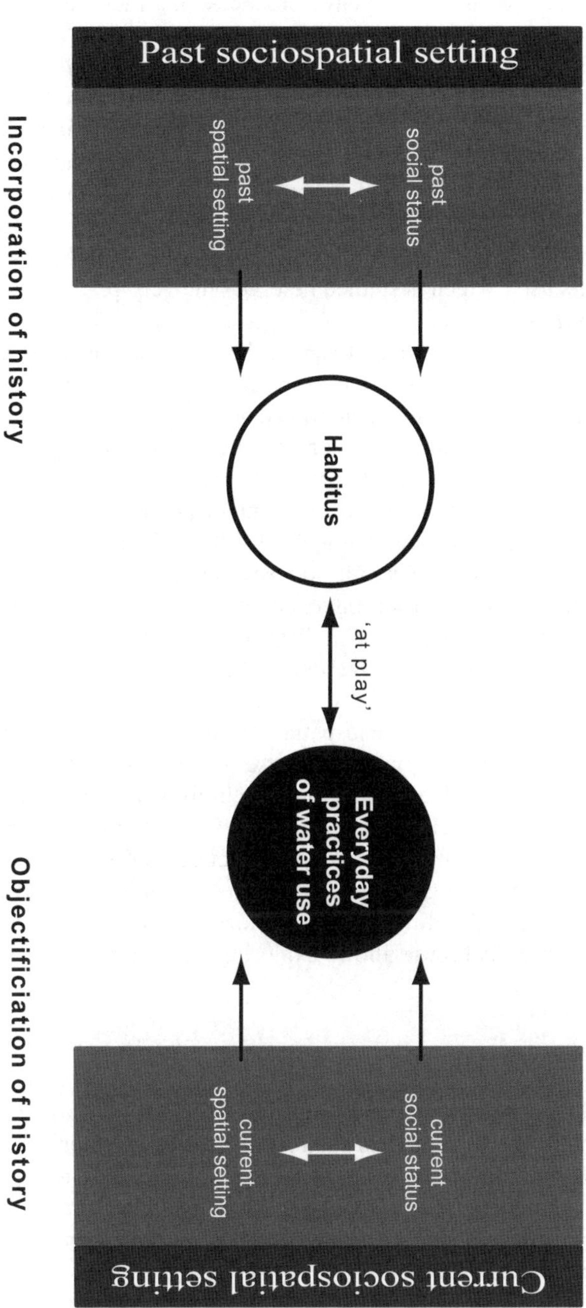

Figure 2.1: Habitus-habitat-practice model. Own elaboration

structures and (everyday) practices. By merging this concept with a relational understanding of space, the products of spatial practice can be understood as a materialization – or more precisely, spatialization – of habitus. And this spatialization applies in dual sense, as figure 2.1 illustrates. On the one hand, today's living conditions may be conceptualized as framing conditions for today's practices, in which habitus is at play – and thus potentially transformed or reproduced. On the other hand, past experiences with living conditions – which the present work calls habitat biographies – may be conceptualized as framing conditions of processes of habitus generation. At the risk of repeating myself I want to stress that it is not habitus itself which is studied here, but the very practices which it is assumed to guide or pre-structure.

The author wants to argue that practices of water use can be analyzed from a sociospatial perspective in as far as habitus is both achieved and enacted under specific sociospatial conditions, in a specific setting which simultaneously serves as a framing condition for practices and a socialization context. Hence the working hypothesis on which the present study is based: the current sociospatial setting (or habitat) frames social practices in the present while at the same time pre-conditioning future practices through a habitus which is maintained or transformed in this very setting. In turn, the sociospatial settings a person experienced in the past are assumed to exert an influence on their present practices of water use through the incorporated set of social dispositions (i.e. the habitus). Studying water use from a sociospatial perspective under these conceptual parameters therefore simultaneously calls for a historical analysis of the individual and collective process of socialization in a given sociospatial setting and for an investigation of current practices. Thus I would suggest to reverse (yet maintain) the habitus-makes-habitat notion, and consider the production of habitus as a thoroughly sociospatial process in the present work. Habitat is, in other words, both a product of spatial practices and a framing condition of everyday practices and processes of socialization. It is in this sense that the generation of habitus has to be studied with respect to the specific sociospatial setting where it (historically) stems from. Such a spatialization of habitus in a dual sense allows exploring water use from a sociospatial perspective.

A Site of Socialization?

This approach was encouraged by previous indications in the Bourdieu-based literature that habitats (in their very quality as social products) might influence the generation of habitus to some extent by tending to reproduce and reinforce social inequalities (see Oßenbrügge 2003: 7). Others grasp the relation between habitus and space under the name of a 'sociospatial habitus', and also pointed towards a reciprocal nature of this relation (see Giglia 2013: 172 ff.). But perhaps more importantly, Bourdieu himself hints towards a potential role of 'physical appropriated' space in the largely under-theorized process of habitus generation (see 2.2): for instance when drafting an incorporation of society through repetitive spatial practices and experiences:

> "the imperceptible incorporation of structures of social order is without doubt taking place to a good part *through permanently* (...) *repeated experiences of spatial distance*, wherein social experiences exhibit themselves; (...) such social structures, naturalized [through the movements of the body] organize societal ones and qualify them." (Bourdieu 2010a: 119; emphasis added)[25]

In other words, the 'symbolic gymnastics of habitus acquisition' are set in a (geographical) spatial context, even in Bourdieu. This idea coincides with Ilse Helbrecht's call for a renewed cultural geography wherein the relation between human beings and the (socially produced) environment should be understood as an interaction rather than presuming that people are passive 'users' of space (see Helbrecht 2003: 165). Always assuming that space is socially constructed to begin with, it has been argued that this very space might also be involved in a pre-structuration of social practices (see for instance Dovey 2005, and for an overview: Parnreiter 2007: 59 ff.). Similar to the workings of habitus, the sociospatial setting might not determine practice but serve as a kind of guideline, or resource, facilitating certain (everyday) practices and complicating others. As Edward Soja has put it:

> "Throughout our lives, we are enmeshed in efforts to shape the space in which we live while at the same time these established and evolving spaces are shaping our lives in many different ways." (Soja 2010: 71)

In a similar vein, some ask whether past spatial experiences have a potential influence on spatial practices, and thus indirectly speculated about some kind of dialectic between habitat and habitus:

> "although it is the habitus which makes the habitat, the former owes its existence to certain social and thus also spatial structures becoming manifest in it. Whether one (...) has grown up in a village, in the outskirts or the city center [...has] left certain marks, condensing into a habitus which in turn has an influence on the way in which the spatial environment is being shaped in the present." (Schroer 2006: 88–89)

Pushing this thought radically further, it may almost seem as if space is somehow able to shape social structure; and this leads us directly to one of the major risks of the present venture – i.e. to follow and at the same time reverse Bourdieu by assuming that the habitus makes the habitat and vice versa, as the working hypothesis does. One of the main challenges any critical attempt of exploring a potential dialectic between social and spatial structures is facing sooner or later is the error of reification, or essentialization. A case in point are the utterances of some scholars of practice theory, in which they seem to place subject-object relations on the same level with social interaction (see e.g. Reckwitz 2002: 253). The results of similar ventures are paradigmatically expressed in the debate over an alleged 'intrinsic logic of cities', which is only mentioned here briefly as it heavily draws on Bourdieu's concept of habitus, attributing it directly to cities[26].

25 The translation from the 2005 edition of *Das Elend der Welt* was performed by the author.

26 In what seems like a reanimation of the reification debate lingering in the field of political geography since the late 19th century controversy between Emile Durkheim, Friedrich Ratzel and others (see Agnew 1994, Barnes/Minca 2012, Minca/Rowan 2015, a fierce debate has developed over the last decade or so on an alleged 'intrinsic logic of cities', particularly in

The present approach does not only operate on a different spatial scale – and is thus, other than the 'intrinsic logic of cities' perspective, aware of inner-urban heterogeneities particularly with respect to living conditions (see Gestring 2011: 46) – it is also, and more crucially, about taking human beings and their practices as a starting point of any sociospatial analysis. Rather than reifying 'the city', the author proposes to spatialize the habitus concept from a strictly subject-based perspective involving the production of space as a practice. In this pursuit, habitus is a set of dispositions incorporated neither by institutions nor by space (be it neighborhoods, cities or countries), but always and exclusively by individual human beings via their own body (see Atkinson 2011: 337, Gestring 2011: 45). In other words, there is no 'habitus of space', only space as a social process, and – simultaneously and hypothetically – as a setting wherein habitus is generated. Without going so far as to conceptualize space as an entity of its own right, the present work thus strives to further clarify a potential dialectical relation between habitat and habitus, which some of the existing geographical literature specifically calls for:

> "Habitus is shaped through the incorporation of certain external material and cultural conditions of being. If we conclude *that these conditions contribute to the formation of each individual's habitus*, we then have to take the conditions of the habitat, the living space, the neighborhoods into consideration" (Haferburg 2003: 67; emphasis added).

In how far these habitats contribute to an incorporation of society (and thus to social formation) is a question which will be pursued in the present work. The production of space in everyday life can thus be conceived as a process of learning through both ordinary and extraordinary experiences within a sociospatial setting – also in the form of 'symbolic gymnastics', as Bourdieu called one of the core elements of the incorporation of social rules through routinized practices. Though space as an ongoing social process clearly does not generate everyday practices, it potentially regulates them:

> "Activity in space is restricted by that space; space 'decides' what activity may occur (…). Space lays down the law because it implies a certain order (…). Space commands bodies, prescribing or proscribing gestures, routes and distances to be covered. (…) this is its *raison d'être." (Lefebvre 1991: 143)*

It is in this pursuit that the Bourdieuian habitus concept benefits from a relational understanding of space. In this context, the present work aims to explore whether habitats exert an influence on social practices – more precisely, to what extent past experiences (which are always also sociospatial in nature) exert an influence on

German-spoken urban sociology and geography (see Berking/Löw 2008, as well as for an overview: Frank *et al.* 2013, and for a critique: Belina/Miggelbrink 2010 and Kemper/Vogelpohl 2011). It draws on a culturalistic interpretation wherein it is possible to "compar[e] the city to a person with a specific biography. Related to it is the idea that cities, too, have their own habitus, or character (…), which becomes the generative principle of life-styles" (Lindner 2003: 46; emphasis added). Attributing urban space with a quasi-human character – "like humans, cities may be recognized by the way they walk" (Berking 2008: 15) – it is based on the idea that such a space-as-social-agent were somehow able to *generate* certain social practices. For a critique of what he calls a "misinterpretation" of Bourdieu, see Gestring (2011).

practices in the present (via a habitus generated under certain conditions). This will be studied empirically through a novel biographical method specifically developed for the present purpose.

Tracing Habitat Biographies

The idea of studying the generation of habitus in the specific sociospatial setting where it is assumed to stem from brings us back to Milton Santos' call to think space and time together and as an ongoing process wherein society is realized (see Santos 1996: 47). A historical-biographical perspective seems apt for an exploration of the generation of habitus if the latter is conceived as a collective product of living conditions evolving from a systematic biography and appropriated throughout the individual life course (see Steinrücke 2006: 68). Applying habitat biographies will allow setting interviewees' practices of water use in relation to their past experiences with infrastructural and housing conditions. Bourdieu put an emphasis on those differences in biographies which represent collective rather than individual experiences: the 'typical biography' of a certain social group, he claimed, reflects their social origin and thus reveals more than an analysis which is based on present social status alone (see Bourdieu 1982: 188). Other habitus-based research on social inequality also stresses the need to understand social formation and the reproduction of inequality as a longitudinal process wherein future social status is at stake, and hence adopt a life course perspective (see Savage 2000: 69). As social status works through the ongoing process of (apparently) individual biographies (see ibid.: 73), this is the precise context in which it should be studied, I want to argue. The spatiality of this process of social becoming is recalled and grasped here by habitat biographies, a method developed specifically for this purpose (see 3.2.3 for details on the method). From a biographical perspective, each interviewee's social background and his or her past places of residence are analyzed as places of socialization, where certain practices of water use were exercised under certain housing conditions and certain conditions of water supply. Such personal narratives on past experiences may show clear lines or ruptures, where today's living conditions may or may not coincide with past experiences. Documenting the spatial setting of these processes of social becoming, this instrument shows the trajectory of each interviewee's personal dwelling biography with respect to location, type of water supply, ownership status and duration of stay. To give an example: it allows tracking whether someone lived in a high-rise apartment with permanent water supply in the past, and to what extent this experience makes a difference today with respect to their practices of water use. These habitat biographies are then translated into a specific graphic iconography, which allows comparing and analyzing them for similarities and differences to construct empirically grounded types reflecting people's past experiences with water supply limitations.

The domestic sphere (or habitat) is understood as an element of spatial practice in the present work, and, at the same time, as a site where social relations are at play (see Haferburg 2007: 60), not least via the actual infrastructural conditions. Based on a combination of the praxeological approach with a relational understanding of space, such an empirically grounded sociospatial approach embraces the urban condition as an ongoing process and strives to unveil its historicity. The following chapter goes into detail on how such a sociospatial approach is translated into an overall, field research-based research design for the present study, including the instrument of habitat biographies.

3. RESEARCH DESIGN

The present study aims to explore the relation between space and everyday practices, taking domestic water use in Mexico City as an example. Its methodological approach is designed to obtain material to clarify this relation, primarily through empirical fieldwork. The set of qualitative methods introduced in the present chapter are considered apt for this purpose, and will also serve future research on everyday practices from a subject-based, sociospatial perspective.

3.1 RESEARCH STRATEGY

If we are to conceptualize urban space as an ongoing process rather than a naturally given entity, a spatialization of Pierre Bourdieu means to infuse his theoretical approach – and the concept of habitus in particular – with a relational understanding of space (see chapter 2 and Schwarz 2009). Which methods are apt for this? With respect to the existing literature, the dialectics of production of space and (re) production of social inequality in the realm of domestic water use – i.e. the link between (everyday) practices of water use and sociospatial conditions – can be studied in particular through levels of domestic water consumption[1], discourses around water, and everyday practices involving water. The present study puts a clear emphasis on the latter: everyday practices of water use, understood as material practices and imaginations. By locating the investigation in the domestic realm, it contributes to a long-running yet narrow stream of down-scaled water research. Rather than producing large quantitative data sets, the author opts for the small-scale, context-sensitive research approach Dena Fam and colleagues call for (see Fam *et al.* 2015). In that sense, the city and the household are sites at the meso and micro scale of what has been identified as a crucial field of ethnological water research: "the culturally meaningful, sensorially active places in which humans interact with water and with each other" (Orlove/Caton 2010: 408). Following a sociospatial research perspective, a qualitative, field-work based approach was chosen for the present study as it seems most adequate to grasp the complexity of urban processes through the provision of a relational understanding. Inquiring as to

1 There is relatively few systematically collected statistical data on actual levels of domestic water consumption which would allow for a correlation with socioeconomic or demographic data – not least due to the prevalence of research strategies based on quantitative data sets on water provision obtained from water utilities and/or model up-scaling, a common confusion and equation of the categories 'domestic consumption' and 'water provision', and methodological obstacles when it comes to qualitative approaches. Actually studying levels of domestic water consumption requires a sophisticated data collection 'on the ground', i.e. directly in the studied households, for instance through water diaries (see Lahiri-Dutt 2015) or a participatory mixed method approach (see Thornton/Riedy 2015).

how the relation between space and social practices works, the present study embodies an explorative approach aiming for a sociospatial contextualization of practices of water use. This is sought through an empirical, actor-based perspective which focuses on subjective experiences which are accessed through personal narratives (and, to some degree, direct observation) in the field. Based upon his readings of Bergson and Proust, Walter Benjamin provides a useful distinction between information and the act of narration, which allows for an active re-appropriation of subjective experience:

> "A story does not aim to convey an event per se, which is the purpose of information; rather, *it embeds the event in the life of the storyteller in order to pass it on as experience to those listening*. It thus bears the trace of the storyteller, much in the way an earthen vessel bears the trace of the potter's hand." (Benjamin 2003: 316; emphasis added)

Consequently, individual and collective narrations are a key element of all three key methods employed here: in individual interviews and focus group discussions as well as the novel method of habitat biographies which will be introduced below. Though distinctively subject-based, the research agenda links social practices back to social structures and urbanization processes. Therefore, practices of water use were explored at and set in relation to one specific sociospatial setting: the domestic realm. (However, this micro scale is by no means regarded as the only relevant setting of urban practice and experience, which always refers to multiple geographies).

In order to allow for some degree of comparability, the diversity of the domestíc sociospatial setting was partly controlled through a pre-selection of interviewees based on the following criteria: steadiness of water supply (permanent and intermittent), housing type (flats and houses) and social status (popular and middle class). As proposed in 2.4, all three aspects were assumed to have a potential impact upon practices of water use.

The fieldwork was set in the boroughs of Iztapalapa and Cuauhtémoc, Federal District. This selection was based on the expectation of similarities in general framing conditions with respect to urban governance (such as the regulatory regime and a single water supplier) on the level of the Federal District as well as on expected differences in water supply conditions between the two boroughs. This expectation also extended to potential differences in levels of water consumption and practices of water use, as indicated in the literature (see for instance Libreros Muñoz 2004: 690, Legorreta 2006: 106). (A detailed introduction to the selected boroughs and studied neighborhoods is provided in 4.2). The present study explicitly sought to challenge the idea of these two boroughs being homogeneous or monolithic entities in terms of infrastructural conditions. It did so by exploring actual experiences with water supply conditions at people's homes and their relation to practices of water use. Though to some extent inspired by 'traditional' comparative approaches in urban studies (see Pickvance 1986, and for a critique Robinson 2011), it hence put an emphasis on exploring and understanding relations more than on a strict explanation of causal relationships. It bridges the inductive-deductive divide (see Denzin 1989: 5 ff.) by employing both theory-driven assumptions (regarding the relation between habitus and habitat, see 2) and allowing for new conceptualizations to evolve from the (empirical material exploring the) experiences of Mexico City's residents with respect to water and everyday life. Their water supply and housing conditions as well as social status provided a starting point for a venture which may be called a sociospatial exploration of everyday practices under (partially) controlled conditions.

3.2 EMPIRICAL METHODS

With respect to the conceptual framework developed in the previous chapter, a range of mostly qualitative methods were employed in the present work to achieve a methodological triangulation (see Denzin 1989: 236 ff.). The studied practices of water use were tackled not only in their materiality but also in their mental-conceptual and imaginary qualities. Alongside theories of production of space, such a multi-dimensional research perspective was also inspired by Anne Huffschmid's and Kathrin Wildner's proposal to 'make spaces talk' through a combination of spatial observations in the ethnographical sense and discourse analysis (see Huffschmid/Wildner 2009: 38). The research design is primarily based on three methods:

- an exploration of domestic water use through **individual interviews**, covering material practices as well as narrations on lived experiences,
- a documentation of past water-related experiences from a sociospatial perspective through **habitat biographies**, and
- an exploration of collective discourses and *imaginarios* through **focus group** discussions.

Other methods included a documentation of each interviewee's sociospatial setting (on the micro scale) through a socio-economic short survey and a protocol of material housing conditions, and a photo documentation of water-related devices in the interviewed households. A mapping of local bottled water economies in two neighborhoods[2] completed the picture.

As argued above, a qualitative research design is of use to explore the interrelations between sociospatial processes and everyday practices. Moreover, a severely limited reliability and accessibility of 'official' statistical data with respect to water

2 This method was motivated by the assumption that local bottled water economies can be seen as crucial framing conditions for domestic practices of water drinking while simultaneously reflecting the local demand for this highly commodified form of potable water. As of 2013, there were no detailed statistical records on the size and composition of this emerging sector in the Federal District. Information was obtained through a series of observatory walks by the author in early 2014 which allowed for a comprehensive mapping of all shops and businesses selling bottled water in two of the studied neighborhoods: the socially mixed neighborhood of San Rafael in Cuauhtémoc and the popular class Ermita Zaragoza housing complex in Iztapalapa. The aim was to document all bottled water sales points, distinguishing vendors of branded bottled water as well as the small local purification plants *(purificadoras)*, and specifically targeting the sale of refillable 20-liter jugs *(garrafont)*, which are clearly targeted at domestic water consumption.

provision and water consumption underpinned the need to generate own empirical material through a fieldwork-based research strategy. According to the literature, much of official Mexican water data is unreliable at best, displaying a variety of limitations including a lack of valid, coherent and updated data sets, a methodological fragmentation, and a limited public accessibility (see Perevochtchikova 2013: 44). This impression was confirmed during fieldwork for the present study when several official requests for water provision and water consumption data from public supplier SACM through the so-called transparency portal *InfoMexDF*[3] did not result in satisfactory access to spatially differentiated data of the required type[4]. Consequently, the main source for a general description of water supply conditions in Mexico City (as provided in chapter 4) was the 2010 *Censo de Población y Vivienda* by the National Institute of Statistics and Geography (INEGI[5]) in combination with a large array of secondary sources – such as governmental reports, a press review[6] and existing literature. More importantly yet, the principal source of choice for an exploration of everyday practices of water use and their sociospatial setting in the present study was empirical fieldwork based on a range of qualitative instruments. The most important methods employed will be introduced in what follows.

3.2.1 Individual Interviews

In accordance with its subject-based nature, individual interviews based on a semi-structured interview guideline were the main source of empirical material for the present study. They combined elements of the half-standardized interview, which aims to reconstruct subjective theories (see Flick 2012: 203) and the focused

3 http://www.infomexdf.org.mx is an online portal channeling all requests for information from public institutions and governmental entities, in accordance with the *Ley de Transparencia y Acceso a la Información Pública del Distrito Federal*. After ten months of inquiry by the author through this portal, a final request for water provision data on the borough level was turned down by SACM upon the justification of there being no macro-metering in the distribution network. The water utility itself declared to have no data whatsoever (!) on water provision and distribution within the Federal District. This claim was contradicted by a visit by the author to one of the redistribution pumping stations in the Federal District, where new in- and outflow meters were installed and reportedly working properly during a field visit in early 2014.

4 The only information of avail obtained from SACM via *InfoMexDF* was a list of neighborhoods subject to water rationing in 2013, and an overview on metered water consumption at domestic taps in 2010. Despite being broken down to the borough and neighborhood level, this data was of little use for a direct correlation with socioeconomic data and/or a GIS-based analysis of sociospatial patterns of domestic water provision and consumption levels in the Federal District, as it excludes unmetered connections (about 35 % of all domestic clients, see SACM 16.10.2013: 67) as well as mixed domestic-commercial contracts and all those households granted a fixed water rate due to rationing.

5 *Instituto Nacional de Estadística y Geografía*.

6 Covering the 2011–2014 period, the online archives of *La Jornada and El Universal* newspapers from Mexico City were consulted for articles on water supply and related topics, as were Mexican newspapers *Reforma* and *Excélsior*, magazines *El Proceso* and *The Economist*, and the homepages of *Animal Político, The Guardian* and *BBC News*.

interview (see Merton/Kendall 1946), with the latter allowing for a greater openness towards the interviewee's own narration. Following Benjamin's concept of experience (see Benjamin 2003: 315 ff.), interview partners were understood as active narrators sharing and interpreting their own experience, rather than passive data providers. People were thus considered experts of the social processes they are involved in (see Gläser/Laudel 2009: 11).

Table 3.1: Structure of interview sample; with steadiness of water supply according to interviewees' own perception. Own elaboration

Steadiness of water supply in neighborhood	Popular class interviewees	Middle class interviewees	Total number of interviews
Permanent	14	21	35
Intermittent	11	7	18
Total number of interviews	***25***	***28***	***53***

In total, 53 individual interviews were conducted by the author in Mexico City between 2012 and 2014 (see Tab. 3.1), covering 12 different neighborhoods in the boroughs of Iztapalapa and Cuauhtémoc, Federal District (see Tab. 4.7 for details on the selected neighborhoods). Ten of these interviews formed part of a pre-test carried out in September 2012. During the main fieldwork period in 2013/2014, eight interviewees from the pre-test were revisited with an updated interview guideline to allow for a full comparability of these interviews in the overall analysis.

As for the selection of interview partners, roughly half were recruited from the borough of Iztapalapa, the other half from Cuauhtémoc. Potential interviewees were contacted via key informants, forming clusters of three to five interviewees per neighborhood. Sampling was based on an a-priori-determination (see Flick 2012: 155 ff.) along three dimensions: social status, housing type and steadiness of water supply. The pre-selection of potential interviewees referred to their social status as roughly pre-defined through their occupation, type of neighborhood and material characteristics of the dwelling. Water supply conditions and housing types – considered the main material-spatial framing conditions for water use on the micro scale – were also taken into account. While only households connected to the public water network were included in the sample, it contains cases with permanent as well as intermittent water supply, and embraces two different housing types (flats and single-family houses).

Though interviewees were not specifically recruited with reference to their gender, the sample also seems to reflect how water-related questions are still mainly considered a female matter, forming part of domestic (care) work. Of 53 interview partners, 39 were female, and another three interviews were conducted with female-male couples. Interviewees were between 30 and 76 years old.

A crucial part of the explorative, sociospatially informed research strategy was to conduct individual interviews by the researcher herself and directly at the interviewee's home. Through some additional effort, it was possible to do so in a vast majority of cases[7]. Conducting the interviews at other places throughout the city might indeed have proven more convenient for the interviewee and the interviewer alike, both in terms of dislocation and scheduling of the interview date. Yet setting up as many interviews as possible at the interviewees' homes was what allowed to take field observations seriously in the sense of a 'reading' of space as an essential part of an ethnographic approach to data collection (see for instance Wildner 2003). This deliberate strategy allowed for a direct, participatory observation not only of each interviewee's housing conditions and water-related devices, but often also of (some) domestic practices of water use. Moreover, travelling through the city in public transport, walking through the streets and visiting homes provided the author with a direct insight into the studied neighborhoods in terms of accessibility, social interactions, and not least, its materiality (see de Certeau 1988b: 91 ff.). Travelling to and being at the interviewee's home allowed the interviewer to immerse herself in the actual sociospatial setting whilst maintaining a stance wherein an active observation in a deliberately open-minded manner is possible. As Georges Perec instructs us:

> "Oblige yourself to write down what has no meaning, what appears to be the most self-evident, the most general, the most unglamorous." (Perec 2014: 85)

Such openness provides the base for an inductive, ethnographical approach on urban space. Though the researcher's perspective is not to be confused with an actual everyday experience, such an approach may nevertheless allow to catch a glimpse of how this habitat is experienced by its inhabitants. The researcher is situated in a position lingering between observer-as-participant and participant-as-observer (see Adler/Adler 2000), with the very act of interviewing people at home and documenting observations serving as a supplement to people's narrations on experiences during the interview itself. Notes on these observations were jotted down in a postscript after each interview. A typical interview lasted between 40 and 90 minutes, and was documented through note-taking and audio-recording. Interview transcripts were analyzed in the Spanish original with the help of MAXQDA software. Nine of 53 interviews were excluded from analysis due to empirical saturation, and other methodological considerations. The analysis therefore drew exclusively upon 44 individual interviews.

After the completion of each individual interview, a short survey covering the household's socioeconomic situation was conducted, which allowed for a quick retrieval of socio-economic and demographic data, for instance with respect to occupation, education, size and composition of the household, and some characteristics of the dwelling. Both education and occupation can be deemed 'classical'

7 All in all, 41 of 53 interviews were conducted at the interviewee's home. To facilitate the participation of employees in particular, interviews were scheduled according to the interviewee's needs, including evenings and weekends. Nine interviews had to be conducted at the interviewees' workplaces rather than at home due to time constraints. In three other cases, the interview took place either in front of the dwelling or in a courtyard , where the author was able to observe the storage facilities and general material conditions of the building but did not enter the dwelling itself.

items for identifying social status. Given Mexico's stratified health system (see Braig 2004), social security affiliation is also a strong indicator of social status. The possession of household goods, the material conditions of the dwelling and the (over)crowding of rooms[8] were also understood as indicators of social status. In addition, the survey included specific water-related items: the perceived frequency of water supply, the availability of some household goods (such as cisterns, pumps, and washing machines), the storage capacity of cisterns as well as the existence and proper functioning of water meters. All in all, the data obtained through the short survey served to document the interviewees' social status as well as housing and water supply conditions at the time of the interview.

In addition, a protocol of material housing conditions was produced whenever it was possible to conduct interviews directly at the interviewee's home[9]. After each interview, impressions of the dwelling and the interview situation were also documented in a short handwritten protocol. Such a documentation of the habitat's materiality not only served to illustrate a household's social status but was also of avail for the elaboration of thick descriptions of typical practices of water use (see Geertz 1973: 9 ff.). Both the documentation of housing conditions and the socio-economic short survey hence served to set practices in a certain context and thus complete the picture as to the sociospatial setting of everyday practices on the micro scale of the home.

Moreover, a photo documentation of housing conditions and water-related items was done if permission was granted by the interviewee. Rather than an 'objective' method of documentation, photography is understood here as a way of empirical data production by the researcher herself (see Harper 2000: 727, Dirksmeier 2007: 86). Following Judith Butler, any alleged objectivity of photographs is an illusion in so far as the act of taking a picture already represents an interpretative act (see Butler 2005: 823). In this sense, photography is understood as a performative act which itself creates realities rather than 'simply' or objectively documenting them. During fieldwork for the present study, pictures of water-related devices, material housing conditions and public spaces in studied neighborhoods were taken by the author whenever appropriate. These photographs provided a direct input for the thick descriptions of water-using practices, and served as a mnemonic device during the interpretation process: together with postscripts and protocols of material housing conditions, the obtained photographs were used during the subsequent process of data interpretation to revive the author's impressions of dwelling conditions and the like.

8 *Hacinamiento* (overcrowding) is defined as an average of more than 2.5 persons per bedroom by INEGI, the Mexico's National Institute of Statistics and Geography (see: *Índice de la calidad de los espacios en la vivienda,* http://www.inegi.org.mx/est/contenidos/espanol/rutinas/glogen/default.aspx?t=cp&s=est&c=10249, accessed 29.07.2014).

9 As already mentioned, this accounted for a vast majority of cases. Just 12 of all 53 interviews were conducted elsewhere, most at the interviewee's workplace.

3.2.2 Habitat Biographies

One particular item during each interview was an open question designed to generate a narration on people's dwelling experiences and respective changes during their life course. These habitat biographies represent a novel empirical approach developed specifically for the present study, and are based on a visualization of each interviewee's individual dwelling history. A first sketch was already done during each interview. These visualizations were later digitalized, enriched with empirical material from interview transcripts, and systematically analyzed in order to identify specific types of habitat biographies.

For the present study, the development of such a novel, specific method to grasp people's dwelling experiences was motivated by Bourdieu's notion of habitus on the one hand, and empirical findings from the pre-test on the other. First, the theoretical considerations which led to a spatialization of Bourdieu through a merger with theories of production of space (as discussed in chapter 2) provided a motivation to systematically explore and document the interviewees' habitats by introducing a historical and at the same time sociospatial perspective. The methodological challenges faced by every kind of empirical habitus research – given that habitus is conceptualized to operate widely in the realm of the unconscious (see Bourdieu 1977: 18; Fuchs-Heinritz/König 2011: 115) – were another contributing factor which called for the development of a new instrument designed to grasp the sociospatial production context of habitus (rather than habitus itself). As an incorporation of society, habitus is constituted through processes of learning in a certain social and, as the author argues, spatial context. In this sense, the habitat biography method was drafted as a way of studying the becoming of habitus in the very sociospatial setting where it (historically) stems from, and to do so from a subject-based perspective. Each interviewee's social background and his or her past places of residence are understood as places of socialization, where certain rules of water use were incorporated under certain housing and water supply conditions. Yet despite referring to Bourdieu's theory of practice and his notion of habitus in particular, (it has to be stressed that) the present study did not strive to explore '*the* habitus of water use'. In this, it differs from 'conventional' habitus research. Here, the empirical material served to identify 'typical' practices of water use, which were then set in relation to sociospatial questions. Therefore, it was crucial to develop an appropriate method for grasping experiences (related to the domestic sphere as the specific sociospatial setting chosen for the present research) from a subject-centered, historical perspective.

Second, apart from these conceptual considerations related to Bourdieu's praxeological approach, the development of the habitat biography method as a way of exploring past habitats was also inspired by topics which emerged during the pre-test as well as a previous study in Mexico City (see Schwarz 2009). When asked for a comparison of water supply conditions in former times in their lives, some interviewees recalled a stark contrast between their childhood experiences with water supply and later places of residence. Others expressed relief over now-improved conditions, or referred to lessons learnt from past limitations, which they felt had prepared them for later hardship. All these accounts called for the development of a specific method to capture these experiences and their relation to current practices in a structured way.

The habitat biography method was developed specifically for the present study, and is based on narratives created at the end of each individual interview. This particular form of biography illustrates a personal history of housing for each of the dwellings an interviewee has lived in from childhood to the present place of residence. Inspired by earlier methods such as genograms – diagrams depicting household constellations and family relationships, a method widely used in genealogical, medical and psychological studies (see McGoldrick/Gerson 1985, Friedman *et al.* 1988, Butler 2008) – the habitat biography is first and foremost a graphical adaptation of the biographical method. Developed by Hans Bahrdt (1975) and others (see Matthes 1983) with reference to Maurice Halbwachs' notion of collective memory (see Halbwachs 1991), it calls for a re-contextualization of personal narrations as past experiences are thought to influence current perceptions (see also Witzel 1985: 238 f.). The habitat biography method thus resembles a topical life history (see Denzin 1989: 189, Cole/Knowles 2001) focused on dwelling experiences and water-related issues in each interviewee's life course. The method is based on assuming an internal perspective in all accounts. It was hence not the researcher's decision whether water supply conditions were qualified as limited, but reflected the interviewee's own perception and narration. For the same reason, supply frequencies at the interviewees' domestic water taps in past and present dwellings were recorded as reported by the interviewee him- or herself rather than as obtained from any official water supply data[10]. This strictly subject-based approach lies at the core of the life history method as Denzin defines it: all practices are studied from the subject's own perspective, which has priority over any seemingly 'objective' definition of the situation:

> "A central assumption of the life history is that human conduct is to be studied and understood from the perspective of the persons involved. (…) it should be noted that because life history presents a person's experiences as he or she defines them, the objectivity of the person's interpretations provides central data for the final report. The investigator must first determine the subject's "own story". In fact, the subject's *definition* of the situation takes precedence over the objective situation because, as Thomas and Thomas (1928: 571–72) have argued: (…) *If men define situations as real, they are real in their consequences*." (Denzin 1989: 183f.)

Upon this basis, the habitat biography method covered past and current housing conditions of the interviewee. Particular attention was paid to the location of the dwelling, the duration of stay, the housing type (ownership and type of dwelling), the type and steadiness of water supply, and the availability of storage devices. During the interview, a first draft of the habitat biography was sketched by hand and later digitalized using specially designed pictograms. Quotes from the interviews were then added to each step in the habitat biography, with an emphasis on comparisons and remarks on water use and water supply in former dwellings. Visualization forms an integral part of the habitat biography approach as it facilitates a systematic

10 Such as lists of neighborhoods subject to water rationing (see SACM 2013) and the like.

comparison of different biographies in search for collective experiences and biographical similarities (see also Bohnsack 2003: 114). It allows for a periodization of dwelling experiences according to processes located on the micro level (biographical experiences, e.g. formation of the first own household) as well as on the meso and macro scale (e.g. urban development), and an easy identification of different types of biographies – for instance those who never suffered limited water supply in any of their dwellings. Such differences were of particular relevance with respect to the research question, as types of habitat biographies formed during the interpretation process were afterwards analyzed in relation to current practices of water use (see 3.2.4).

3.2.3 Focus Group Discussions

In contrast to individual interviews, focus group discussions directly aim to reveal collectively shared meanings and rules (see Kitzinger 1994). The method is characterized by a form of "non-directive interviewing" (Hennink 2007: 5) – essentially, it is about the interaction between participants themselves (see Myers 1998: 106 f.). It is the relative autonomy ceded to the participants regarding the topics discussed which allows for a reproduction of a collective narrative. The value of the method thus lies in the visibility it gives to participants' own experiences and views, as they "negotiate their own agenda" (Merryweather 2010). Through a dynamic conversation dominated by the group rather than the researcher, the method is able to grasp both collectively shared practices and their meanings. For this reason, focus group discussions are one of the methods central to current qualitative empirical research on habitus and class formation (see for instance Reay 2005, Wills *et al.* 2011). In the present work, focus group discussions were used to gain insight into collective narratives regarding implicit logics of water use, in particular the imagined landscapes of water supply introduced in chapter 5.4. To ensure comparability across several focus groups, a rather open topic guide with five thematic dimensions was employed, designed to generate a detailed narration. Additional material of a potentially controversial nature was employed to foster group interaction (see Kitzinger 1994: 106 f.). Along with several water-related statements, this included two pictures chosen to provoke a debate. The method of photo elicitation – a deliberate use of photographs in the interview process – was developed in the context of visual anthropology (see Collier/Collier 1986), and is praised for its ability to animate the narrative flow and unearth emotions and memories (for an excellent overview on the method and its applications, see Harper 2002).

Four focus group discussions with a total of 46 participants were conducted in Mexico City by the author in January 2014. Groups had between 10 and 14 participants, which were recruited to form a relatively homogeneous group. To facilitate the discussion of shared experiences, focus groups are typically constructed homogeneously along categories such as gender, age, social status or ethnicity (see Merryweather 2010). While it seems that a strongly heterogeneous group composition tends to inhibit group interaction due to a lack of trust, leading to superficial statements or participants' failure to contribute (see Myers 1998: 89), homogeneous focus groups are considered to be more apt to represent (or reproduce) collectively shared discourses (see Bohnsack 2007: 373). For the present study, the recruitment of participants for the focus group discussions was based on social status, and, as far as possible, place of residence, type of water supply and housing type. Given this *a-priori* determination of the sample, a certain degree of shared experiences, for example regarding water supply conditions, could be assumed. Access to the field was gained through facilitators – established members of the local community who supported the recruitment process. Together with an appropriate location for the discussion, such an approach seems crucial for the method's success as it boosts people's commitment to participate in and actively contribute to the group discussion (see Kitzinger/Barbour 1999).

3.2.4 Methods of Interpretation

The interpretation of the empirical material for the present study mainly drew upon qualitative content analysis (see Mayring 2010) and the construction of empirically grounded types (see Kluge 2000). Two interconnected reading perspectives provided the leitmotif during the reading and interpreting of the interview material. The first follows the question how practices of water use are related to the sociospatial setting at the time of the interview. The second refers to the relation between practices and habitat biographies, i.e. past experiences in the domestic realm. (Again, it should be stressed that the micro scale is understood here not as an exclusive but rather as one specific sociospatial setting amongst others).

First, a preliminary system of categories was derived from the interview guideline, adding subcategories stemming from the pre-test material. This system of categories served as a starting point for an analysis of the interview material. Apart from the preliminary categories, all codes were developed inductively, that is, derived directly from the interview material (see Mayring 2010: 75). In-vivo codes were employed to strengthen the actor-based perspective (see Kelle 1995: 370). Using MAXQDA 10 software, a preliminary code system was deployed for coding all ten pre-test interviews. Advancing through all 44 interview transcripts, the code system was adapted continuously, with codes constantly being linked back to the research question through both above mentioned reading perspectives: current sociospatial settings and habitat biographies.

In a next step, the resulting code system provided the base for the identification of sets of typical practices of water use. Thematic codes stemming from the analysis of the interview material were used as relevant dimensions of analysis, starting with the main categories Water Supply, Drinking Water, Hygiene, Storage, and Habitat biographies. In this context, reconstructing the inherent logic of water supply which forms part of people's imaginarios (see 5.4) required a special assembling act. These imagined landscapes were identified from the interview material related to questions on intermittence and reasons for differences in supply conditions throughout the city. This material was analyzed for patterns, identifying the interviewees' imagined landscapes of supply. Later on, these landscapes were enriched with collective narratives evolving from the focus group discussions.

Next, the habitat biographies generated as part of the interview process were analyzed with the aim of forming empirically saturated types. Such types result from a combination of empirical analysis and theoretical knowledge (see Kluge 2000). Past experiences with water supply conditions, and in particular limitations in water supply, provided the relevant analytical dimension in the present case. To increase comparability, a rough periodization was also introduced. A timeline marks events on the micro level (biographical perspective), and the 1985 earthquake as a city-wide event marking a rupture in urban development as well as (often) in people's biographies. Parallels in habitat biographies stemming from similar experiences with water supply conditions were then identified to allow for the creation of several empirically saturated habitat biography types with respect to water supply limitations (see 6).

Finally, the identified sets of practices of water use were brought in relation to the research question and theoretical framework. In order to produce an empirically grounded typology of water use with respect to past and present sociospatial conditions on the micro scale, and hence make sense of the imagined landscapes of water supply as well as the other sets of water-using practices, they were analyzed for specific patterns and empirical regularities according to current sociospatial conditions on the one hand, and habitat biography types on the other. The findings are presented through thick descriptions of four typical practices of water use (set in relation to sociospatial conditions at the time of the interviews) in chapter 5, and by a crossing of selected practices of water use with two specific types of habitat biographies in chapter 7.

Before turning to the empirical findings from chapter 5 onwards, the following chapter provides an overview on today's water supply conditions in Mexico City from a city-wide, infrastructural perspective, thus introducing the general framing conditions for domestic practices of water use.

4. THE URBAN LANDSCAPE OF WATER SUPPLY AND WATER CONSUMPTION IN MEXICO CITY

The modern infrastructural ideal of a centralized, universal water supply has never been brought to realization in most urban areas around the world where some part of the population is usually left unserviced or underserviced in one way or the other (see Gandy 2008). Mexico City – or more specifically the Federal District with its roughly 8.8 million inhabitants, where around 97% of the urban population have access to piped water in the dwelling or on the plot (see CEPAL 2011) – seems relatively well-serviced in contrast. Nevertheless, the water supply situation is disparate, with the Federal District resembling "a mosaic in which the highest [water] provision and quality are located in the West and the city center, while the biggest deficits in quantity and quality are to the East and in the margins where the urban sprawl invades the conservation area" (Consejo de Evaluación del Desarrollo Social del Distrito Federal (CEDS) 2010: 94). In the words of the United Nations' Environmental Programme, water poses "one of the most complex challenges which [Mexico City] faces today, and the biggest obstacle for urban development" (Programa de las Naciones Unidas para el Medio Ambiente (PNUMA) 2003: 100). The relation between urban development and domestic access to water infrastructure is of a more complex nature, however. The boroughs of Iztapalapa and Cuauhtémoc, which were selected as research areas for the present study, reflect this complexity.

The densely populated borough of Iztapalapa, to the South-East of the Federal District, is infamous for its supply interruptions and low water quality. In the second half of the 20th century, the agricultural areas surrounding the former villages of Iztapalapa (many of which date back to pre-Hispanic times) were successively converted either by informal urbanization, industrialization or by the construction of large housing complexes and transport infrastructure. Over the last five decades, peripheral areas of Mexico City in particular were developed formally and informally by and for all kinds of social classes despite the lack of public infrastructures, including water. The ex-post introduction of infrastructural networks is common practice in these neighborhoods. Iztapalapa is one of the poorer boroughs of the Federal District, lacking sufficient technical, educational, cultural, recreational and transport infrastructure for its population of 1.8 million.

The centrally located Cuauhtémoc borough, in contrast, represents the direct opposite, having been urbanized over a period of several hundred years and in some parts dating back to the colonial period. Constituting the political, cultural and (traditional) economic center of this metropolis of over 20 million inhabitants, this area is supposed to have better public services and infrastructural conditions than many other parts of the city. Home to government offices, specialized retail, run-down colonial *vecindades* now partially being upgraded, late 19th century buildings as well as modernist housing complexes such as Tlalteloco, and new apartment blocks and offices being thrown up in the last decade, its population is socially mixed. After an ongoing population decline in the second half of the 20th century, a redensification policy seems to have stabilized the number of its inhabitants since the year 2000. However, it is disputed whether the subsequent construction boom also led to a deterioration of public services, including water supply (see Fuerte Celis 2013).

Which are the main features of the landscape of water supply in Mexico City? Given that water supply frames possible use, this chapter will give a general overview on sources and characteristics of water supply and water distribution in the Federal District (4.1), followed by an introduction to the two research areas Iztapalapa and Cuauhtémoc (4.2).

4.1. SOCIOSPATIAL PATTERNS OF WATER SUPPLY IN MEXICO CITY

What springs to the eye immediately in any report on the topic is the way questions of water and waste water infrastructure are organized in Mexico City. Not only is water imported in considerable volumes from external sources but at the same time, an even larger amount of water is discharged from the valley in the form of untreated sewage. This is nothing less than an entirely human-made "paradoxical hydraulic cycle. Extreme environmental degradation is (...) inimical to this city, whose natural destiny is to be a lake, situated as it is in a closed basin (...) Successive drainage of the Mexico City valley and the consequent need for external water supplies have been achieved through a series of increasingly ambitious engineering feats over five centuries" (Connolly 1999: 61). Through an increasingly complex, highly regulated socio-technological system, Mexico City receives water from two basins in the Pacific divide, which, after mixed together with water from its own aquifer, is used and then emitted as waste water to the Gulf of Mexico divide (instead of many, see Peña Ramírez 2012: 164). Consequently, an ongoing overexploitation of its own aquifer as well as other basins, and the export of polluted water are the main characteristics of what Perlo Cohen calls the *hydropolitan region* of Mexico City (see Perló Cohen/González Reynoso 2005). The colonial draining of Lake Texcoco and the construction of canals to prevent inundations of the city center laid the foundation for today's drainage system. Three large-scale drainage systems[1] are now at work in Mexico City, the first of which was built as early as in the 17th century, and a fourth deep drainage system is currently under construction: the *Túnel Emisor Oriente*. The ongoing expansion of this entire system seems to result from a combination of urban growth, infrastructural path-dependency and soil subsidence as a consequence of the overexploitation of the aquifer beneath the city. Huge pumping stations have been installed to secure the flow of waste water, as soil subsidence has

1 These are the 17th-century *Tajo de Nochistongo*, the *Gran Canal de Desagüe* from the late 19th century, and the *Sistema de Drenaje Profundo,* which was constructed from 1967 onwards (for details, see Connolly 1999, GDF/SACM December 2004, Legorreta 2006).

overturned gravity by neutralizing the decline necessary for a proper functioning of pipelines.
Less than 10% of Mexico City's waste water undergoes treatment, and most of it is discharged through these tunnels and pipelines towards an agricultural region in neighboring Hidalgo state where it is used for irrigation of food crops such as corn and vegetables, much of which are later re-imported to Mexico City (see Legorreta 2006: 56 ff.). In 2008, a total of 27 plants for the treatment of municipal waste water[2] operated in the Federal District, but they were reportedly only running on roughly half of their installed capacity of 6.48 m^3/s (see Peña Ramírez 2012: 62). The treated water is mainly being used for the irrigation of public greenspaces, and a small share is used directly for aquifer recharge. Since 2010, a large new waste water treatment plant with a treatment capacity of 35 m^3/s is being built in Atotonilco, Hidalgo state, by a consortium dominated by Mexican industrial tycoon Carlos Slim[3]. According to CONAGUA, this privately operated mega plant will heavily improve treatment levels, increasing them from 6% to 60% of Mexico City's entire waste water (see CONAGUA 2015).

Mexico City's Ruptured Hydrological Cycle

Mexico City's water problems seem widely socially produced, as the natural conditions in general are not unfavorable. First of all, the local climate is rather humid, with average annual precipitation comparable to that of Hamburg, Germany. Rainfall is distributed quite unevenly throughout the year, however, with most of it occurring in the rainy season during the summer months. Mexico City has a subtropical highland climate, with marked differences between day and night-time temperatures but very moderate seasonal variations in temperature. Average low temperatures are at 11°C and average high at 24°C. The metropolis receives an average 846 mm of rainfall per year, most of which falls in the rainy season between June and September. Mexico City's air is driest between February and April, when air humidity falls below 50% and the lack of rain and increased levels of evaporation impose a need for irrigation of urban green spaces and private gardens alike (all climate data for Tacubaya station, Federal District, see Fig. 4.1).

The prospects for future water availability are hardly a source of hope. In recent climate change scenarios for Mexico City it is estimated that by 2050, water availability will decrease between 10–16% in the Cutzamala system, 12–17% in the Lerma system and 13–17% in the Mexico City aquifer (see Escolero Fuentes *et al.* 2009: 120 ff.). This could to be caused by a rise in average temperatures (and with that, evaporation) rather than a decrease of precipitation. An increase in strong rainfall events and, as a result, lower infiltration and aquifer recharge (as well as more flood events) could also occur (see ibid. 116). In a similar vein, other climate change

2 In contrast to industrial waste water treatment, which lies in the responsibility of the industrial water users themselves.

3 The same company is also involved in the construction of the *Túnel Emisor Oriente,* which will directly feed the Atotonilco waste water treatment plant.

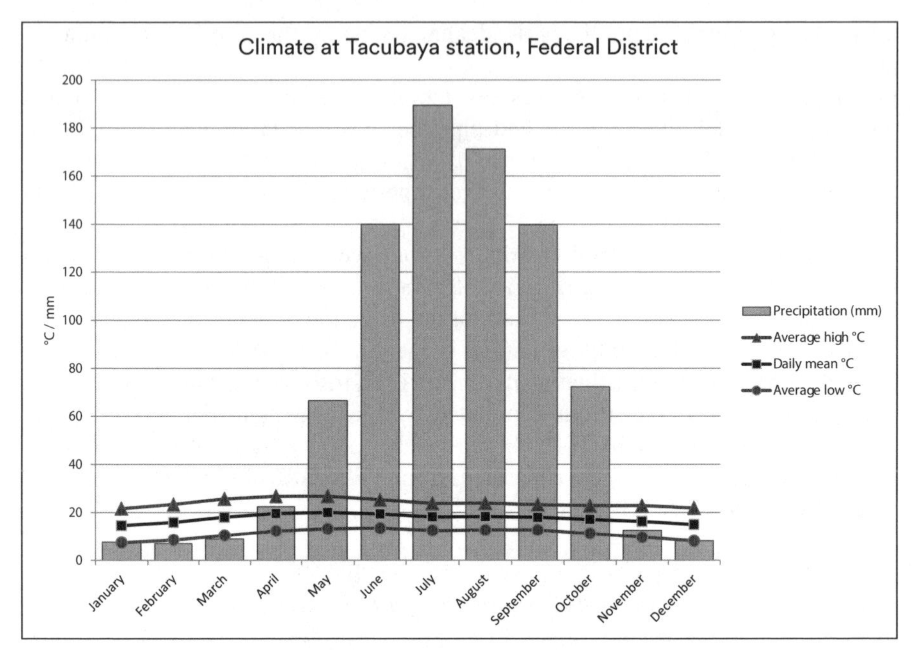

Figure 4.1: Climate: Average precipitation and temperatures in the 1981–2000 period at Tacubaya station, Federal District. Source: Own elaboration based on CONAGUA, http://smn.cna.gob.mx/observatorios/historica/tacubaya.pdf, accessed 10.01.2015

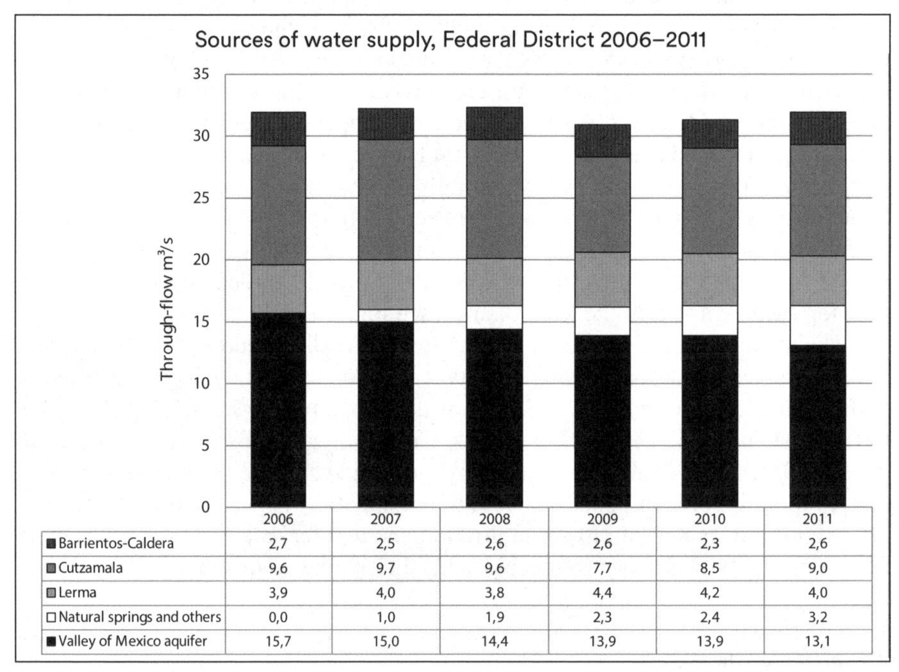

	2006	2007	2008	2009	2010	2011
Barrientos-Caldera	2,7	2,5	2,6	2,6	2,3	2,6
Cutzamala	9,6	9,7	9,6	7,7	8,5	9,0
Lerma	3,9	4,0	3,8	4,4	4,2	4,0
Natural springs and others	0,0	1,0	1,9	2,3	2,4	3,2
Valley of Mexico aquifer	15,7	15,0	14,4	13,9	13,9	13,1

Figure 4.2: Water sources used by SACM water utility to supply the Federal District, 2006 to 2011, in m^3 per second. Own elaboration based on GDF/SACM October 2012: 37

scenarios for 2050 indicate rising average temperatures (between +1 and +2°C) as well as changes in precipitation between +5% and -15%, depending on the scenario (see Arredondo Brun 2007: 6–7).

This would perpetuate the today's situation, as the hydrological cycle of the valley of Mexico is already ruptured and water recharge strongly reduced. According to official figures, only about 20% (or 260.9 million m^3) of all rain water is infiltrated and thus available for recharge of the aquifer, while 21.5% (or 282.4 million m^3) is surface run-off, which in this case goes to its most part directly down the drain and is exported to Hidalgo State without contributing to local rivers or water bodies. Of the 1,313 million m^3 of average annual precipitation in the Federal District, another 57% is consumed by evapotranspiration (753.5 million m^3), and roughly 1.3% evaporates directly (own calculations based on GDF/SACM December 2004: 4). Reduced infiltration is a result of extensive urbanization in most of the Federal District. Once surface run-off enters the drainage system, it is mixed with waste water and pumped out of the valley, thus no more being available for recharge. Mexico City's water shortage is hence fundamentally human-made rather than natural.

4.1.1 The Multiple Sources of the Federal District's Tap Water

Where does water for supplying the Federal District's taps originate? In 2011, 31.9 m^3/s of water were provided to the Federal District, of which roughly two thirds stemmed from sources owned and operated by the public water utility Sistema de Aguas de la Ciudad de México (SACM). With 41%, the biggest share was obtained via deep wells from the Valley of Mexico aquifer, and some 2% from surface waters – the Magdalena river and some of the last remaining natural springs (for detailed volume data, see Fig. 4.2). Despite the city being located in the closed Valle de México basin, a considerable share of its water is nowadays being imported from three adjoining catchment areas (Lerma, Cutzamala and Tula). Two of these major water extraction systems outside the Federal District are operated by SACM: the Lerma river system, providing 13% of the total water through-flow, and the Chiconautla system with 8%. The remaining third of water originated from sources run directly by the federal water commission CONAGUA, mainly the Cutzamala system providing 28% of the total through-flow (see GDF/SACM October 2012: 36 and Fig. 4.2).

Despite impressive volumes of water being imported from catchment areas outside of the *Valle de México* basin, its own aquifer remains the most important source of water for Mexico City. Wells operated mainly by SACM[4] contributed 13.1 m^3/s to the Federal Districts water supply in 2011 (see Fig. 4.2 and GDF/SACM October 2012: 37), though these volumes decreased continuously over the

4 Rights for the operation of commercial wells are granted for the industrial sector, including the beverage industries, whereas the operation of private domestic wells is not permitted. For obvious reasons, there is no detailed information on the amount of wells operated clandestinely.

last years. The number of wells operated today by SACM in the Federal District[5] varies according to source: between 542 wells in 2013 (see Instituto Nacional de Estadística y Geografía (INEGI) 10.04.2013: 2) and as much as 976 wells, without further specification of year (see GDF/SACM October 2012: 56). In addition, experts estimate that over 2,200 unlicensed wells operate in the Mexico City valley, 900 of which are located in the Federal District (see Peña Ramírez 2012: 172).

Concerns with respect to an overexploitation of Mexico City's aquifers have been expressed by experts for quite a while. According to estimations, water extraction more than doubles recharge in the four aquifers of the Mexico City basin, putting them under strong pressure (see Peña Ramírez 2012: 66 f. and 171). In the Federal District alone, annual overexploitation of the local aquifer amounted to 111 million m^3 – the annual groundwater recharge of 513 million m^3 is outgrown by an extraction of 624 million m^3 (see SACM 2012: 120). Mexico City's water stress is hence mainly the result of an ongoing logic of ever-increasing water extraction, which started as early as the mid-19th century. In order to keep pace with the rapid industrialization and urbanization in the Valley of Mexico, a "Plan for Immediate Action" recurring to water extraction from newly explored deep wells was announced in 1974 and was expanded subsequently. Today, four boroughs of the Federal District, amongst them Iztapalapa, receive part of their water from deep wells according to this plan (see CONAGUA 2009a: 121).

Notwithstanding its effects, the logic of expanding the water offer through technological fixes continues to be hegemonic. Apart from distracting from a debate over just distribution and access to water, this logic (which is materialized in long-lasting, large-scale water and drainage infrastructures) also has undeniable and highly visible material impacts upon the urban texture of Mexico City. Not only is the underground perforated and permeated by pipes, conduits and sewers as in so many other cities, it is also sinking. Various experts state that as a result of overexploiting the Valley of Mexico aquifer, the ground level in several parts of the city are continuously subsiding (see Breña Puyol 2007: 87, Legorreta 2006: 110 ff.). Water was extracted from a growing number of artesian wells from the mid-19th century on. Their number grew together with the urban fabric, as new urbanizations where equipped with own wells, so that in 1950, around 700 deep wells operated in Mexico City. In consequence, ground levels in particular in the inner city, located on the clayey soil which forms part of the former bed of now-drained Lake Texcoco, subsided as much as 10 meters during the 20th century alone (see Peña Ramírez 2012: 172; Legorreta 2006: 110 ff.).

Since a draining of several local lakes was begun in the 17th century, a similar logic of grand technical engineering has been pursued through large infrastructural projects such as the deep drainage (1967) and the Tunél Emisor Oriente, under construction. Paradoxically, these projects are a reaction to soil subsidence, while they themselves further aggravate the lack of aquifer recharge and thus ultimately proliferate further subsidence. This both heightens flood risk and, in combination with frequent seismic activity, aggravates the risk of damage to urban infrastructure

5 Wells under industrial and commercial use are not included in this calculation.

networks such as public transport, water and drainage (see National Research Council *et al.* 1995: vii; GDF/SACM December 2004: 80). The effects of ground subsidence are now clearly visible in many parts of the metropolis, where residential buildings as well as heavy-built monuments (such as the Metropolitan Cathedral in Zócalo square) are sinking and suffering structural damage. When it comes to water supply, subsidence-induced material damage to the primary and secondary distribution network is likely to result in elevated water loss through leakage, as well as possible groundwater contamination through leaking drainage and gasoline pipelines (see Legorreta 2006: 123). Ironically, the very overexploitation of the Valley of Mexico City aquifer provided the main motivation for an import of water from other aquifers (see ibid: 114), mainly from Lerma and Cutzamala.

As the first external source of water for Mexico City, the Lerma system was established in 1951. It feeds on subterranean water from the river Lerma aquifer near Toluca, a city of two million inhabitants some 60 kilometers west to the Federal District. Due to heavy extraction in the first decades after its installation, the Lerma aquifer was declared as depleted as early as the 1990s, and on at least one other occasion in 2003. Subsequently, the through-flow to be sent to Mexico City was temporarily reduced, yet the Lerma system continued to provide 4 m^3/s to the Federal District in 2011 (see GDF/SACM October 2012: 37).

The Cutzamala system, installed three decades after the Lerma system, was the first water source for the Federal District which is not run by the local authorities but rather the Mexican federal water commission CONAGUA[6]. It provides water from a number of dams located at a distance of more than 100 kilometres from Mexico City in the lower-laying region of Valle Bravo. The Cutzamala system is one of the biggest water supply systems worldwide, supplying around 485 million m^3 of water per year (data for 2008, see CONAGUA 2011: 70). Planned as a much-needed additional source of water for the growing industrial metropolis in the mid-1970s, it started providing water from the Villa Victoria reservoir in 1982. The Valle de Bravo reservoir was added in 1988, and the Vaso de Colorines and Chilesdo reservoirs in 1993 (see GDF/SACM December 2004: 13–14). Covering a complex geography with a large and sophisticated system of seven reservoirs, three mountain-piercing tunnels, six large pumping stations, the Los Berros purification plant and a number of elevated storage tanks and large pipelines, the Cutzamala system is designed to move large volumes of water over a distance of 126 km and an elevation of 1,100 meters. The lowest point is at Presa Colorines (1,628 meters above sea level), and the highest at an oscillation tower at 2,701 meters, near the Villa Victoria reservoir, from where water flows with gravity towards Mexico City, located at 2,300 meters above sea level (see CONAGUA 2011: 70). The energy consumed by operating the Cutzamala system's pumps was 1.37 TWh in 2012 alone (see CONAGUA 2013: 82).

6 The Comisión Nacional del Agua (CAN, later renamed CONAGUA) was founded in 1989 in the realm of neoliberal opening of the Mexican water sector. Prior to this, the Cutzamala system was operated by the *Comisión de Aguas del Valle de México (CAVM)* (see GDF/SACM December 2004: 13).

Today, the Cutzamala system makes up for 28% (i.e. 9 m^3/s) of the Federal District's total water sources, serving 11 of all 16 boroughs of the Federal District (including Cuauhtémoc and Iztapalapa), in addition to 13 municipalities in Mexico State (see CONAGUA 2011: 71). Water from the Cutzamala system arrives to the Valley of Mexico through the Analco-San José tunnel system. From there on, it is distributed through a two-armed mega aqueduct: the northern branch *(Macrocircuito)* leads towards several municipalities of the State of Mexico, and the southern branch, called *Acuaférico,* towards the Federal District. At a length of 33 kilometers and a diameter of 3 to 5 meters, the southern branch is shorter yet at a projected 25 m^3/s has more than double the through-flow capacity of the northern branch (see Legorreta 2006: 87). The first sections of its pipelines, driven to volcanic rock at a depth of over 90 meters, were taken into operation in 1988. Entering the capital state near the new business district of Santa Fe to the West, the southern branch of the aqueduct serves parts of the Federal District, forming an incomplete ring as it stretches along the Ajusco Mountains to the south-eastern borough of Xochimilco. Initially, the two branches of the macro-distribution system were to be joined to form a circular aqueduct, roughly surrounding the former limits of Mexico City's urbanized area. In 1997, the Interamerican Development Bank promised a credit of 288 million USD for the construction of this last section of the Acuaférico (see Nascimento 1997). But this plan was suspended in 1998 by the Federal District's government and the projected last 10 kilometers, leading towards Milpa Alta where they were to join the Macrocircuito, have not been built. Local resistance against a further expansion of the Cutzamala system in the communities where water was to be extracted and sent towards Mexico City – and in consequence CONAGUA's inability to provide an increased water through-flow necessary for feeding such a potentially enlarged aqueduct – are cited as reasons for this suspension (see CONAGUA 2010: 83). Experts have also warned that in case the Acuaférico were completed, it would incite an urbanization boom in the peri-urban agricultural zones and natural reserves to the South of the Federal District. Ironically, this could particularly involve the elevated parts of the Ajusco – an area protected for its importance for rain water infiltration and evapotranspiration (see Legorreta 2006: 87). However, a possible future expansion of the Cutzamala system and its potential benefits for those parts of Mexico City experiencing considerable limitations in water supply have been a recurrent topic in the political discourse and the media (see for example Romero Sánchez/Camacho Servín 17.03.2012: 35, and Cruz 21.12.2012).

Amongst Mexico City's population, the Cutzamala is probably the best known water source, as temporary supply suspensions periodically affect large parts of the city. During these *cortes de agua,* water provision through the Cutzamala system is partly reduced or cut off completely during a period of 24 to 72 hours, with resulting supply limitations displaying a marked spatial differentiation. Via the local media, government officials and the urban water utility justify these supply suspensions by maintenance works in the Cutzamala system. Hopes are that a third supply tunnel, currently under construction for an estimated 403 million USD, will make an end to water supply suspensions due to maintenance works (see Conan 13.03.2013, Robles 14.11.2013).

Nevertheless, maintenance is not the only problem the complex Cutzamala system is facing. Due to hydrological variations, in particular a reduction in annual rainfall, and exploitation at unsustainable levels, the amount of water stored in most of the reservoirs feeding the Cutzamala system has continuously decreased since 2003 (see CONAGUA 2011: 72). Amid a drought in 2009/2010, which led to historically low water levels in several reservoirs feeding the Cutzamala system (see for instance Robles 09.04.2009), water extraction from some of the main reservoirs was temporarily suspended (see CONAGUA 02.09.2009: 1). At that time, water levels were down to 44% of their total installed capacity (ibid.), and it had been as low as 30%, near their limit of operability, in April 2009 (Robles 09.04.2009). In consequence, an emergency program for the Federal District, which aimed at securing "minimal supply during the year" was hastily implemented by SACM, reducing water provision in "a considerable number of neighborhoods (…) through cuts in the provided through-flow between 6 a.m. and 6 p.m. from Monday to Saturday" (CEDS 2010: 174). A similar situation of water shortage recurred in 2013, when CONAGUA and SACM agreed to reduce the Federal District's supply from the Cutzamala system by 10% during the entire year in order to allow storage levels in the three main reservoirs to recover. The Villa Victoria, Valle de Bravo and El Bosque reservoirs reported their lowest levels since the 2009 drought period, running at only half their capacity[7], and the entire system at 60% (see Bolaños 23.01.2013, Macías 03.02.2013). Measures like this are likely to become more common in the future, as climate change scenarios for Mexico City suggest a negative impact on water availability. As existing water sources in and near Mexico City are more and more exploited, extractable volumes are already decreasing. This leads to a situation where water provision in the Federal District is still on an elevated average level (327 liters[8] per capita and day in 2007, see Jiménez Cisneros *et al.* 2011: 54), yet shows a negative trend. Over the last two decades, the average amount of water provided to the Federal District has decreased continuously, from 35 m^3/s in the mid-1990s (see Departamento del Distrito Federal, Comisión de Aguas del Distrito Federal May 1994: 5) to roughly 32 m^3/s in 2011 (see GDF/SACM October 2012: 37).

In the light of this development, governmental reactions are mixed: on one hand, there are proposals to reduce demand, increase the distribution network's efficiency (not least by repairing leaks) and employ new water sources from within the Mexico City basin – in particular the exploration of new wells, an artificial recharge of the aquifer, and the reuse of treated water (see GDF/SACM October 2012: 92). On the other hand, an expansionist logic remains hegemonic, heading for new and more distant water sources for Mexico City. There are plans to develop a

7 These reservoirs were at 44% of their installed capacity on May 21, 2013 (see CONAGUA 22.05.2013), and water levels had risen to no more than 55% of total capacity by August 20, 2013, at the height of the rainy season (see CONAGUA 21.08.2013).

8 This refers to overall water provision, not only domestic but also non-domestic and industrial use, and includes losses.

new water source for the whole metropolitan region, which is supposed to provide as much as 8m^3/s from 2018 on, and is to be developed in a joint effort by the Federal District, neighboring Mexico and Hidalgo states, and CONAGUA (see GDF/SACM October 2012: 106). The Federal District's share of this project shall be fed from sources in the Tecolutla and Amacuzac basins as well as the Temascaltepec River, the latter a controversial extension of the Cutzamala system (see CONAGUA 2012: 56). In fact, all three basins have been discussed as possible future sources for Mexico City since the early 1990s (see Legorreta 2006: 91 ff.). Nevertheless, such an expansionist logic has been contested by a number of social movements, most prominently in areas of water extraction (see Gómez Fuentes 2009 and Campos Cabral/Ávila García 2013).

4.1.2 The Distribution of Tap Water

After obtaining it from the complex set of sources described above, water is processed in purification plants in order to make it apt for domestic and industrial use. A total of 36 purification plants operated in the Federal District in 2008, producing 33.46 m^3 of potable water per second – and their number had grown to 50 in 2013 (see CONAGUA 2009b: Anexo A-128; SACM 16.10.2013: 5). In 2012, SACM announced plans to install more water purification plants in order to provide water of a better and even potable quality in the near future (see GDF/SACM October 2012: 103). Due to the regional scale of its water sources, the biggest water purification plant serving the Federal District is not even located within its jurisdiction but in neighboring Mexico state: the Los Berros plant forming part of the Cutzamala system, with an installed purification capacity of 20,000 liters per second (see CONAGUA 2009b: 102).

To provide a population of 8.8 million, plus the industrial and commercial sector, the Federal District has one of the largest water distribution networks in the world, consisting of 567 kilometers of aqueducts and numerous elevated water tanks, as well as 1,273 kilometers of primary and roughly 12,000 kilometers of secondary network pipelines (see GDF/SACM October 2012: 56). Notably, the water supply networks serving the Federal District and those serving the adjoining urban municipalities in the State of Mexico – which house more than half of Mexico City's population – are not physically interconnected in any way. This appears to be a result of longstanding political tensions between these two entities dating back to at least the 1950s (under the reign of Federal District's mayor Uruchurtu) and hence predating the strongest period of industrialization and rapid urban expansion of Mexico City.

As the Federal District forms part of one of the largest metropolitan regions in the world, it is hardly a surprise that the biggest share of water is assigned to domestic users. Official data is not devoid of some blurs[9], yet the general picture seems clear: In 2011, domestic users accounted for 44% of water consumption in the area

9 These numbers are somehow inconsistent when comparing with those reported for the 2001 to 2007 period (based on SACM data), when losses where at 32%, industrial and commercial use at 17% and domestic use at 51% (see Jiménez Cisneros *et al.* 2011: 158), while the same source sees losses reduced to 30% in 2008/2009.

serviced by SACM, whereas 21% were consumed by the industrial and commercial sector[10], and no less than the remaining 35% were accounted for as water losses (see GDF/SACM October 2012: 38). The literature distinguishes between real and apparent losses in water supply systems. Real losses involve overflows and leakage both in the public network and the private realm (up to the user's water meter), while apparent losses are an effect of faulty meters and irregular connections (see Alegre et al. 2007: 127). Comparing several sources, it is clear that information on leakage and other forms of water loss in the Federal District often remains blurry as to the type of water loss, and estimations on the share of water lost during distribution oscillate between 30% and 37% of the Federal District's total through-flow (see Jiménez Cisneros et al. 2011:158, GDF/SACM December 2004:90). It is claimed that repairing and substituting broken mains and pipes, fixing leakage and inhibiting unaccounted-for water connections could increase the total water through-flow in the Federal District by roughly 13 m^3/s without adding any new water sources (see CEDS 2010: 178). However, despite considerable efforts by the Federal District's water utility – which repaired an average of 26,000 burst pipes per year and replaced 1,200 kilometers of the secondary distribution network in the 2003–2011 period, in addition to renewing water mains and valves (see GDF/SACM October 2012: 57) – water losses apparently remained quite stable over the last decade, if the above-mentioned estimates are to be trusted.

Distribution: Spatial Differentiation in Water Provision

There is no publicly available information on the state of the distribution network and leakage on the borough level, but when it comes to the distributed volumes, considerable differences between boroughs become clear. The Federal District's average water provision per capita and day decreased from 362 liters in 1997 to 327 liters in 2007 – down 9.7%. This can hardly be explained by population growth, which was 0.3% in the same period (see Tab. 4.1). As a consequence of an overall reduction in water availability over the last two decades, SACM's water provision also diminished on the borough level. A strong divide between the central and western parts of the metropolis and its eastern areas in terms of water supply has already been diagnosed by various authors (see for instance CEDS 2010: 94). In the 2007 dataset, differences between boroughs are striking, with the central borough of Cuauhtémoc receiving 480 liters per capita in 2007 – twice the volume a resident of Iztapalapa was provided in average. Iztapalapa belongs to a group of boroughs which received less than 250 liters per capita and day in 2007, along with the more peripheral boroughs of Milpa Alta, Tláhuac, Tlalpan and Xochimilco. Receiving more than 400 liters per inhabitant and day, Cuauhtémoc formed part of the other end of the spectrum, together with the widely affluent

10 Whereas the relatively low amount of remaining agricultural activities – such as in the peripheral boroughs of Tláhuac and Milpa Alta – seemed to be fed by local wells rather than the urban water utility's piped water.

boroughs of Benito Juárez and Miguel Hidalgo, and the sociospatially rather heterogeneous ones of Magdalena Contreras and Cuajimalpa[11] (see Tab. 4.1).

Table 4.1: Comparison of water provision per inhabitant, per borough of the Federal District, 1997-2007. Source: Jiménez Cisneros et al. 2011: 54, based on Dirección General de Construcción y Operación Hidráulica: Plan Maestro de Agua Potable del DF, 1997–2010

Borough	Water provision: Liter/capita/day (WP)		Population (P)		1997–2007: Variation in %	
	1997	2007	1997	2007	WP	P
Álvaro Obregón	431	391	682,900	716,992	-9.3	0.5
Azcapotzalco	323	326	457,400	424,998	0.9	-0.7
Benito Juárez	463	455	371,800	362,530	-1.7	-0.2
Coyoacán	317	312	659,400	630,004	-1.6	-0.5
Cuajimalpa	686	525	141,600	181,897	-23.5	2.5
Cuauhtémoc	491	480	543,600	530,035	-2.2	-0.2
Gustavo A. Madero	347	343	1,259,400	1,189,747	-1.1	-0.6
Iztacalco	318	317	419,200	393,516	-0.3	-0.6
Iztapalapa	269	238	1,714,600	1,847,666	-11.5	0.7
Magdalena Contreras	460	414	217,400	233,102	-10.0	0.7
Miguel Hidalgo	491	478	366,600	358,063	-2.7	-0.2
Milpa Alta	343	231	83,400	122,887	-32.7	3.9
Tláhuac	247	177	263,100	359,431	-28.3	3.1
Tlalpan	286	249	563,400	616,716	-12.9	0.9
Venustiano Carranza	329	337	488,200	445,827	2.4	-0.9
Xochimilco	270	214	341,700	416,012	-20.7	2.0
Federal District	**362**	**327**	**8,575,697**	**8,831,430**	**-9.67**	**0.29**

Metering and Domestic Water Consumption

For sure, such water provision data provides clear indications of spatial differences in water distribution – though it does not say much about actual domestic consumption. A consumption of 250 to 280 liters of water per capita and day on average is estimated for the Federal District (see Animal Político 30.03.2014). However, this calculation seems to be based on the entire water provision (rather than consumption alone), divided by the number of inhabitants – with no distinction being made between domestic, non-domestic and industrial use, as well as including physical water losses. The only way to establish how much of the total water provision is assigned to the domestic sector is an approximation based on metered consumption.

11 There is no proper explanation for the strongly elevated levels of water provision in Cuijamalpa, the borough to the West where the Cutzamala system enters the Federal District. Some sources point towards high levels of water loss (see PNUMA 2003: 102).

As an interface between (public) water provision network and consumption at the domestic tap, wateWr metering (and prepayment meters in particular) is discussed in the literature as a strong symbol of water sector privatization and exclusion from emerging 'premium network spaces' (see Graham/Marvin 2001: 298, Narsiah 2011). From a more neoliberal perspective, it is argued that metering data has the potential to serve as a means of lowering consumption levels through economic and psychological incentives, and of raising environmental awareness (see Shirley 2002, Jensen 2008).

As part of a push towards an opening of the Mexican water sector to private investment, concessions for water metering, billing and several other tasks of the Federal District's public water utility were handed over to multinational consortiums since the mid-1990s (see Wilder/Romero Lankao 2006). In consequence, metering strongly increased as water meters have been installed for some 64% of SACM's domestic clients[12] over the past two decades, according to official data (see SACM 16.10.2013: 67). Controversially discussed 'prepayment' meters (see Jaglin 2008) are unknown to Mexico City. An overview on consumed volumes of the past months appears on domestic water bills for comparison, and in 2013, each of SACM's roughly 1.1 million domestic clients with water meters consumed an average 31.3 m^3 of tap water per two-month billing period (i.e. 521 liters per day). Average consumption by non-domestic and mixed users was considerably higher, with 70 m^3 and 99 m^3 respectively (see SACM 16.10.2013: 68). However, these volumes are not to be confused with per-capita consumption, as the number of water clients provides no information on the actual number of persons served by each domestic water connection. In Mexico City, an individual water meter does not necessarily represent a single dwelling or individual household, but may also be shared, serving several families or even an entire building. Older mechanical meters were being successively replaced by digital water meters in some of the households interviewed for the present study in 2013, particularly in the city center. Domestic clients themselves had to cover installation expenses, and some sensed a loss of ownership as data from these devices is inaccessible to the user, requiring a special terminal for read-out.

In the present work, water metering is not taken into focus, as functioning meters were installed in 31 of all 44 interviewed households[13], yet 19 of them payed a fixed water rate and another six were generally and officially exempted from payments due to faulty water provision in the first place (see below). All in all, water metering did not feature prominently in the interview material, and seemed to be of limited relevance to everyday users in Mexico City.

Based on an average daily consumption of 557,472 m^3 of tap water on domestic contracts with water meters in the Federal District in 2010, it is possible to set a

12 The Federal District's water utility issues three types of contracts: to domestic and non-domestic (i.e. industrial and commercial) users, and to so-called "mixed users" – for buildings which combine residential and commercial functions. A dwelling with a grocery store or workshop on the ground floor, as common in many of Mexico City's neighborhoods, would be a typical example.

13 The remaining 9 interviews were excluded from in-depth analysis due to their limited value (see chapter 4.2).

(calculative) per-capita consumption of 63 liters of metered water per day and resident of the Federal District (own calculation based on SACM 2014 and INEGI 2010). However, this can only be a first and very rough approximation, as the population served through contracts for mixed use, through unmetered connections, those granted fixed water rates due to rationing, and all those without any access to piped water are all missing from this picture. Focusing on everyday practices of water use, a quantitative analysis is hardly the focus of the present work, yet further research on sociospatial patterns of water provision, distribution and consumption in Mexico City is needed without doubt – and better access to data on public water provision is one of its necessary preconditions. Reliable, updated information on water distribution in the Federal District, and water provision on the borough level and below, is hard to come by, but it seems as if marked differences in the volumes provided to different parts of the city present a continuing trend. Power relations, political decisions and technological issues are intertwined in this matter, which becomes most obvious in terms of supply frequencies and the effects of temporary supply suspensions throughout the metropolis.

4.1.3 Frequencies of Tap Water Supply

At first glance, domestic access to water seems almost universal in Mexico. Over the past three decades, access to tap water rose steadily, to roughly 89% of all dwellings in the country by 2010. Mexico City as the capital and a focal point of urban and industrial development reached almost universal access to tap water by the mid-1990s. By 2010, 97.5% of all dwellings in the Federal District had a water tap installed either in the house or on-site – that is, outside the dwelling but within the premise (see Fig. 4.3).

Most dwellings in the Federal District had in-house water taps in 2010[14], and around one-tenth of dwellings disposed of water taps on the premise (see Tab. 4.2). Supply through water tankers (1.4% of dwellings), or water from wells and local surface water are the only supply types where no water network is installed. In addition, some 15,000 dwellings (0.6% of all) in the Federal District obtained water exclusively from public hydrants – a supply type which had been very common in the past, in particular in recently urbanized areas. More differences emerge when comparing types of access on the borough level (see 4.2), revealing local characteristics of supply.

Having access to tap water as the primary type of water supply does not necessarily make this a household's one and only source, however. Despite an almost universal presence of water taps in the Federal District's dwellings, understanding the actual nature of access to tap water experienced by its inhabitants requires a

14 That year, 96.6% of dwellings had either in-house or on-plot access, according to this source. Differences in shares result from changes in the definition of "inhabited dwelling" by INEGI in 2010 in contrast to earlier census. While the earlier definition included all types of accomodations used for dwelling purposes (such as mobile homes, shelters and other improvisations, and inhabited constructions not designed for housing), the more recent one excludes them.

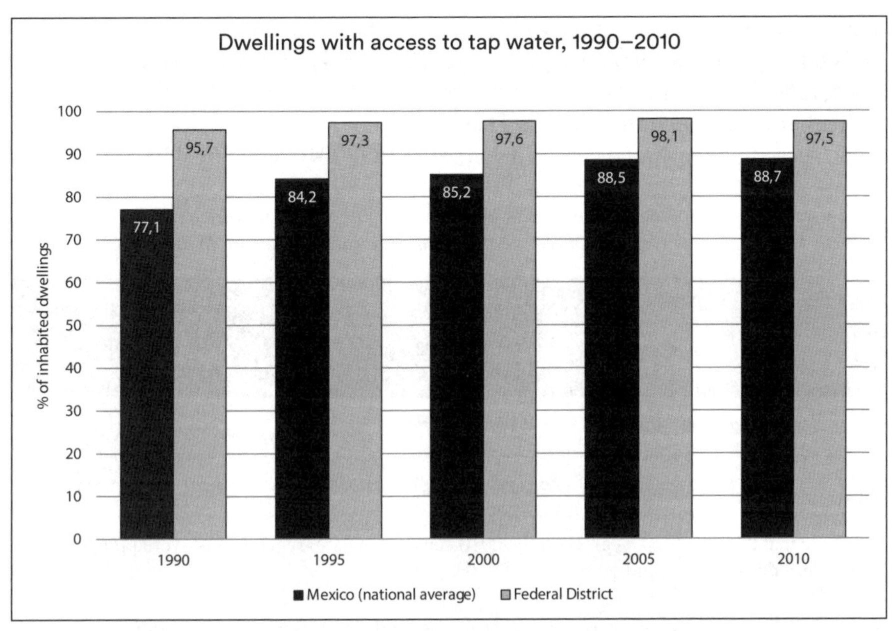

Figure 4.3: Coverage of tap water in inhabited dwellings * *of the Federal District and national average, 1990–2010. Sources: INEGI Censo de Población y Vivienda 1990, 2000, 2010 and Conteos 1995, 2005*

* Includes both in-house and on-plot water taps. An inhabited dwelling is defined by INEGI for this data set as: "Vivienda particular que en el momento del levantamiento censal tiene residentes habituales que forman hogares. Incluye también cualquier recinto, local, refugio, instalación móvil o improvisada que estén habitados".

Table 4.2: Type of water supply in inhabited dwellings in the Federal District, 2010. Own elaboration based on INEGI 2010

Type of access to water	Number of inhabited dwellings served	% of inhabited dwellings served
Federal District (total)	2,386,605	100.0%
Tap water in the dwelling	2,085,363	87.4%
Tap water outside of dwelling but on-site	227,476	9.5%
Public tap or hydrant	14,974	0.6%
Tap water obtained from other dwelling	5,978	0.3%
Water truck (pipa)	33,238	1.4%
Water from well, river, lake, stream or other source	4,078	0.2%
Unspecified	15,498	0.6%

more profound analysis. As we will see, a water tap in the dwelling or on the premise is no guarantee for a permanent and reliable supply, given widespread intermittence, rationing schemes and low water pressure. As much as 18% of all dwellings with an in-house tap in the Federal District did not receive water permanently[15] in 2010, according to official census data (see INEGI 2010 and Tab. 4.3).

Table 4.3.: Frequency of water provision in inhabited dwellings with in-house or on-plot piped water in Mexico and the Federal District, 2010. Own elaboration based on INEGI 2010

	N° of inhabited dwellings with piped water	Frequency of water supply (% of these dwellings)				
		Daily	Every 3rd day	Once or twice a week*	Sporadic	Not specified
Mexico (nationwide)	25,360,800	73.04%	14.77%	8.22%	3.59%	0.38%
Federal District	2,367,139	82.04%	8.2%	6.55%	2.89%	0.31%

* Note that this category, as given in the 2010 census, is incongruent as it is overlapping with the "every third day" category

Spatial differences seem to persist within the Federal District in the average amount of water provided per inhabitant, to begin with. Moreover, spatial disparities in distribution are also expressed in water supply frequencies. When distinguishing supply frequencies, as documented by census on the borough level, it becomes clear that they also vary strongly between different parts of the Federal District[16] (see Fig. 4.4).

Non-permanent water supply is more common in the peripheral boroughs to the South, such as Milpa Alta, Tlalpan, Xochimilco, and Cuijamalpa and La Magdalena Contreras to the West. Marked differences in supply frequencies also exist between the two boroughs which form the focus of the present study. More than one third (or 163,000) of all dwellings in the densely populated borough of Iztapalapa had no permanent supply despite having water taps installed on the plot or in the house, whereas in Cuauhtémoc it was only 3% of all dwellings. Water was provided only once or twice a week to roughly 67,800 dwellings in Iztapalapa (see INEGI 2010; for details see 4.2). When it comes to the steadiness of water supply, both boroughs hence represent the upper and lower end of the spectrum.

15 Supply reductions applied by the water utility as a means of disciplination are another possible reason why even households equipped with a water tap may need to recur to additional types of supply. In the first ten months of 2013 alone, SACM has suspended or partially restricted water supply due to non-payment in over 34,000 cases (see SACM 16.10.2013: 66).

16 However, a more fine-grained analysis would reveal that water supply frequencies in these areas are far less heterogeneous as the borough map might indicate. In the borough of Coyoacán, for instance, non-permanent water supply is mainly an issue in the neighborhoods of Santo Domingo and Santa Ursula – precisely those areas product of popular urbanization are subject to SACM's rationing policy (see SACM 2013).

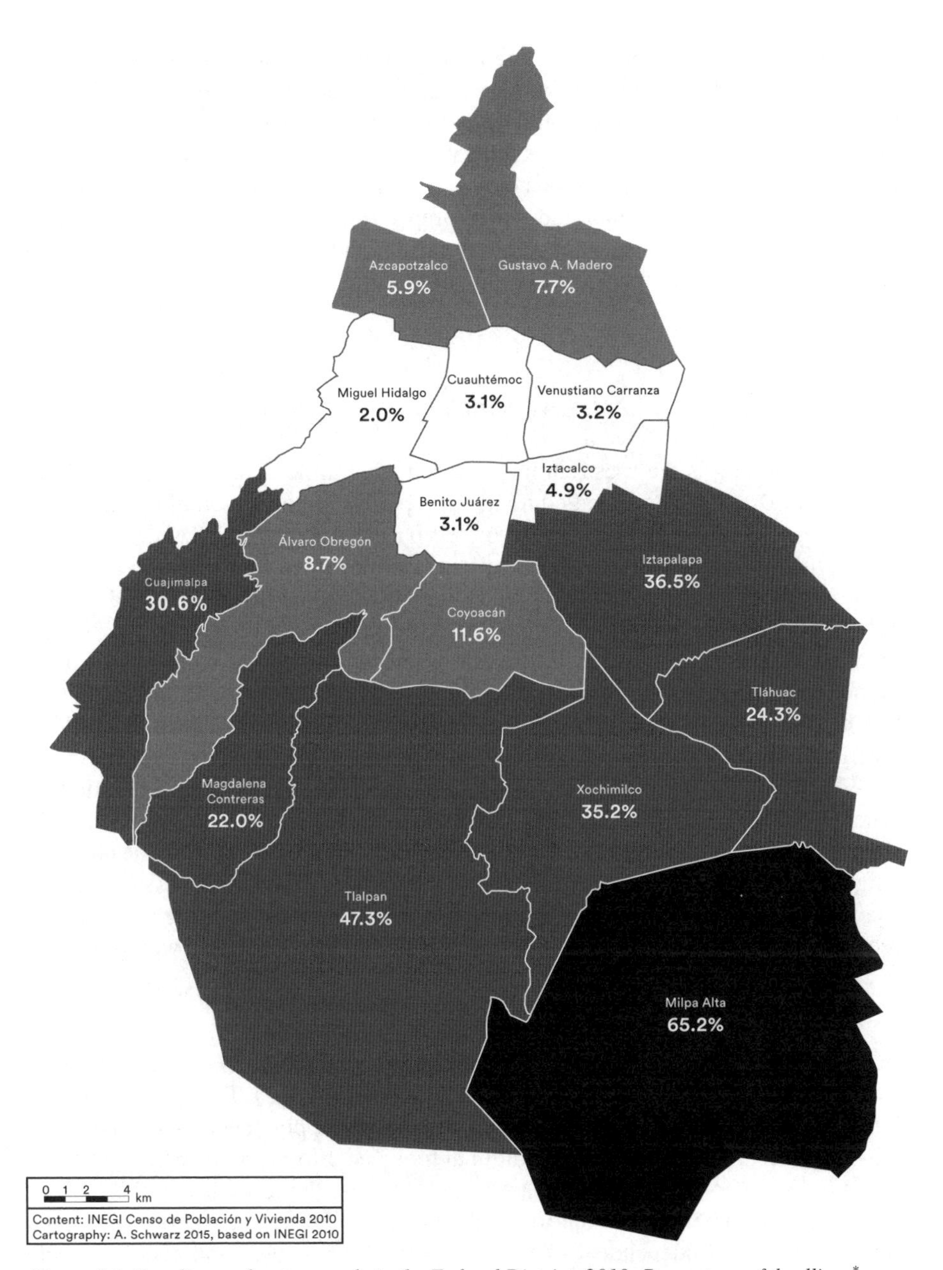

Figure 4.4: Steadiness of water supply in the Federal District, 2010: Percentage of dwellings per borough receiving non-permanent water supply. Own elaboration based on INEGI 2010*

* Inhabited dwellings disposing of piped water, either with water tap in-house or on plot.

Clearly, more than the mere presence of a domestic water tap, the main question continues to be one of distribution: "Today, it is not the presence of the [water supply] network but the flow of this liquid which establishes differences between distinct areas of the metropolis." (Rubalcava/Schteingart 2012: 32). Three types of limitations in water supply frequencies on the neighborhood level can be distinguished in Mexico City: rationed water supply formally scheduled by the water utility *(tandeo)*, unplanned supply interruptions and irregularities (intermittence and low pressure), and periodic supply suspensions *(cortes de agua)* affecting a wider area.

Rationing

A direct rationing is the most common reason for non-permanent water supply in the Federal District. The so-called *tandeo* is a planned interruption of water supply according to a rationing scheme elaborated by the water utility SACM in accordance with the local government year by year. Employees of each of the 16 boroughs are responsible for the technical implementation by manually opening and closing valves in the secondary distribution network (see GDF 01.11.2007: 262) – though this task might be taken over by private companies with the recent renewal of private concessions now also including the management of water distribution (see CEDS 2010: 96). According to official figures, 278 neighborhoods in the Federal District were subject to water rationing in 2013 (see SACM 2013) – of these, 93 were located in the borough of Iztapalapa (and none in Cuauhtémoc). The rationing scheme laid out in the water utility's annual *Programa de Tandeo* sets seemingly rigid standards, assigning specific days and durations of supply to the affected neighborhoods. However, the situation on the ground turns out to be less schematic. Maria from the neighborhood of Miravalle in Iztapalapa, for instance, told me she receives water only once a week – but there is no fixed schedule she would rely on:

> *It is not the same schedule every week, nor the same day – it varies. This week, for instance [water] came on Monday.* (Maria E2: 62, Iztapalapa) [17]

In spite of this, her neighborhood is officially scheduled by SACM to be supplied twice a week, on Tuesdays and Thursdays (see SACM 2013). The impacts of such rationing on the household level and of unreliable supply schemes in particular, creating insecurity and requiring additional time and efforts, are discussed in 5.3.

As a means of easing the impact of rationing on domestic users, the Federal District's government has taken to a policy of financial compensation. It decreed fixed water rates (independent of actual consumption levels), and even complete suspensions of domestic water bills for some (but not all) neighborhoods experiencing limitations in water supply. The latter, *condonación de pago,* is a measure which allows for a complete exoneration from payment in case water supply is irregular

17 All translations of empirical material from Spanish to English are by the author. Interviewees' names were anonymized; quotes indicate the interview number (E2) and paragraph numbers from transcripts.

(see GDF 2012: 274, §14). As a kind of social benefit, it is designed to be granted to poor urban areas with particularly strong limitations in water supply (see Jefatura de Gobierno del Distrito Federal 05.04.2013: 3–6). In most cases, this refers to sporadic supply frequencies (rather than rationing) and/or very low water quality. The condonación is granted directly by the mayor of the Federal District, based on a list provided by the public water utility (see GDF 2013: 17, §44). Through this mechanism, inhabitants of 52 Iztapalapan neighborhoods were exempted from paying their 2008-2013 water bills (see Jefatura de Gobierno del Distrito Federal 05.04.2013: 3–6). The same year, an additional 252 neighborhoods were granted fixed water rates for receiving rationed water supply (see Gobierno del Distrito Federal, Secretaría de Finanzas 15.03.2013). The borough of Cuauhtémoc – where water supply is not subject to official rationing – was not amongst them, but 60 neighborhoods in Iztapalapa. The strong presence of the latter borough in these water policies reflects both its problematic water supply situation and it being a stronghold of the governing PRD party.

In contrast to official rationing schemes, intermittence and low pressure go mostly unaccounted for. Intermittence refers to spontaneous interruptions of water supply which are not (formally) scheduled by the water utility. Be it due to in-house rationing, spontaneous disruptions or low pressure in the public network – even residents of neighborhoods officially listed as receiving permanent water supply may experience supply interruptions.

Temporary Supply Suspensions

As already mentioned, CONAGUA resorted to partial suspensions of its Cutzamala system as an emergency measure during extended periods of drought in 2009 and 2013. This led to a reduction of the throw-flow provided to the Federal District's SACM, resulting in lower supply levels down the line (see for instance CEDS 2010: 174). But another policy has a wider and more noticeable impact on the Federal District: the periodic supply suspensions CONAGUA applies several times per year in order to give the Cutzamala system maintenance. On the occasion of these *cortes de agua,* water inflow from the Cutzamala system towards Mexico City is partially or totally suspended over a period of several days. In 2012 and 2013, there had been eight total and six partial supply suspensions respectively due to maintenance, according to CONAGUA's press releases.

In their annual rhythm, the application of *cortes de agua* seems decoupled from local climatic conditions. CONAGUA usually conducts those maintenance works on the Cutzamala system which lead to supply reductions during long weekends and holidays such as Easter and the Day of the Death. The timing is based on the assumption that water demand decreases as most industrial production is paused and those who can afford it leave the city for vacations. Water inflow from the Cutzamala system towards Mexico City (including those parts of Mexico State served by the Macrocircuito) is usually reduced by 10% on these occasions (see CONAGUA 30.01.2013). On the ground, the result is a temporary but often complete interruption of supply in several parts of Mexico City. Typically, a large number of boroughs of the Federal District are affected on the occasion of these cortes, experiencing either a reduction of through-flow or complete supply disruption lasting up to three or four days. These effects tend to last longer than the officially announced suspension period, as it takes some time to refill the distribution network's mains, pipes and tanks, and restore full service, according to SACM (see Romero Sánchez 07.04.2012). As water supply is reduced more in some boroughs than in others, this leads back to questions of distribution.

The relation between a reduced or suspended inflow from the Cutzamala system, and supply suspensions on the borough level is not as straight-forward as it might seem. Even Venustiano Carranza and Tláhuac – boroughs which do not receive water from the Cutzamala system at all – have been subject to limited supply during *cortes*[18]. As the example of the so-called *megacorte* during the 2013 Easter holidays shows, the effects on the population also strongly depend on the total duration of a supply suspension, and the season. They are more severe to the end of the dry period, that is, the Mexican late winter and spring. From November to March, the Federal District receives a monthly precipitation below 20 mm on average. With no rain, air quality in the city is worst in this period. Levels of particulate matter – stemming from cars, industries, and those parts of the dried-out beds of the former lakes which have not yet been urbanized – are elevated, and the air is drier than usual. Private and public green spaces require irrigation. This coincides with rising daytime temperatures. In late March 2013, these overall conditions combined with an exceptionally long and comprehensive supply suspension during the Easter holidays and resulted in what media termed a megacorte, representing the "greatest water scarcity faced by the inhabitants [of the affected boroughs] during the last ten years" (Royacelli/Cruz 28.03.2013). CONAGUA had officially announced a supply suspension on its Cutzamala system lasting from Thursday March 28th to Saturday March 30th: first, a complete suspension over 55 hours, and then a reduction to 40% of through-flow for another 12 hours. This was motivated by major maintenance works and the installation of an additional pipeline[19], which is supposed to help to avoid these very same supply suspensions due to maintenance works in a near future (see CONAGUA 26.03.2013 and 31.03.2013). According to press reports, water supply in the metropolis was severely affected and could only fully be restored in Iztapalapa, Cuauhtémoc and the other 11 affected boroughs of the Federal District after a week (see El Universal 30.03.2013).

Given that the Cutzamala system is not the only source the Federal District relies on, providing roughly 28% of total through-flow in 2011 (see GDF/SACM October 2012: 37), it remains an open question why parts of the city experience total supply suspensions as an effect of these maintenance works. The controversy over institutional responsibility for supply suspensions is often taken to the public (see for instance Animal Político 02.11.2012, Espino Bucio/Cruz López

18 Only three of all 16 boroughs were reportedly not affected: Gustavo A. Madero, Milpa Alta and Xochimilco (see Espino Bucio/Cruz López 02.04.2013). The latter two run a number of wells in their jurisdiction which serve part of its population directly.

19 The first step of installation of the so-called third line of the Cutzamala system.

02.04.2013, Animal Político 30.03.2014). It is a demonstration of disagreement between federal institutions (represented by CONAGUA) and the Federal District's government (represented by SACM) wherein each side is eager to pass the buck to the other. On two occasions in 2012, for instance, CONAGUA published a press release claiming that "the maintenance of the Cutzamala system does not induce a lack of water supply to the Federal District" (CONAGUA 22.08.2012). It also pointed out that distribution and areas of suspension lay within the responsibility of Mexico City's institutions, not those on the federal level (see Animal Político 30.03.2014). Undoubtedly there has been a lack of coordination and cooperation between the Federal District and its neighboring states Hidalgo and Mexico, specifically in terms of water policies encompassing the entire metropolis (see for instance Eibenschutz Hartman 1997, CEDS 2010: 88, Peña Ramírez 2012). Meanwhile, the government of the Federal District resorts to explaining the disparate landscape of water supply by technical reasons, in particular a differentiated infrastructural set-up and insufficient water availability (see for instance GDF/SACM December 2004: 80). In the domestic sphere, limitations in water supply volumes and frequencies impose a need for private storage strategies and create a dependency on additional types of supply. Such off-grid types of domestic water supply – water tankers and bottled water – are discussed below.

4.1.4 Quality of Tap Water

The quality of tap water is another factor with a strong potential of limiting its usability. Different types of use require different levels of water quality, with drinking-water per se calling for the highest standards. The World Health Organization defines safe drinking water as one that "does not represent any significant risk to health over a lifetime of consumption, including different sensitivities that may occur between life stages. (...) Safe drinking-water is required for all usual domestic purposes, including drinking, food preparation and personal hygiene" (WHO 2011: 1–2). The quality of raw water from which potable water is produced is assessed for its chemical composition and microbiological safety. As for Mexico City, a study from 1997 found that 40% of raw groundwater (prior to chlorination) tested at 40 wells in the city did not meet drinking water standards set by the Mexican government in terms of microbial contamination (see Mazari Hiriart *et al.* 2000: 98). There is a lack of more recent data on microbial aspects, but it is clear that dissolved minerals in groundwatér in some parts of the basin exceed the norms set for drinking water. In some parts of the borough of Iztapalapa, for instance, groundwater showed elevated levels of both dissolved minerals (iron and sodium) and of ammonia nitrogen, the latter linked to the infiltration of untreated waste water (see GDF/SACM October 2012: 46). An overexploitation of the local aquifer and resulting soil subsidence also has a negative impact on water quality, again particularly in eastern parts of Mexico City (see Mazari Hiriart *et al.* 2000: 99) and GDF/SACM December 2004: 80). After extraction, raw groundwater is processed in a number of water purification plants and (depending on the location) sometimes mixed with water from other sources before being distributed to the domestic users. SACM, the Federal District's water utility, assures that most water entering the network is of potable quality – but official data on tap water quality in Mexico City is hard to obtain (see for instance Jiménez *et al.* 2011: 71). Apart from detailed and up-to-date borough data on residual chlorine, there is no comprehensive water analysis information publicly available from SACM. On its homepage, we learn that as of October 2012, levels of residual chlorine[20] and bacteriological analysis were "satisfactory" for 99% and 84% of all tests in the Federal District – that is, within the thresholds defined by the Mexican norm for drinking water quality (NOM-127-SSA1-1994) (see SACM October 2012). The question whether tap water in Mexico City is actually potable is disputed amongst experts. Existing studies struggle to establish straight-forward links between tap water quality and health risks in Mexico City, given both a lack of data on tap water quality and the potential influence of a number of other everyday risk factors[21]. For the Federal District, some 96% of all tap water was defined as potable by the water utility in 2013 (see SACM 16.10.2013: 6), but other sources estimate that around 40% of tap water in Mexico City is not suitable for human consumption (see Oswald Spring 2011a: 499). As of April 2014, 10 out of 16 boroughs in the Federal were provided with "100% potable water" according to SACM. At the same time, a total of 60 neighborhoods throughout the city's other boroughs were provided water of limited quality[22] but still deemed fit for consumption without health risks (see SACM 14.04.2014). Despite all this, the common sense amongst most of Mexico City's inhabitants seems to be that tap water should not be drunk without further pre-treatment (see 5.1).

Independent of the quality of raw water, a number of additional factors further down the line – from the effectiveness of purification over processes of distribution and supply frequencies to domestic water handling – exert a potential influence on the quality of tap water (see for instance Mazari Hiriart *et al.* 2000: 98). For instance, it has been argued that there is an interrelation between water quality and supply frequencies. Intermittent water supply, common in parts of Mexico City, creates underpressure in pipes potentially facilitating an infiltration of pollutants:

> "Leaks in the distribution system are a major cause of concern for both water quality and water supply. When the soil is permeated by sewage from leaking sewers or from other sources [...] leaky pipelines will be infiltrated with contaminated water when pressure is low. According to

20 As for the importance of residual chlorine levels for water quality monitoring: "The standard levels of chlorine (0.2 milligrams/liters) maintained in the distribution system as it reaches the customer's tap are not sufficient to inactivate microorganisms that may have entered the pipelines. The value of maintaining a chlorine residual is to prevent the growth of slime in the system and, more importantly, to be a marker as to whether recontamination may have occurred. Recontamination can use up the chlorine residual. Therefore its absence is a cause for concern" (National Research Council (1995: 48).

21 For instance, it could not be established whether increased rates of parasite-induced intestinal infections such as Giardiasis, Cryptosporidiosis and Amoebiasis in children younger than one year in Mexico City are linked to unsafe drinking water conditions or due to other exposure (see Sánchez Vega *et al.* 2006).

22 Principally due levels of iron and manganese exceeding the limits set by the Mexican drinking water norm.

the Federal District's water quality laboratory, neighborhoods that experience more frequent interruptions in service have poorer quality water compared to neighborhoods with a constant supply." (National Research Council 1995: 46)

While a non-permanent domestic water supply in Mexico City is thought to be a risk factor for diarrheal infections in young children (see Cifuentes *et al.* 2002), it was not studied which role the quality of tap water itself played in this. A lack of hygiene as a result of limited water availability and other domestic water handling, or an infiltration of pollutants in the distribution network are possible causes for a decrease in water quality in areas supplied intermittently. What is more, as a non-permanent water supply imposes a need for domestic storage, domestic tanks become a potential source of microbiological contamination, in particular due to a dissipation of residual chlorine during longer storage (see WHO 2011: 104). In addition, water stored in uncovered tanks also bears a risk of airborne microbial (and other) contamination.

Mexico City's water problem is one of supply not quality, according to a biologist working for SACM quoted recently in the local press (see Montalvo 27.03.2014). However, given a lack of reliable and up-to date water quality data on different spatial scales, it remains difficult from an everyday perspective to understand if and where tap water is potable, and to evaluate in how far and at which point during distribution and storage water quality might be affected. As with supply quantities, authorities again seem to be shifting responsibilities around. In 2012, the Federal CONAGUA declared water quality in the Cutzamala system as good and apt for human consumption, shifting responsibility for a possible contamination to local water utilities and their distribution network (see CONAGUA 16.07.2012). The Federal District's water utility SACM, in turn, passed responsibility for limitations in water quality over to the domestic user – turning domestic storage tanks into a contested ground between the public and private realm.

4.1.5 Off-Grid Supply by Water Tankers

Apart from water supply through the public water network, there are several other modes of water supply in Mexico City. In some peri-urban areas, water transport by donkeys and self-installed low-tech networks connected to local springs and streams are employed (see for instance López/Hernández Lozano 2013) – yet the two predominant types of off-grid water supply are water tankers and bottled water. In contrast to public water supply, these can be defined as externalized and commercialized types of water supply in the sense that they are widely unrelated to SACM's centralized water distribution network and in part run by private companies. According to official figures, about 2% of the Federal District's dwellings relied entirely on decentralized supply, as they did not receive any water from the public water network (see 4.1.3). However, in the light of widespread intermittences, rationing and limited water quality, it is actually a majority of the Federal District's population who resorts to some decentralized type of water supply in addition to tap water. This is more than obvious when it comes to bottled water consumption as a substitute of potable water (see also 5.1).

Not only during the dry period are water tankers a common sight in Mexico City's streets. Typically equipped with a capacity of 10 m^3, the so-called *pipas* have not only become a symbol of insecure supply and rationing, but also of corruption and struggle over water. All in all, 1.4% of the Federal District's dwellings (over 33,000 in total) were supplied exclusively by water tankers in 2010, as they lacked access to piped water (own calculation based on INEGI 2010). But many other residents also resort to water provision by tanker trucks as an additional source, given that water is supplied intermittently in many neighborhoods, and that the through-flow is being suspended periodically during *cortes*. Meant to secure access to water for all[23], and alleviate the effects of rationing and temporal supply suspensions for domestic users, a public tanker service is run by the Federal District, and there is also a growing private sector.

Over 400 public water tankers are operated by the boroughs, all of which are fed at public filling stations. The borough of Iztapalapa alone ran 135 public tankers and between six and ten filling stations *(garzas)* in 2013. Most filling stations have exclusive wells fed by the local aquifer, independent of the drinking water system, but at least one station in Iztapalapa is connected to the Cutzamala system (see SACM 2005: 43, El Universal 02.05.2013, and author's interview with tanker driver, January 2014). Serving both public buildings (such as schools and hospitals) and domestic users, a typical public water tanker runs between four and seven journeys per day on a regular basis – though this number can rise to up to eleven journeys in the hotter and drier months, between mid-February and May. In the borough of Iztapalapa, this adds up to a calculated daily supply of roughly 1,735 m^3 in the main season (see El Universal 02.05.2013) – not even counting those trucks owned and run by private subcontractors. Usually, each journey covers a number of households, as one tanker's load is shared either through filling a collective cistern in one of the Federal District's numerous housing complexes, or smaller, individual storage tanks[24] (see 5.3). The amount of water distributed by truck surges in emergency situations such as the temporal supply suspension during the 2013 Easter holidays: as there was no piped water in large parts of the city for at least three consecutive days, public tankers distributed at least 9,470 m^3 of water in 947 journeys on a single day to the affected population in the Federal District, of which 452 journeys were in the borough of Iztapalapa (see Royacelli 29.03.2013). These numbers highlight the need for additional sources of bulk water in the light of limitations of the network-bound supply. The commercialization of off-grid water is more obvious in Mexico City's growing private water market, yet water distributed by truck tends to come at a cost in any case. Officially, the flotilla of the Federal District's water

23 For instance, the *Ley de Aguas del Distrito Federal* defines that no public water network is to be installed in irregular settlements, but that they be supplied by tankers in order to secure the inhabitants' constitutional right to water (see CEDS 2010: 97–98).

24 In the poorest and most recently urbanized areas, it is common to use barrels lined up along the closest accessible road instead of a storage tank installed in the house. Water is then carried to the house in buckets.

tankers provides water to the affected population free of charge but there are widespread reports of bribery. As Dolores, a resident of one of Iztapalapa's many housing complexes, explained:

> *Supposedly [public water tankers] are for free, but they are not free of charge [...] as a donation is requested, whatever anyone is able to give.* (Dolores E8: 235, Iztapalapa)

The usual bribe for a 10,000 liter-truck is 50 to 100 Mexican peso (2.85–5.70€[25]), and the common euphemism *"para el refresco"* (Regina E12: 286) conceals this practice as a tip for the driver. These informal payments seem to be higher in the city center (where less water shortages occur in the first place). Jorge from Cuauhtémoc said they gave a tip of 300–400 Mexican peso (17–22€) to the truck driver for filling the collective water tank shared in his housing complex of three buildings (see Jorge E17: 79–81). In particular when demand is thriving during daylong suspensions in the dry period, these unofficial charges for a public and supposedly free service are on the rise (see for instance Royacelli/Pantoja 31.03.2013). During periodic supply interruptions, extra payments to the drivers of public water tankers are a way of accelerating delivery. Interviewees in a housing complex in Iztapalapa paid around 300 Mexican peso (around 17€) per delivery to an employee of the borough in order to get priority access to water delivered by a public tanker, and avoid the official waiting list[26]. In other parts of the city, the waiting period for a public truck during a long supply suspension in early 2013 could reportedly be cut from three days to some two hours for a payment of 500 to 700 Mexican peso (31–44€) (see Royacelli/Pantoja 31.03.2013). It is common for interviewees to stress that these payments are voluntary, and that tanker drivers did not ask for them (see Silvia E26, Teresa E49, and Marta E51, all from Iztapalapa). Just how crucial an additional water supply by trucks becomes during emergencies, when households run out of stored reserves is not only highlighted by tanker drivers and public employees taking advantage and requesting unofficial payments. Politicians and local representatives also see the chance to gain influence and votes, and social conflicts evolve as people stand in line for hours to register on the waiting list for public water tankers. In the densely populated borough of Iztapalapa, where water supply is more erratical, and supply suspensions have a particularly strong impact, some even resort to hijacking public water tankers as a way of securing access to water. On two unrelated occasions, this was witnessed by a tanker truck driver employed by the borough of Iztapalapa, and by Marta, a local resident:

> *It was indeed impressive to stand in line and watch the water truck arriving [...]. You had to wait in line [for several hours]. And then you see [water trucks] descending the bridge when people from [a close-by housing complex] hijack the tankers and take them elsewhere. At gunpoint, that is. [...] Several guys, I don't know, maybe five or ten, [...] they climb the truck [...] and threaten the driver, gun in hand.* (Marta, E51: 216–220, Iztapalapa)

25 All conversions from Mexican peso to € are based on exchange rates of the date of inquiry, obtained at http://www.oanda.com.

26 Source anonymized to protect the identity of the informant.

We had a hijacking of water tankers […] during a night shift. […] People climbed our truck in a place where several water trucks were waiting, and they made us drive to a different part of the same housing complex. We better not ask for trouble – we didn't even step out to get the hose, they served themselves. Even our bosses tell us: Don't get involved in conflicts, it's dangerous. Better do what people tell you to. (Author's interview with tanker driver, Iztapalapa, January 2014)

Buying bulk water from private vendors is a way to avoid this competition over the public water truck service and the waiting lists. It comes at an elevated cost – households supplied by private water trucks pay up to 40 times as much as SACM's tariffs for piped water (see Tab. 4.4). A survey conducted by the author in January 2014 showed that private water sellers were charging between 900 and 1,400 Mexican peso (50–78 €) for 10 m³, depending on the area of delivery and type of water[27]. The actual source and quality of water resold by private water vendors often remains unclear – some claim to provide water in potable quality, and to obtain the water from "certified wells", others from natural springs.

Table 4.4: Domestic water supply by public network, public and private truck: Comparison of actual tariffs, 2013/2014. Own elaboration

	Charges for 10 m³ of water	Source
Home delivery by private water vendor to Cuauhtémoc borough	50–78 €	Author's survey, January 2014
Typical bribe or unofficial payment for home delivery by public water truck, to be shared by all households served	2.85 €–8.50 € in Iztapalapa; 17–22 € in Cuauhtémoc	Author's interviews, 2013 and 2014
SACM water tariff for metered domestic user connected to public water network*	1.88 €–8.50 €	GDF 2013: 93–96

* Fixed rate for first 15 m³ of water consumed bimonthly, according to sociospatial IDS class (see 4.2).

Who uses truck water? In the Federal District, both upper and lower class settlements are formally negated access to public infrastructures if they were built in conflict with urban legislation or in conservation areas. As an effect, residents of unserviced areas pay much more per m³ of water – be it through purchasing water from private vendors or bribes for the public tanker service (see CEDS 2010: 99). In the present study, in contrast, supply by tankers was exclusively employed as an additional source, as the fieldwork was conducted in areas where all dwellings disposed of piped water. In the first place, it was a means of securing water availability for domestic purposes in the case of emergencies. In the borough of Iztapalapa, both inhabitants of housing complexes and of individual dwellings with

27 Phone and online price requests were placed to eleven different private water sellers in different parts of Mexico City (boroughs of Álvaro Obregón, Benito Juárez, Venustiano Carranza, Iztapalapa and Tlalpan), requesting a delivery of 10 m³ via a water truck to a domestic ground floor cistern in the San Rafael neighborhood in the borough of Cuauhtémoc.

installed storage tanks usually resorted to public water tankers during longer supply suspensions. In the borough of Cuauhtémoc, it was reportedly almost impossible to obtain water from public tankers during emergencies, with private water vendors representing an exclusive option for those who can afford their services. On a much smaller scale, additional water supply by truck was used on a regular basis in some better-off inner-city households in the face of elevated levels of domestic consumption which coincided with limited supply. In all three cases, they relied on private water vendors, which were organized by the concierge. This way, an intermittent water supply becomes almost imperceptible to residents, as Eva's example (E22) illustrates. In her apartment block in the neighborhood of Cuauhtémoc, close to Paseo de la Reforma, the concierge orders water tankers several times per week – whenever water levels in the building's cistern (with an installed storage capacity of 1.4 m^3 per flat) fall below a certain threshold. Thus, an estimated 77 m^3 of water are supplied by truck every week to serve the demand of the inhabitants of 22 flats. As becomes clear, private vendors here serve a better-off clientele, who understand it as a means of increasing their independence from public supply and its shortcomings while securing elevated levels of domestic water consumption. As was demonstrated elsewhere (see Swyngedouw 2004: 58), maintaining the status quo of a limited public water supply lies in the interest of private water vendors, who often resort to highlighting its unreliability. Bottled water vendors, in contrast, are heavily capitalizing on widespread mistrust in tap water quality.

4.1.6 Bottled Water: Commercializing Mistrust in Tap Water

In Mexico, the commodification of water is most obvious in the widespread consumption of bottled water, which is usually sold in refillable 20-liter plastic jugs. Since the mid-2000s, "Mexico leads the world in bottled water consumption" as a beverage industry report put it (Rodwan 2010: 14), ahead of countries such as Italy and the United Arab Emirates. On average, every Mexican consumed 243 liters of bottled water per year and capita in 2010 – more than twice as much as in the late 1990s (see Fig. 4.5).

The global bottled water market is expanding rapidly[28], in particular in the so-called emerging economies. According to estimations by the beverage industry, Mexico consumed 12% of the bottled water sold globally in 2011 (see Rodwan 2012: 18). The same year, bottled water sales in Mexico had a value of 9 billion USD, and analysts projected them to grow to 13 billion USD by 2015 (see Castano 22.02.2012). In 2013, almost two thirds of Mexico's population drank only this type of water (see INEGI 09.04.2013: 12). This large and still growing market is dominated by transnational beverage companies: Danone with its Bonafont brand

28 This trend is clear, even though more detailed market data are quite hard to come by – the yearly Global Bottled Water Market Report for instance, which is published by the US-based Beverage Marketing Corporation and lists volume data by country, five-year growth projections and industry trends, is available for a 4,995 USD fee (http://www.beveragemarketing.com/reportcatalog.pdf, accessed 03.03.2015).

holds 26% of the market and is in fierce competition with Coca-Cola's Ciel (23%), followed by PepsiCo with its main brands *Electropura* and *e-Pura,* holding a 13% market share. PepsiCo is particularly strong in sales of the 20-liter refillable jugs called garrafones which make up 85% of the Mexican bottled water market, but sales in the more profitable smaller single-serve containers are growing (see Castano 22.02.2012). Branded bottled water now provides such a profit in Mexico that it has even become attractive for product piracy – refilling brand bottles with simple tap water, as the Mexican federal office for consumer protection PROFECO warns (see Huerta Mendoza March 2004: 41). This lead to the introduction of a range of security measures in packaging, such as sophisticated holograms on seals. In addition to global brands, a growing number of small enterprises pushes into the Mexican bottled water market, particular on the micro level. Most of Mexico City's so-called *purificadoras* purify and bottle water in small shops and feed a local market on the neighborhood level. Their no-name product comes at prices lower than those of the large brands of the beverage industry – and it can be assumed that it is not included in the market reports cited above. Actual bottled water consumption levels are thus likely to be significantly higher than what is indicated in figure 4.5. In the Federal District, lower-income households seem to be the main clients of these enterprises, which are a common sight in poorer neighborhoods of Iztapalapa yet much less so in inner-city areas (see 5.1.3).

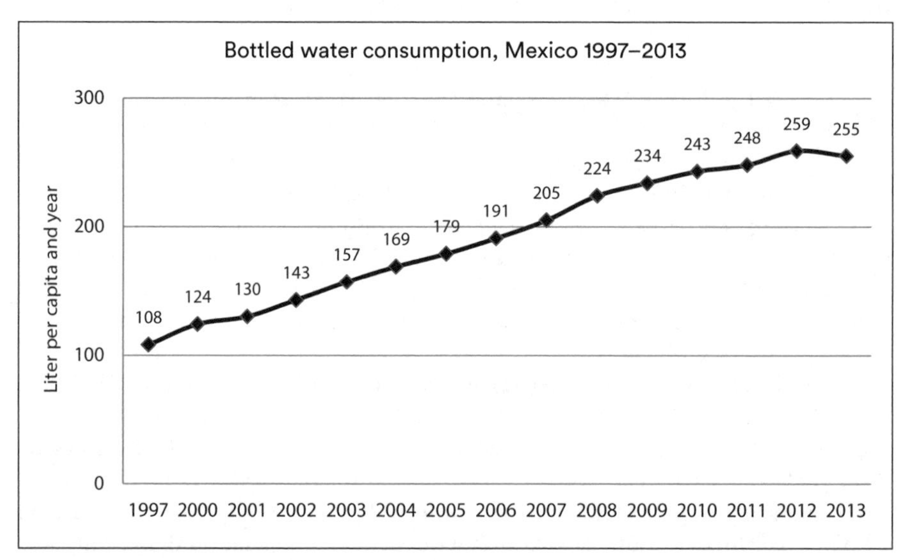

Figure 4.5: Bottled water consumption in Mexico per capita and year (no data available for 1998–1999). Own elaboration based on International Bottled Water Association 2003, 2006, 2007; Rodwan 2009, 2010, 2011, 2012, 2013, 2014.

Bottled water is particularly ubiquitous in the urban context, with Mexico City's per-capita consumption more than twice the nationwide average. A recent study found that in 2011, as much as 87% of the Federal District's households consumed bottled water regularly, amounting to an average of 391 liters per capita and year (see Monroy/Montero 22.01.2014). Similar to the increase all over the country, the consumption of bottled water in Mexico City has been constantly on the rise in the last decades. 77% of the Federal District's households drank bottled water in 2008 (see Jiménez Cisneros *et al*. 2011: 260), and as early as the year 2000, a study in the borough of Xochimilco found that one quarter of households resorted to bottled water for drinking (see Cifuentes *et al*. 2002: A 622). In the borough of Iztapalapa, with its notoriously low tap water quality and widespread rationing, as much as 95% of residents rely exclusively on bottled water for drinking (see Monroy/Montero 22.01.2014). However, the elevated levels of bottled water consumption indicate that it is not only a much welcomed source of potable water per se. In addition to drinking it, people actually use bottled water for a variety of purposes, as we will discuss later.

Rather than an act of conspicuous consumption, the predominance of bottled water in Mexico City reflects what is deemed a "near universal distrust for the quality of the tap water" (Tortajada 2006: 15). What is this mistrust based on? Limitations of tap water quality such as color or smell are perceptible in various Mexico City neighborhoods – but that is not the only motivation for purchasing bottled water. It seems as if widespread notion of tap water being undrinkable is also rooted in collective memory. The devastating 1985 earthquake in Mexico City in particular seems to be a crucial moment when it comes to the question of potability of tap water and the population's trust in it (see chapter 5.1). Facing enormous damage to water mains and the drainage network as well as an unknown number of corpses trapped in the rubble of collapsed buildings, the authorities took to recommending not to drink tap water in the aftermath of the quake. It can be assumed that this experience, together with further warnings against tap water consumption due to cholera outbreaks in the early 1990s (see Peña Ramirez 2012: 80), opened a pathway for bottled water consumption in the long run. It seems as if trust in the potability of tap water didn't come back for most of Mexico City's inhabitants. Tap water was boiled or filtered in most of the interviewed households until the early 2000s, when bottled water vendors and small purification plants selling non-branded water started to appear all over the city.

In Mexico and elsewhere, the bottled water industry heavily capitalizes on public fear over tap water quality and the bad image of public water supply, not least by highlighting an alleged purity of bottled water (see Wilk 2006: 317). Such a sales strategy lies in the very heart of the industry's concept of bottled water as an ideal commodity, much "unlike tap water, which may be exposed to a variety of conduit materials during its fairly constant time of travel to the consumer" (Bryan 2001: 122). Some have gone so far as to call it "the ultimate artefact of social alienation: an individual container of water which, drawn directly from the earth, circumvents the risk of any connection with a wider social or physical environment" (Strang 2004: 217). As Hamlin argues, individualized solutions such as bottled water consumption and home purification highlight above all mistrust in municipal authorities responsible for water supply.

> "The capabilities of domestic water purification devices vary enormously, as does the quality of the product sold by the bottled water industry [...]. Indeed, these responses say less about our need for water we can trust than they do about the institutions we trust." (Hamlin 2000)

It seems as if this trust is now increasingly being projected on transnational beverage companies and their products. Margarita, the owner of a small shop, whom I interviewed in Iztapalapa speaks the mind of those who prefer a renowned brand over bottled water from a local purification plant: *It's a higher quality, therefore it is more expensive* (Margarita E1: 397). This reflects an argument made by the bottled water industry itself, pointing towards higher prices as an alleged indicator of quality: "the perception that bottled water meets or exceeds the standards of tap water (...) is supported by the fact that people commonly pay a higher unit cost for bottled water than for tap water" (Bryan 2001: 122). Bottled water hence is hardly a feasible substitute for a reliable drinking water supply. According to a 2011 study, it cost up to 40 times as much as a liter of tap water, and households in the Federal District spend an average of 229 Mexican peso (around 14€) each month on bottled water (see Monroy/Montero 22.01.2014). The present study found that both bottled water prices and consumption levels showed a spatial differentiation throughout the city (see 5.1.3).

Ironically, the quality of bottled water, which is its major selling point, goes widely uncontrolled in Mexico City. Norms for commercial water purification have been set up by the authorities, and the treatment process in small-scale purification plants usually involves multiple processes such as mechanical filtration, activated carbon filtration and ultraviolet radiation in combination with chlorination, ozonation and water softening. Yet doubts over the proper implementation of these norms have risen, as both experts and the press lately call for an increase in the control of purification plants and warn of the unpredictable quality of bottled water in the city (see Contreras 02.04.2013 and Monroy/Montero 22.01.2014). In the end, the quality of bottled water depends on a whole range of factors in addition to the purification process itself, including proper handling, storage and transport of the final product as well as the kind of supplies used for production (see WHO/FAO 2007: 17–21 for details). Though the provenience of some of the more exclusive brands of mineral water are announced in a way that resembles a good wine, the origins of most bottled water sold in Mexico City remain vague as they are usually not labelled at all. That accounts not only for the local purification plants which have sprung up everywhere in the poorer parts of the metropolis over the last decade, but also for bigger brands. Water tankers registered in Mexico state can be observed delivering water to the small purification plants producing "no-name brands" in Iztapalapa. The only aspect that can be verified by those who buy bottled water from one of the major brands is the place where the bottle was filled, while the origins of this water are mostly kept undeclared. The impressive volumes of water bottled by the three transnational beverage companies operating in Mexico are mostly extracted from wells operated by the industry itself, but their locations and water tariffs paid are seldom subject to public scrutiny. What is sure is that they extract considerable amounts of water from the local aquifer. Market leader Danone, for instance, runs two large bottling plants exclusively for *garrafones* in Mexico City, one in Vallejo and one in Texcoco[29].

To wrap up, it is essentially the power of a brand and a promise of purity on the intersection of nature and technology, as one author put it (see Wilk 2006: 310) that is trusted by Mexico City's residents when it comes to bottled water quality, while actual sources and purification processes are less questioned. Efforts and costs related to bottled water consumption and other sources of potable water used by Mexico City's residents will be discussed in detail in 5.1., together with sociospatial patterns of bottled water consumption.

4.2 CONTEXTUALIZATION: WATER IN IZTAPALAPA AND CUAUHTÉMOC

Why study domestic practices of water use in the Federal District, and more specifically, in the boroughs of Iztapalapa and Cuauhtémoc? As the present study opted for a contrasting comparison, two boroughs in the Federal District were selected for investigation due to deep urban and political differences in the wider metropolis. Mexico City with its 22 million inhabitants (known as *Zona Metropolitana de la Ciudad de México*) has long since expanded beyond the Federal District, incorporating a large number of municipalities in neighboring Mexico state as well as one in Hidalgo state. Due to the particular and outstanding political position of the Federal District[30], urban development often took a different path there than in the neighboring municipalities. This is also true for infrastructural policies, including water supply. To put it in a nutshell, a comparison of domestic water supply and practices of water use in two boroughs of the Federal District rests within the same political and institutional framework as there is only one public water utility responsible for supply. In the municipalities of Mexico State forming part of Mexico City, in contrast, water is provided under different conditions by a number of water utilities, both public and private.

Cuauhtémoc and Iztapalapa, the two selected boroughs of the Federal District, represent a contrast in their status in the urban hierarchy, the overall quality and reliability of water supply and in the discourse evolving around it. They share similarities, though. Unlike some of the more peripheral parts of the metropolis where water infrastructure is lacking and water supply precarious, both selected boroughs have an almost universal coverage of tap water. In the inner-city borough of Cuauhtémoc, 98.4% of all inhabited dwellings dispose of water taps either in the dwelling or on the plot, and some 97.3% in Iztapalapa – both slightly above the

29 With an average production capacity of 2,400 jugs per hour (http://grupodanone.com.mx/conocenos/plantas.aspx, accessed 03.03.2015), the calculated output of each plant would be between 211 and 384 m³ per day – assuming an eight-hour shift where exclusively the 11- to 20-liter jugs these plants are equipped for are filled.

30 The Federal District as the national capital had long been directly dependent on and funded by the federal government, and became independently governed by a directly elected mayor only in the late 1990s.

Federal District's average. Yet, limitations of water supply in terms of quantity and quality are present in both (see Tab. 4.5).

Table 4.5: General water supply situation in the boroughs of Cuauhtémoc and Iztapalapa in comparison with the Federal District as a whole. Own elaboration based on INEGI 2010, SACM 2013

	Cuauhtémoc	Iztapalapa	Federal District
Number of residents in 2010	531,831	1,815,876	8,851,080
Number of inhabited dwellings in 2010	167,781	453,471	2,386,605
Number of inhabited dwellings with tap water in the dwelling or on-site, 2010	161,951 (in-house), 3,227 (on-site)	382,141 (in-house), 63,479 (on-site)	2,312,839
% of inhabited dwellings with tap water in the dwelling or on-site, 2010	98.4% total 96.5% (in-house), 1.9% (on-site)	97.3% total 84.2% (in-house), 13.9% (on-site)	96.9% total 87.4% (in-house), 9.5% (on-site)
Number of neighborhoods with officially rationed water supply, 2013	0	93	278
Number of neighborhoods with dispensed water bills*, 2013	0	52	52
Number of neighborhoods studied (place of residence of interviewees)	6	6	12

* *Condonación de pago* is a policy which allows for the complete dispensation of water bills if water supply is limited.

Perhaps like no other one, the borough of Iztapalapa to the South-East of the Federal District represents the transition from rural to urban periphery during the second half of the 20th century. The former lake bed makes up a large part of the borough, which also encloses the Sierra de Santa Catarina with several volcanoes, reaching elevations up to 2,800 meters above sea level. The land surrounding the 16 *pueblos originarios* (villages dating back to the Aztec rule) had been mainly under agricultural use up until the 1960s, when the area became one of the main destinations of rural-urban migration from poorer states of Mexico, and its industrialization was started shortly afterwards. After a fast and often unplanned popular urbanization, it has undergone a process of consolidation, including the ex-post introduction of infrastructure which is a characteristic of these kind of settlements (see Horbarth Corredor 2003). Today, roughly 90% of the borough is urbanized, with large social housing complexes and consolidated *colonias populares* beside large swaths of land under industrial, commercial and infrastructural use. The remaining areas are inapt for human settlements and under natural preservation. With a population of over 1.8 million, this borough houses almost three and a half times the inhabitants of the borough of Cuauhtémoc (see Tab. 4.5). Iztapalapa is the most densely populated borough of the Federal District, with a mixed but mainly poorer population. The common discourse identifies Iztapalapa as a marginal, socially deprived area of informal origin. Despite its name (*Place where the waters cross* in Nahuatl), which refers to its former location on the southern banks of Texcoco lake and the formerly widespread *chinampa* agriculture, the borough of Iztapalapa is notorious today for its water supply problems. Tap water quality is often poor, domestic water supply is severely rationed in many neighborhoods, bottled water consumption is ubiquitous and water a highly politicized issue.

The borough of Cuauhtémoc, in contrast, represents the mere center of Mexico City, and the cultural and political capital of the country. It was built on the ruins of the pre-Hispanic city of Tenochtitlán, destroyed by the Spanish colonists after the fall of the Aztec capital in 1519. Most of its built-up area is located on the former bed of Lake Texcoco, whose soft clay soil has sunk rapidly since the late 19th century as a result of strong exploitation of the Mexico City aquifer. This ground subsidence damaged not only buildings but also and in particular subterranean infrastructures, including water and waste water. Besides the colonial city center, vast residential areas established around the turn of the 20th century, large modernist housing complexes, and a considerable amount of commercial, hotel and office space, the borough of Cuauhtémoc is home to the Federal as well as the city government. Since 2000, urban policies encourage a redensification of the inner-city neighborhoods, and new residential buildings have sprung up in many areas. At the same time, the colonial city center, which was declared a UNESCO World Heritage site in 1987, became a hub of urban renewal. The borough is diverse and mixed in both material conditions of urban space and socio-economic status of its population. There are no official rationings scheduled by the water utility, and the common discourse identifies no problems with water supply in this part of the metropolis. Nevertheless, there are reports of increasing intermittence as a result of redensification policies in combination with a lack of proper investment in water infrastructure (and other public services) (see Fuerte Celis 2013).

The boroughs of Iztapalapa and Cuauhtémoc are adequate study areas as they serve as paradigmatic examples of the Federal District's pattern of sociospatial differentiation, which still is very much organized along a center-periphery logic. In general terms, Mexico City's better-off tend to live to the West and South-West of the city center, whereas the eastern parts of the metropolis are more marginalized (see for instance Duhau/Giglia 2008: 185; CEDS 2011: 48). The so-called social development index (IDS, for *Índice de Desarrollo Social)*, which forms the base for social policies in the Federal District, provides an insight to the divergent sociospatial conditions in the two boroughs studied here. It is based on a multidimensional conceptualization of poverty based on the satisfaction of basic needs (see Boltvinik 1997, 2010). Ten indicators from the 2005 census are employed to cover a range of topics, from housing quality, access to water, drainage, and electricity over household goods to education and access to health services and social security (see CEDS 2011: 23 ff.). According to this index, a strong social differentiation between central and more peripheral areas of the Federal District can be identified. A vast majority of Iztapalapa's population lives in neighborhoods characterized by a low to very low IDS. In contrast, neighborhoods subject to these least developed conditions seem entirely absent[31] in

31 However, the method is unable to detect poverty on a more fine-grained scale, such as in the overcrowded and often informal *ciudades perdidas* and in individual buildings.

the borough of Cuauhtémoc, where the overall conditions are more heterogeneous and neighborhoods display a medium level of sociospatial development according to the IDS (see Fig. 4.6).

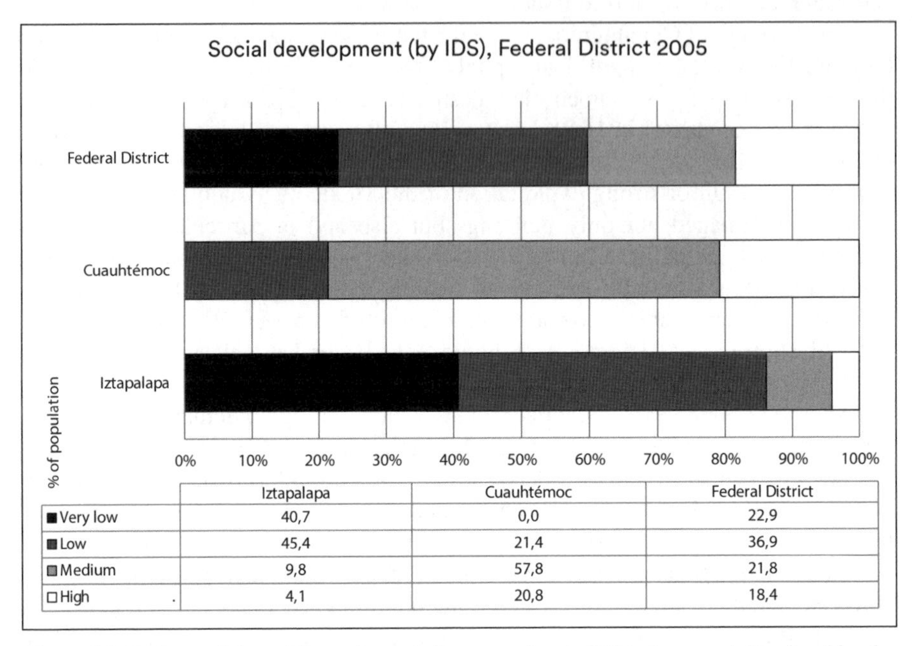

	Iztapalapa	Cuauhtémoc	Federal District
■ Very low	40,7	0,0	22,9
■ Low	45,4	21,4	36,9
■ Medium	9,8	57,8	21,8
□ High	4,1	20,8	18,4

Figure 4.6: Socio-spatial conditions of population according to IDS index on neighborhood level for Iztapalapa and Cuauhtémoc in comparison with the Federal District, 2005. Own elaboration based on CEDS 2011: 50

Since 2010, the Federal District has implemented a new water-tariff policy not only based on an increasing block tariff and cross-subsidies but also aiming for a redistributive effect (see CEDS 2010: 115 ff.). To this end, it is based on a sociospatial classification employing the IDS in combination with indexes on household income and property value (see CEDS 2011: 14 ff.) and Gobierno del Distrito Federal, Secretaría de Finanzas 2010). This extended IDS index[32] is calculated on the block level, and the classification is usually disclosed on domestic water bills. For the present study, this index was employed for the pre-selection of neighborhoods where interviewees were to be recruited. All six studied neighborhoods in Iztapalapa were listed with an average extended IDS being low or very low (popular), as were all the blocks where Iztapalapan interviewees resided[33]. Of the neighborhoods studied in Cuauhtémoc, three had an average extended IDS classified as medium or high, same as all interviewed households from that borough with the exception of those residing in the Tlatelolco housing complex (see Tab. 4.7).

32 Officially called *Índice de Desarrollo e Infraestructura de la Ciudad* (see CEDS 2010: 110).

33 Upon the author's request, one third of all 53 interviewees provided water bills indicating the classification by the extended IDS. All other blocks were identified based on maps published by Gobierno del Distrito Federal, Secretaría de Finanzas (n.d.a).

Water Supply in Iztapalapa and Cuauhtémoc

Considering the urban water supply situation in the Federal District, Iztapalapa and Cuauhtémoc represent opposite ends of the spectrum. The only point in common is a near universal access to water infrastructure in the sense that nearly all inhabitants had a water tap installed either in their house or at least on site (see Tab. 4.6).

Table 4.6: Type of domestic water supply received by inhabitants of Cuautémoc and Iztapalapa in comparison with the Federal District, 2010. Own calculations based on INEGI 2010 as quoted in Jiménez Cisneros et al. 2011: 63

Type of access, 2010	Federal District	Cuauhtémoc	Iztapalapa
Tap water in the dwelling	86.54%	96.76%	84.07%
Tap water outside of dwelling but on-site	10.15%	1.96%	14.16%
Public tap or hydrant	0.70%	0.07%	0.59%
Tap water obtained from other dwelling	0.27%	0.09%	0.12%
Water truck (pipa)	1.59%	0.03%	0.70%
Water from well, river, lake, stream or other source	0.19%	0.05%	0.03%
Unspecified	0.55%	1.04%	0.33%

When it comes to provided volumes, supply frequencies, rationing or bottled water consumption, however, the boroughs are strikingly different. First, the kind of access to tap water is astonishingly homogeneous when comparing the two boroughs and the Federal District as a whole. The only remarkable difference is the nearly universal presence of water taps in all dwellings of the borough of Cuauhtémoc (most likely an effect of denser built-up) while the situation in Iztapalapa is very close to the general situation in the Federal District, with a considerable share of its population disposing of a tap on their plot but not inside the dwelling. Despite roughly comparable shares in all other types of access, this meant that in Iztapalapa, some 12,000 people were provided only by water trucks, and another 10,000 obtained it from public hydrants in 2010, whereas this applied to only a few hundred residents in the borough of Cuauhtémoc.

As discussed in 4.1.2, levels of water provision differ widely between the Federal District's boroughs, while a decrease of provision levels is the overall trend. In this panorama, the borough of Iztapalapa forms part of the less affluent and more peripheral parts of the city which continue to receive the lowest amount of water per inhabitant. In 2007, 238 liters were provided per capita and day – a reduction of 12% since 1997. Notably, Iztapalapa's population grew by less than 1% in the same period, which points to a strong reduction in the amount of water assigned to this borough rather than population growth as a reason for this decrease. Forming the city center, the borough of Cuauhtémoc with its many non-domestic and often privileged users – from tourists in hotels to governmental offices and stores – was provided twice as much: 480 liters per inhabitant in 2007. Apart from high overall volumes, variation in water provision in the 1997–2007 period is also less dramatic in Cuauhtémoc – a reduction of 2% in water provision, while its population remained stable (all data see Jiménez Cisneros *et al.* 2011: 54).

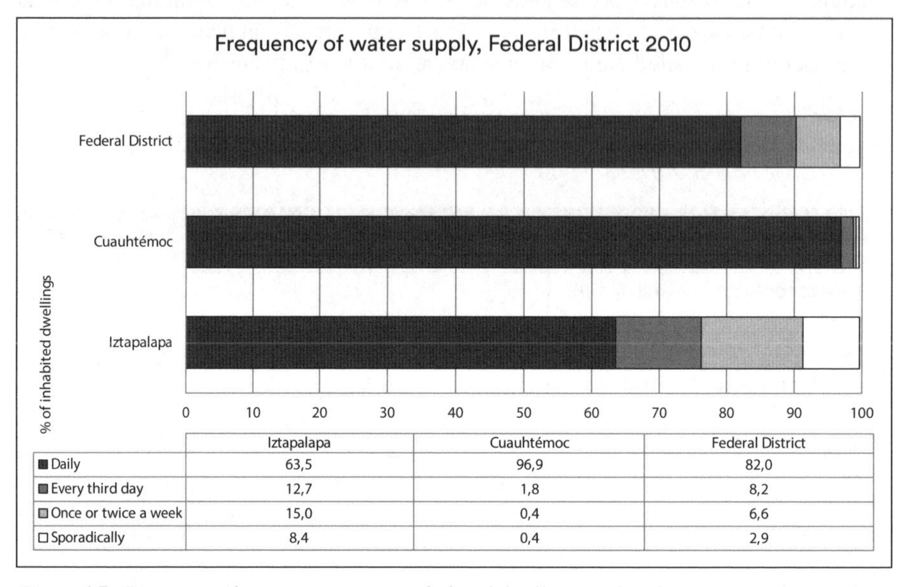

	Iztapalapa	Cuauhtémoc	Federal District
■ Daily	63,5	96,9	82,0
■ Every third day	12,7	1,8	8,2
■ Once or twice a week	15,0	0,4	6,6
□ Sporadically	8,4	0,4	2,9

Figure 4.7: Frequency of water provision in inhabited dwellings with in-house or on-plot piped water in Cuauhtémoc and Iztapalapa in comparison to the Federal District's average, 2010. Missing shares to 100% are listed as ‚unspecified'. Own elaboration based on INEGI 2010

More striking differences between the two boroughs begin to emerge when it comes to supply frequencies. Water supply is more stable in the borough of Cuauhtémoc than in Iztapalapa. At the time of the 2010 census, it was permanent in all but 4,575 dwellings (less than 3%) in Cuauhtémoc, according to official figures. In contrast, more than one third (or 163,000) of all dwellings in Iztapalapa did not dispose of permanent supply despite having water taps installed on the plot or in the house. In many buildings, water is provided only once or twice a week, and the share of dwellings with some kind of non-permanent supply are well above the Federal District's average (see Fig. 4.7). The main reason is that the water utility's policy of rationing is applied in the borough of Iztapalapa (and elsewhere) but not in Cuauhtémoc. In 2013, some 93 neighborhoods were subject to rationing in Iztapalapa, most of them located in the south-eastern fringe of the borough and the Sierra de Santa Catarina (see SACM 2013). The number of Iztapalapan neighborhoods subject to rationing has sharply risen over the last decade, from 36 neighborhoods in 2005 (see SACM 2005: 121).

Water rationing in the Federal District mainly seems to be a result of differential distribution. Limited water availability and poor conditions of the local distribution network are repeatedly cited as reasons for insufficient supply in Iztapalapa (see SACM 2005: 121 ff.). The borough relies entirely on water extracted from around 80 local deep wells operated by SACM (see ibid: 44 ff.). However, water extracted from the local aquifer by deep wells is also redistributed to neighboring boroughs Iztacalco, Venustiano Carranza and Tláhuac (see ibid: 42). Ironically, large parts of Iztapalapa also lack water during the periodic supply suspensions related to the maintenance of the Cutzamala system[34], even though the borough is not (officially) a beneficiary of Cutzamala water. Inner-city Cuauhtémoc, in contrast, feeds mostly on external sources, namely the Lerma system supplying the western and central parts of the borough, the Chiconautla serving its northern parts, and several aqueducts providing water from Chalco, Xochimilco and Xotepingo. There are also six wells operated by SACM, of which one feeds the large modernist housing complex of Tlatelolco (see SACM 2009: 25 ff.). With no rationings scheduled, the entire borough is supposed to receive a permanent supply. But there are reports of low pressure in the water provision network and intermittence in a number of neighborhoods, reflected in the roughly 3% of dwellings found to be supplied non-permanently in the 2010 census (see Fig. 4.7). These limitations are reported in particular from neighborhoods near Avenida Reforma and other areas where the construction of offices and residential buildings has boomed[35] during the last decade. Intermittence is reported mostly from the neighborhoods of San Rafael, Cuauhtémoc and Roma Norte, which form part of the present study (see SACM 2009: 71–72). A combination of an increase in population – and in consequence, a higher water demand – with a lack of investment in the maintenance and renewal of water infrastructures is blamed for the deterioration of water supply conditions in these areas (see CEDS 2010: 98, Fuerte Celis 2013: 175).

As for tap water quality, information is relatively scarce, but tap water tends to be of lower quality in Iztapalapa in comparison with the borough of Cuauhtémoc (see for instance CEDS 2010: 155). According to the latest water quality analysis published by public water utility SACM and dating back to April 2014, all studied neighborhoods in Cuauhtémoc and most in the borough of Iztapalapa were supplied with "100% potable water" complying with the Mexican drinking water norm NOM-127-SSA1-1996. However, two of the Iztapalapan neighborhoods studied here were amongst the 60 neighborhoods of the Federal District where tap water did not comply with this norm (see Tab. 4.7).

The Selected Neighborhoods

By changing scales and concentrating on six neighborhoods in each of the two boroughs, the present study explores the relation between domestic practices of

34 Furthermore, the interview material indicates that the effects of temporary supply suspensions had a longer duration and stronger effect in neighborhoods of Iztapalapa than in the borough of Cuauhtémoc – though the local context down to street and building level also seemed to exert an influence. Even during the 2013 Easter holidays, when CONAGUA suspended the entire inflow from the Cutzamala system, the borough of Cuauhtémoc was expected to be less affected than other, more densely populated boroughs (see Bolaños 28.03.2013).

35 Under what came to be known as the *Bando Dos* policy, the Federal District promoted a redensification of the four central boroughs from the year 2000 on.

Table 4.7: Overview on 12 neighborhoods studied empirically between 2012 and 2014. Sources: Interviews E1–E53, SACM 2009: 71–72, SACM 2013, SACM 14.4.2014, Gobierno del Distrito Federal, Secretaría de Finanzas 15.03.2013: 10, Jefatura de Gobierno del Distrito Federal 05.04.2013: 5, Gobierno del Distrito Federal, Secretaría de Finanzas n.d.a, Gobierno del Distrito Federal, Secretaría de Finanzas n.d.b

Borough	Neighborhood	Type of urbanization	Predominant housing type	Extended ISD index (interviewees blocks)	Type of water tariff applied in 2013*
Iztapalapa	Miravalle	Colonia popular	Single-family houses	Popular	Exempt from payment
	El Manto	Colonia popular and small housing complexes	Single-family houses and flats	Popular; Low	Fixed water rate
	Ermita Zaragoza	Housing complex	Terraced houses	Popular	Fixed water rate
	Fuentes de Zaragoza	Housing complex	Flats	Low	Fixed water rate
	Chinampac de Juárez	Housing complex	Flats	Popular	Regular
	Presidentes de México	Housing complex	Flats	Low	Fixed water rate
Cuauhtémoc	Paulino Navarro	Inner city	Flats	Low	Regular
	Roma Norte	Inner city	Flats	Medium	Regular
	Tabacalera	Inner city	Flats and single-family houses	Medium	Regular
	San Rafael	Inner city	Flats	Medium	Regular
	Cuauhtémoc	Inner city	Flats	High	Regular
	Nonoalco Tlatelolco	Housing complex	Flats	Low	Fixed water rate

Neighborhood	Steadiness of water supply, according to SACM	Steadiness of water supply, interviewees' perception	Water quality according to SACM
Miravalle	Rationed: Supplied Tuesdays and Thursdays for 24 hours	Non-permanent, supply once a week	Water of limited quality** but "apt for consumption without health risks"
El Manto	Rationed: Supplied 5 hours per day	Intermittent, supply several times a week (E13, E29) or some hours per day	100% potable
Ermita Zaragoza	Rationed: Supplied 6 hours per day	Permanent; low pressure (E50)	100% potable
Fuentes de Zaragoza	Rationed: Supplied 6 hours per day	Permanent; low pressure (E51)	Water of limited quality (see above)
Chinampac de Juárez	Permanent	Permanent	100% potable
Presidentes de México	Rationed: Supplied 6 hours per day	Permanent but low pressure	100% potable
Paulino Navarro	Permanent	Permanent; low pressure during day (E18)	100% potable
Roma Norte	Permanent	Permanent	100% potable
Tabacalera	Permanent	Permanent, sometimes low pressure during day	100% potable
San Rafael	Permanent, reports of intermittence***	Permanent but low pressure	100% potable
Cuauhtémoc	Permanent, reports of intermittence	Intermittent, supply 4–5 hours per day	100% potable
Nonoalco Tlatelolco	Permanent	Permanent	100% potable

* Note that in areas where regular water tariffs apply, dwellings without water meters pay fixed water rates according to the IDS status of their block.

** Principally due levels of iron and manganese exceeding the limits set by the Mexican drinking water norm (see SACM 14.04.2014).

*** Intermittence was reported in over 1,200 cases in the San Rafael and Cuauhtémoc neigbourhoods in 2009 (see SACM 2009: 71–72).

water use and questions of space and social status. Table 4.7 provides an overview on the twelve neighborhoods covered by empirical fieldwork. The selection covers a range of different modes of urbanization, water supply conditions and housing types. The neighborhoods studied represent three basic modes of Mexico City's habitat production: popular urbanization, housing complexes, and the heterogeneous space of the inner city.

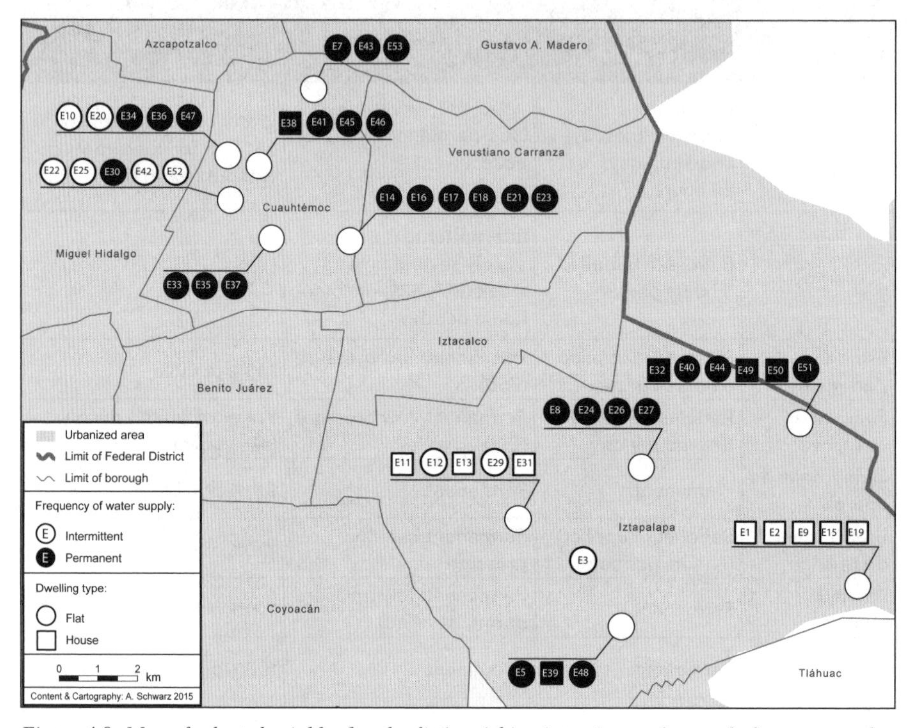

Figure 4.8: Map of selected neighborhoods, distinguishing interviewees by supply frequency and dwelling type. Own elaboration

A total of 53 residents from the twelve selected neighborhoods were recruited for individual interviews – between three and five persons per neighborhood (see Fig. 4.8). Interviewees were preselected based on their social status, housing type, steadiness of water supply and last but not least starting with existing field contacts from previous research (for research design, see 3). The material obtained from 44 of these interviews provided the base for an in-depth analysis; empirical findings are presented in the following chapters. With the exception of Miravalle and El Manto, a majority of interviews (33 out of 44) were conducted with flat dwellers. Of the eleven interviewed house dwellers, only one was a resident of the borough of Cuauhtémoc. This roughly reflects the composition of the housing stock, which is clearly dominated by houses in Iztapalapa and by flats in Cuauhtémoc. As a consequence of its historically strong processes of popular urbanization, Iztapalapa's most common dwelling type is the single-family house, making up as much as three quarters of its housing stock in 2010[36]. In contrast, flats make up around 80% of inhabited dwellings in the borough of Cuauhtémoc and these shares have not changed much over the past two decades (see INEGI *Censo de Población y Vivienda* 1990, 2000, 2010 and Fig. 4.9).

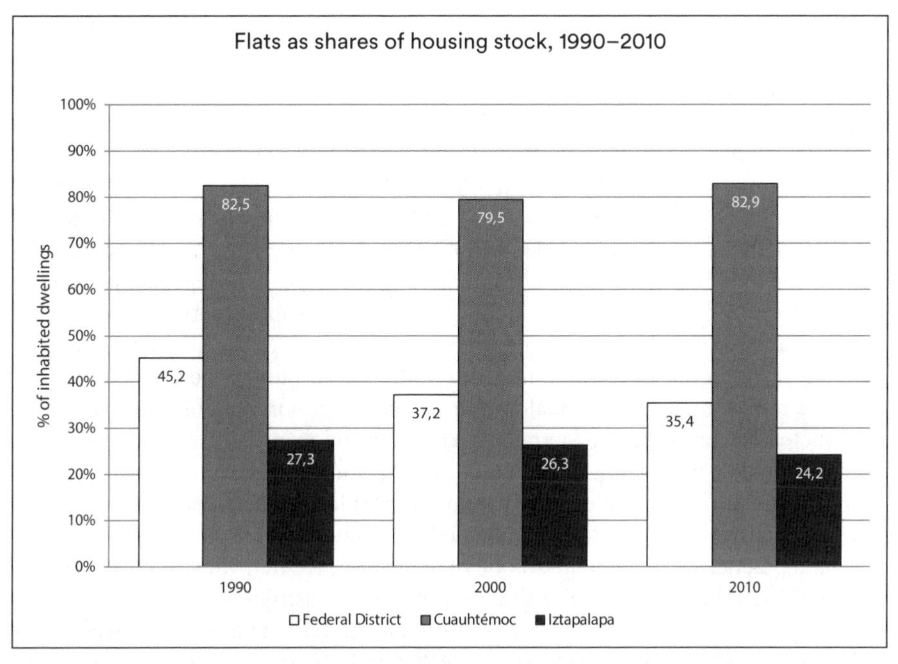

Figure 4.9: Predominant dwelling types: Percentage of housing stock made up of flats and single rooms in Cuauhtémoc, Iztapalapa and the Federal District, 1990–2010. Own elaboration based on INEGI Censo de Población y Vivienda 1990, 2000 and 2010

Whereas *popular urbanization* was the dominant mode of Mexico City's urbanization in earlier decades, creating about half of its dwelling stock, its dynamic has now been dwarfed (see Duhau/Giglia 2008: 185). This is mainly the result of a change in housing policies including a reform of the loan system in the mid-1990s, which closed down the sources of land for popular urbanization by formalizing the commodification of collectively owned agricultural land in the ejidos (see for instance Cymet 1992). In consequence, apartment buildings are now providing a growing part of the dwelling stock in the centrally located boroughs of the Federal District, while the periphery is coined by prefabricated terraced houses in settlements paradoxically called conjuntos urbanos. Despite a decrease in the share of flats in the Federal District, where it sank from 37% to 35% of inhabited dwellings from 2000 to 2010 (see Fig. 4.9), this dwelling type continues to be predominant in the inner city areas of the Federal District. In fact, apartments are being constructed

36 Refers to inhabited dwellings, with the remaining share made up of *cuartos de azotea*, and no data available.

in considerable numbers in many of these neighborhoods, where at the same time, water supply conditions tend to be subject to increasing limitations, as interview material shows. Hence there is a need to understand whether there are specific water using practices of flat dwellers, and how they cope with eventual supply limitations.

First, the neighborhoods of *Miravalle* and *El Manto,* both in Iztapalapa, are the product of popular urbanization (instead of many, see Schteingart 1989, Gilbert/Varley 1991, Horbarth Corredor 2003 on informal urbanization). This kind of self-built habitat, the so-called colonias populares, was the dominant mode of urbanization in Mexico City and many other cities during the better part of the second half of the 20th century. Colonias populares were home to more than half of all inhabitants of the Metropolitan Region of Mexico City (ZMCM) in 2000 – housing some 9.2 million people in total (see Duhau/Giglia 2008: 177). Serving as the main source of housing for the population excluded from the formal housing market, it is coined by an inversed mode of spatial production – starting from the dwelling and introducing collective infrastructure ex-post, including the water supply network. As Duhau and Giglia put it, this is a negotiated space, both in terms of the absence of formal, state-led planning in the initial phase and in the use of public and collective space (see ibid.: 329 ff.). As a typical example, the neighborhood of *Miravalle* was constructed on the slopes of Sierra de Santa Catarina from 1985 on, when the first residents purchased small plots of land – hitherto under agricultural use – from a local *fraccionador*. Land titles were eventually regularized between 1994 and 1998, and public infrastructure was implemented subsequently (for an in-depth analysis of spatial practices in the neighborhood of Miravalle, see Schwarz 2009). Today the settlement is relatively consolidated in terms of infrastructure, but water supply remains rationed to once a week and is of such low quality that domestic water bills in Miravalle have been suspended by the Federal District's government since as long as 2008. The process of consolidation on the dwelling level seemed stalled in several of the interviewed households.

El Manto is an older and more heterogeneous neighborhood near the Cerro de la Estrella and the pre-hispanic villages of Iztapalapa. Popular urbanization started in the mid-1970s and was later combined with small industries and commercial use as well as smaller housing complexes. The neighborhood shows a stronger grade of consolidation and is more centrally located than Miravalle, and the interviewees' socio-economic status tended to be slightly higher. Water supply is intermittent but frequencies vary strongly between different streets and parts of the neighborhood.

Second, the selection includes a number of housing complexes – the major mode of housing production dedicated to those parts of Mexico City's population employed in the formal sector. A product of formal planning implemented by private developers and public institutions alike, housing complexes form the second largest part of the housing stock of the metropolitan region. They are providing accommodation for 15% of its population, or roughly 2.6 million residents in total (see Duhau/Giglia 2008: 177). These housing complexes are generally dominated by flats but some include a mix of different housing types. Their origins are also more divergent than it might seem at first glance.

Nonalco Tlatelolco is one of the most prominent and iconic of these settlements, housing over 10,000 residents and a full range of public infrastructure including hospitals, schools and shops. Finished in 1964, this neighborhood close to the colonial city center is one of the pioneering large-scale modernist housing complexes of Mexico City. It became infamous as the site of brutal repression against the student movement in 1968, and again in 1985 when several high-rises were badly damaged or even collapsed during the earthquake. Tlatelolco was built for a middle class clientele, although it predated the introduction of the two most important institutional lines for housing credits – namely the INFONAVIT (private sector employees) and FOVISSSTE (public sector employees) – in 1972. Tlatelolco disposes of an own well and its water supply is permanent even during the periodic supply suspensions, as an enormous central cistern is supposed to serve large parts of the complex. It thus functions on an insular logic, as several of the surrounding, older and more popular neighborhoods such as Guerrero are subject to supply limitations that are unfelt in Tlatelolco.

The settlement of *Ermita Zaragoza* was built by the government in the early 1970s on the margins of Iztapalapa and Nezahualcoyotl to house families which were being resettled from other parts of the Federal District, either from informal dwellings or from locations at risk. Paula (E32) for instance, moved in as a child, after being evacuated as her family's self-built dwelling in a former mining zone in the western Federal District had collapsed. In consequence, Ermita Zaragoza's original residents were more popular class than middle class, and not necessarily employed in the formal sector. In contrast to other housing complexes, Ermita Zaragoza has a relatively low density, as the main housing type are duplexes and small terraced houses, which both offered the opportunity of subsequent expansion on a small plot. Water infrastructure was implemented from the very beginning, and supply only deteriorated over the last years as water pressure is low in some parts of the settlement and water quality varies. The original houses had no cisterns or other storage facilities.

Located more towards the central plains of Iztapalapa, close to the Central de Abastos, *Chinampac de Juárez* is another popular class housing complex similar to Ermita Zaragoza but with housing blocks of six small flats each. It was established in 1990, after residents organized in the urban social movement UPREZ[37] had occupied the construction site of a private developer where construction had stalled as a result of the crisis years of the 1980s. UPREZ was eventually able to obtain the unfinished blocks of Las Frentes, as this settlement is known to its residents, and to make them inhabitable in collective work sessions called faenas, which are common during the consolidation process of colonias populares. As a result, this particular housing complex provided accommodation to families otherwise excluded from the formal housing market, such as domestic workers like Dolores (E8). Water supply is permanent in this area, which had a formally introduced water supply network from the very beginning.

37 *Unión Popular Revolucionaria Emiliano Zapata*, part of the influential *Movimiento Urbano Popular* – for an introduction, see Moctezuma Barragán (2012).

Also located in Iztapalapa, the neighborhoods of *Presidentes de México* and *Fuentes de Zaragoza* are the result of the reform of federal housing policies in 1994. Finished in 1996/1997, they were constructed by private developers and contain a considerably lower number of dwellings than the other housing complexes studied. The flats were constructed for a lower middle class clientele, namely formal sector employees who were granted loans via INFONAVIT and FOVISSSTE. Fuentes de Zaragoza is in walking distance to the above mentioned complex of Ermita Zaragoza and effectively surrounded by neighborhoods with a lower socio-economic status. Water supply is permanent but of such a low quality that the Federal District's government grants fixed water rates to its residents. The Presidentes de México neighborhood, near Metro station Constitución de 1917 in Iztapalapa, is actually home to a number of housing complexes, and there are consolidated colonias populares in adjoining streets. Water supply had been permanent and of good quality but began to deteriorate around 2010, when residents experienced more frequent supply suspensions, low pressure and a lower water quality.

The third type of neighborhoods studied here are located in the inner city, which covers Cuauhtémoc and three neighboring boroughs. In contrast to housing complexes and colonias populares, these neighborhoods are characterized by a larger heterogeneity of land use and inhabitants. Founded between the late 19th century and the 1940s as expansions to the colonial city center, they are now a contested by a range of different actors and between residential and commercial use (see Duhau/Giglia 2008: 239). Eventually, non-residential land use became dominant in some of these neighborhoods due to a loss of population during the last decades of the 20th century. As of the year 2000, the inner city was home to no more than 6.6% of the Metropolitan regions population – 1.1 million in total (see ibid: 177). However, that number has likely increased in the meantime as a result of the Federal District's redensification policies introduced in the same year. Contrasting dynamics can be observed in the five neighborhoods which were studied, as they combine diverse, heterogeneous modes of urbanization.

The neighborhood of *Cuauhtémoc* in the borough of the same name is an impressive example of these disputed dynamics. This traditionally better-off neighborhood was founded between 1882 and 1900 (for an overview on Mexico City's processes of urban expansion, see Streule forthcoming), and is located next to Avenida de la Reforma, where a recent construction boom has put forth a large number of office high rises such as iconic *Torre Mayor*. Height limitations are practically inexistent, and the strong redensification seems to create a fierce competition over urban services and infrastructure. Residents in middle-class apartment blocks and vertical condominiums, or the respective janitors and domestic workers, reported a growing intermittence and insufficient water pressure.

Not far from here, the neighborhood of *Tabacalera* lingers on the margins of the colonial city center and the first expansions realized from the 1870s onwards. Limitations in water supply similar to the Cuauhtémoc neighborhood – low pressure and intermittence, which appeared during the last decade or so – are reported from here as well as close-by *San Rafael*. The latter was drafted as bourgeois *fraccionamiento* in 1881 and has become home to a socially heterogeneous population as well as shops and small workshops over time. As an effect of the Bando Dos policy, the neighborhood is now subject to a dynamic process of upgrading and redensification, which is most visible in the construction of apartment blocks with a comparatively high density and dedicated to the middle class. It seems as if the urban water supply network does not catch up with the pace of development, and supplied quantities seem to differ from street to street, or even plot to plot, in some cases. Originally founded as a fraccionamiento in 1902, the *Roma Norte* neighborhood is very similar in its origins, heterogeneity and ongoing construction boom, but water supply seems relatively stable in contrast.

Situated to the south-eastern limit of the borough Cuauhtémoc, close to Venustiano Carranza, the neighborhood of *Paulino Navarro* has a considerably lower density than the more centrally located areas. Its streets are lined by often quite run-down two- to three-storey houses, which mostly seem to date back to the 1930s but are combined with others representing the bourgeois style of the turn of the 20th century, and some more recent apartment blocks. Today, Paulino Navarro displays a very fine-grained social heterogeneity, where one end of its tree-lined side streets is made up of workshops and run-down but crowded *vecindades*. Then, only some blocks further, small blocks of new flats can be observed side by side with *Porfiriato* residences refurbished as middle class single-family houses, including small front yards. The dynamic of upgrading which can be observed in San Rafael and elsewhere has not yet fully reached this area and a number of interviewees have lived here for several decades despite being tenants and not homeowners[38]. Water supply is reportedly permanent, and some of the poorer interviewees even took to drinking tap water without any pre-treatment.

To summarize, Mexico City's water supply shows a spatial differentiation not so much in access to water taps as in supply frequencies and water quality. Almost everybody has a water tap today; but especially in terms of supply frequencies, the Federal District's water supply system, far from providing a universal service, resembles an infrastructural archipelago, as Bakker (2003) put it. How do people cope with these conditions? What are the consequences for everyday life? The following chapters will explore everyday practices of water use in Iztapalapa and Cuauhtémoc in relation to living conditions (chapter 5), and in relation to past experiences, introducing a historical perspective (chapters 6 and 7).

38 Across the entire social spectrum, around 72% of Mexico City's residents are homeowners (see Rubalcava/Schteingart 2012: 90). However, tenants are more common in the inner city. For more than a century, the colonial city center, for instance, provided simple (and often overcrowded) tenant housing for a considerable part of the urban poor in converted buildings formerly inhabited by the colonial elites.

5. PRACTICES OF DOMESTIC WATER USE IN MEXICO CITY

Performed under conditions defined not least by both housing conditions and the specific characteristics of a neighborhood's infrastructures as expressions of a wider logic of supply, an exploration of domestic practices of water use calls for a close reading of material spatial practices. The production of urban (everyday) space – here meaning the dwelling or home – is a process intrinsically intertwined with infrastructural conditions, as material dwelling conditions are adapted to both expected and experienced conditions of water supply. The installation of a domestic storage tank is only one example. Infrastructural features are linked to the service quality, that is, the reliability of supply with respect to water pressure and supply frequencies, as well as the quality of the water. As elaborated in chapter 4.1, tap water quality is not only linked to the sources and purification process but also to the technical-material set-up of the supply network itself (e.g. its age, size and material).

From drinking and cooking to cleaning and watering plants, water is essential to a wide spectrum of domestic practices. Previous studies have found that hygiene-related use accounts for the largest share of Mexico City's domestic water consumption. According to a 2011 survey[1], 62% of tap water consumed by the Federal District's domestic sector was used for personal hygiene (including flushing of toilets), and 23% for doing laundry. Just 2% was used for food preparation, 1% in watering gardens, and less than 0.3% for drinking (see Monroy/Montero 22.01.2014). This not only indicates that tap water is not a source of drinking water for most of Mexico City's residents but also reveals the importance of hygiene-related water use reflected in the spread of basic water-using household devices. Micro data from the latest Mexican census shows that more than three quarters of all inhabited dwellings in the Federal District had a washing machine in 2010, and 81% had a shower. At the same time, 86% of those inhabited dwellings that dispose of a toilet[2] are equipped with a water tank for flushing (see INEGI 2010). It can be assumed that at least the two last-mentioned installations stand for higher domestic water consumption levels in comparison to the simple and low-tech application of water by buckets and smaller containers which is common in households without proper showers and flushing tanks. Apart from some scattered data, detailed information on what water is used for in Mexico City's households, and how much of it, is quite scarce. However, when compared to older reports it seems as if hygiene-related water use has lost some ground over time. In 2002, as much as 73% of

1 The *Encuesta sobre consumo y percepción del agua en los hogares del Distrito Federal, 2011* was conducted by researchers from UAM-Iztapalapa. and covered 689 households in all 16 boroughs of the Federal District. Preliminary results were presented in a press conference in January 2014. Missing shares of water use were reported to be employed for other purposes.

2 In 2010, about 2% of all inhabited dwellings in the Federal District did not dispose of a toilet (see INEGI 2010).

domestic tap water consumption in the Federal District was attributed to hygiene-related use (personal hygiene and flushing toilets). A much lower 14% was employed in doing laundry, and the rest in unspecified other types of use (see INEGI 2005: 99). Yet most of these reports provide no methodology which would enlighten the reader on how such data was obtained. Leaving these number games behind, the present chapter turns to concrete practices of domestic water use and explores their links to sociospatial conditions. Based on empirical field work in Mexico City, it focuses on the following practices in the realm of domestic water use: drinking, hygiene and cleaning, storing water, and imagining urban water. In the following, each of these four sets of water-related practices are introduced and then set in relation to current housing and infrastructural conditions.

5.1 DRINKING: INGESTING WATER

Nobody drinks [tap] water like that.
(Ada E1: 146, Iztapalapa)

Drinking water constitutes the smallest share of human water consumption, yet it is essential for health and survival. In this context, adequate water quality is of crucial importance – more than for any other domestic use. When washing dishes, doing laundry or taking a shower, water will touch the human body – or some of the artefacts in direct contact with the body – only on the surface. To drink water, then, is a much more physical act of water use as it actually enters the human organism through the act of ingestion. Drinking could hence be understood as the body-related water use *par excellence* whereas other domestic practices may also be fulfilled with a spectrum of more 'technical' or process-oriented types of water with lower quality standards. The perceived and experienced quality of tap water is a crucial point when it comes to judging whether it can be drunk – or used for dental hygiene, for instance – or is only feasible for purposes where water is not in direct contact with the human body. As the present and following chapter will show, a difference is often made between water used for external or more 'technical' domestic purposes – such as taking showers and doing laundry – and potable water actually deemed fit for ingestion.

In the case of Mexico, it is remarkable how bottled water has become the most used type of water for drinking over the last decades. Extremely elevated rates of bottled water consumption characterize those who inhabit its biggest metropolis. In 2008, around three quarters of all households in the Federal District used bottled water for drinking. Most of the remaining quarter relied on domestic tap water pre-treatment of various forms (such as the use of filters or disinfectants), whereas less than 5% of all households took to drinking tap water without any measures of purification (see Fig. 5.1).

A representative survey from 2011 found that bottled water consumption in the Federal District had risen to 87% of all households (see Monroy/Montero 22.01.2014). Today, bottled water sellers and small purification plants are a ubiquitous sight throughout the city. This is hardly surprising when taking into account how Mexico City's population perceives the quality of tap water. A widespread mistrust in tap water, supply limitations, and aggressive marketing strategies for bottled water could all be blamed for this development. In addition, one event in particular was deeply engraved in the collective memory of this city: The 1985 earthquake. Besides far-stretching political implications[3] – not least related to the hegemonic PRI party's ability to provide public services and organize the production of urban space – it also had a strong impact on water supply and its perception in the short and long term. De facto, it laid the base for a rather persistent way of imagining Mexico City's tap water as being of an unpotable quality, which seems to be perpetuated even some three decades after the initial event.

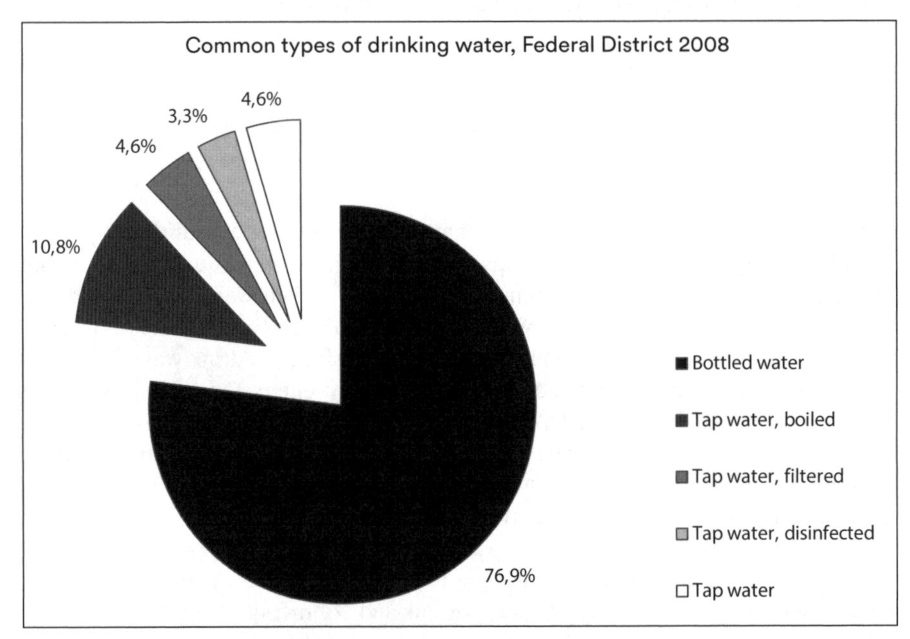

Figure 5.1: Type of water used for drinking in the Federal District's households, 2008. Adapted from: Boltvinik/Figuera Palafox 2010 based on INEGI 2008: Encuesta Nacional de Ingreso y Gasto de los Hogares, as cited in Jiménez Cisneros et al. 2011: 260

3 In particular the Federal District's government under Ramón Aguirre Velazquez was unable and unwilling to respond to the earthquake's devastating effects in an adequate way, which led to a strong civilian movement and grassroots organizations taking responsibilities in many neighborhoods. The PRI, ruling Mexico since the aftermaths of the revolution, was already losing ground in the capital, but its handling of the earthquake response was what set the base for an eventual toppling of its rule and the democratization of the Federal District's political landscape by fostering urban social movements such as CONAMUP (see Davis 1994: 280 ff. and Moctezuma Barragán 2012). In the long run, it resulted in the first direct elections of the Federal District's mayor in 1997.

Don't Drink Tap Water! The 1985 Earthquake

On the morning of September 19, 1985, an earthquake with a magnitude of 8.1 on the Richter scale shook Mexico City, followed by several strong aftershocks. Reliable numbers on casualties have never been published, but it is estimated that between 7,000 and 26,000 persons[4] died as a result of the earthquake, and huge numbers were injured when a great number of buildings, both residential and commercial, were reduced to rubble (see Monsiváis 2005). An estimated 30,000 dwellings were destroyed and another 70,000 damaged (see Ramírez Cuevas 11.09.2005). The city center and the adjoining north- and southeastern areas, built on soft lacustrine clay of the former lake bed of Lago Texcoco, were particularly prone to damage. Urban infrastructure throughout the metropolis was heavily affected, and practically all public services, including water, waste water, electricity, telecommunications, gas and public urban transport, instantly became dysfunctional. As for the water supply network, "extensive damage [was caused] to the buried transmission and distribution lines" (Ayala/O'Rourke 1989: 4). One of the main aqueducts, located in Tláhuac, was severely damaged and had to be suspended. Altogether, around 170 water mains in the Federal District's primary system and between 5,100 and 7,200 in the secondary distribution system were damaged and required repairs (see American Life Lines Alliance April 2001: 19; Ramírez Cuevas 11.09.2005).

As a result, the public water supply was disrupted, and up to 3.5 million inhabitants of the Federal District and another 1.8 million in Mexico State were left without running water – in total, one third of Mexico City's population at the time (see Ayala/O'Rourke 1989: 4). Some emergency water distribution was realized using water tankers, but the coverage of this service was apparently limited. Many residents recurred to fetching water from public taps in the aftermath of the quake – such as Felipe, then ten years old:

> *We lived in Neza[hualcoyótl] and we didn't have water because the pipes broke. We had to walk [about ten] blocks to (...) line up at a water hydrant. It took us about six hours to get a bit of water. (...) waiting in line and when they finally filled (our containers) we went back (...). There were many people, everybody was without water.* (Felipe E24: 215–219)

As of October 2, 1985, tap water supply had not been restored yet in at least ten neighborhoods in the Cuauhtémoc borough (including Centro, Doctores, Roma Norte, Roma Sur, and Paulino Navarro) and some 17 in Iztapalapa. Eventually, water provision through the damaged Southeastern aqueduct was restored to 7.1 m^3/sec by the end of October, almost reaching the pre-earthquake 7.6 m^3/sec. However, repairing all damages to the distribution network took several months (see Ayala/O'Rourke 1989: 4), and routines of water use were severely disrupted by the supply suspensions. This experience changed people's perception of water supply, as Lucia, who was living in the borough of Cuauhtémoc at the time, recalls:

4 The lower numbers stem from the Mexican Federal government, the higher ones from United Nations Economic Commission for Latin America and the Caribbean (ECLAC), according to Ramírez Cuevas (11.09.2005).

We were in the Doctores [neighborhood] in 1985 during the earthquake (...). We had to walk four blocks to the water trucks. (...) There wasn't even [water] to take a shower nor for the toilet, there was nothing. (...) It was containers and buckets and all that we could fill once the tanker trucks arrived. Yes, that taught me to appreciate water because it was a terrible effort. (...) It took about three months until it got fixed. We went to the [public] steam baths (...) as there was literally not a drop of water at home, nothing. People came to donate [bottled] water for drinking. (Lucia E40: 252–260)

A lack of running water over a period of several weeks and the possible contamination of tap water due to ruptured pipes in combination with health risks related to corpses entrapped in the rubble of collapsed buildings meant an increased risk for the spread of an epidemic such as cholera or typhoid fever. A contemporary study by the Mexican Ministry of Health found that the bacteriological and chemical quality of tap water[5] in the Federal District was affected, in particular during the first four weeks after the quake (see Valdespino Gómez *et al.* 1987: 414 ff.). The implications of this cannot be underestimated. David, who lived in central Iztapalapa and Pablo, who lived in a residential area of the Miguel Hidalgo borough in 1985, both recall the earthquake being the moment after which a pre-treatment of tap water became a common domestic practice.

Because of 1985 we remain distrustful of tap water quality, boiling it. (David E11: 16)

It all began with 1985, when there was this quake here in Mexico City. The pipes were affected and the authorities recommended to not consume [tap] water. And it stayed that way – at least you keep it in mind when you have had to experience this at an early age. (Pablo E34: 174)

In effect, as water got contaminated directly in the distribution system in the aftermath of the quake, the population was officially warned against the consumption of untreated tap water:

"Even as water treatment plants had maintained adequate levels of chlorine (…), fractures (…) in the distribution network explain the low frequency of [tap water] samples apt for human consumption (…). Even shelters, hospitals and schools had serious deficiencies in drinking water supply and quality (…), which is why the activities of [public] health education that were broadcast via the mass media played a key role for the guidance of the public. By this means, the population was informed about the need for boiling or disinfecting drinking water." (Valdespino Gómez *et al.* 1987: 419)

Even in those areas where tap water supply was not disrupted during the first weeks after the earthquake, according to the interview material it seems to have become widely perceived as undrinkable as an effect of the warning issued by the government – and this perception would result to be quite persistent (see 5.1.1). This way, the 1985 earthquake not only had an immediate impact on the everyday life of most residents of Mexico City – it also represented a crucial moment inducing a change in the way its public urban services and the entire political system were perceived.

"The September 1985 earthquake threw millions onto the street and wrought widespread infrastructural destruction. Equally important, the quake (…) shook the faith of many in the Mexico City government, the PRI, and the Mexican political system as a whole by exposing

5 The study by Valdespino Gómez *et al.* 1987 covered both households still provided by tap in the aftermath of the earthquake, and those relying on domestic storage tanks supplied by tankers.

> the inability of local and national politicians to manage the city's most basic services in a time of disaster." (Davis 1994: 281)

Whether it is drunk directly from the tap, filtered or boiled prior to consumption, or stems from bottles – the question emerges why people drink this kind of water in particular. From a spatial point of view, the relation between the practice of water-drinking and the material spatial setting is analyzed for infrastructural conditions, more specifically, the steadiness of water supply and the perceived tap water quality. Presumably, a non-permanent water supply rhythm (and thus lower water availability) could influence on the amount of bottled water consumed in a household. Furthermore, there are claims that non-permanent supply patterns are a possible source of contamination as low pressure in the empty water mains might facilitate the infiltration of polluted water, especially in water distribution systems with problems of leakage (see National Research Council *et al.* 1995: 46).

Regarding the quality of tap water, there are strong differences between Cuauhtémoc and Iztapalapa, with the latter frequently tagged as one of the boroughs with worst tap water quality in the Federal District by residents and the local press alike (see Mora 21.05.2013, Royacelli 12.01.2013). Apart from leaking pipes, the problem has been linked to a low quality of raw water from the Iztapalapan aquifer (see GDF/SACM October 2012: 37):

> "Water extracted from wells in this area exceeds the permitted limits of substances such as iron, which causes a color like coffee (...), along with manganese and ammonium nitrate causing bad odor." (Martínez 13.02.2009)

Despite the purification process applied by the water utility, these effects are still observable at domestic taps in various neighborhoods in Iztapalapa (see SACM 14.04.2014). Recently, some governmental programs were announced which will allegedly bring universal coverage with (literally) potable water for the Federal District by 2018 by increasing the number of purification plants in the distribution network (see GDF/SACM October 2012: 102), but it is too early to evaluate the actual implementation and success of these efforts. Moreover, as the interview material reveals that (perceived) water quality also tends to vary on a very small scale, drinking water practices are not only compared between the two boroughs but also for each of the studied neighborhoods.

There is a general lack of reliable and up-to-date information on tap water quality in the Federal District as publicly available data from the water utility itself is scarce (see 5.1.4). For this very reason, the city's residents themselves need to evaluate the quality of water running from domestic taps – a task usually fulfilled by relying on everyday perceptions rather than a chemical analysis of water quality. It can be assumed that particularly the practice of drinking is strongly framed by such a perceived tap water quality. In the present study, smell, taste and appearance of tap water were introduced as possible indicators of water quality by the interview guideline – in accordance with the so-called acceptability aspects (see WHO 2011: 219 ff.). They are also employed in health studies, where both the perceived color and taste of tap water are identified as indicators for a possible risk of water-borne diseases (see Cifuentes *et al.* 2002). A yellowish or brownish color, visible floating particles, or an unpleasant smell reminiscent of eggs, fungus or chlorine were reported by interviewees as signs of limited water quality. Where tap water is deemed unfit for human consumption, it is either pre-treated to make it potable, or substituted by bottled water. In effect, bottled water is omnipresent in Mexico City's homes – the advantage for the present study being that in contrast to tap water, the actually consumed volumes are relatively easy to measure. As water is usually purchased in the 20-liter refillable plastic jugs called *garrafones* in regular intervals, it can be assumed that people keep track of the volumes of bottled water consumed in their households. Data obtained from the household interviews hence allows identifying bottled water consumption practices, comparing consumption levels on a per-capita level, and analyzing them for links to the steadiness of supply and the perceived tap water quality in the neighborhood.

5.1.1 Drinking from the Tap?

Is tap water potable? In sharp contrast to announcements made by the water utility[6], most residents of Mexico City would not hesitate to answer this question with a clear 'no'. As in many places around the world, the occasional tap-water drinker is the exception rather than the rule. A 2011 survey found that in no more than three[7] of the 16 boroughs, tap water was deemed potable by a significant share of the population – including Cuauhtémoc, where roughly 26% of households drank tap water without any kind of pre-treatment (see Monroy/Montero 22.01.2014). As for the present study, in spite of strong differences in perceived tap water quality between the boroughs of Cuauhtémoc and Iztapalapa, tap water is thought to be inapt for drinking by almost all interviewees – they either take to tap water pre-treatment (5.1.2) or bottled water consumption (5.1.3). As a result, no more than five of all 44 interviewees took to the tap as a source of drinking water – two of them filtering or disinfecting tap water prior to consumption, one drinking directly from the tap, and another two combining the consumption of untreated tap water with bottled water (see Fig. 5.2).

Surprisingly, those who drank tap water without any pre-retreatment all lived not only in the same borough but also in the same neighborhood: Paulino Navarro. Of all six neighborhoods studied in the borough of Cuauhtémoc, Paulino Navarro is clearly the most 'popular' or poorest, in terms of its population and material appearance. Water supply in the area is reportedly permanent but of low pressure at times. People took a pragmatic approach to judge tap water quality, citing the absence of observable impacts on their health – such as 72-year old Oscar, who has lived in the neighborhood since 1985:

> *If I don't get ill and my children haven't gone ill either, it means that the [tap] water is good.* (Oscar E14: 80, Cuauhtémoc)

6 According to which currently no more than 4% of tap water coverage (see SACM 16.10.2013: 6), or some 60 neighborhoods (see SACM 14.04.2014) do not receive water of potable quality.

7 The other two boroughs (both with 25% of respondents consuming untreated tap water) were the peri-urban Milpa Alta to the south, where there is still a huge amount of agricultural land use, and Iztacalco, between Iztapalapa and the city center, which has a number of own wells.

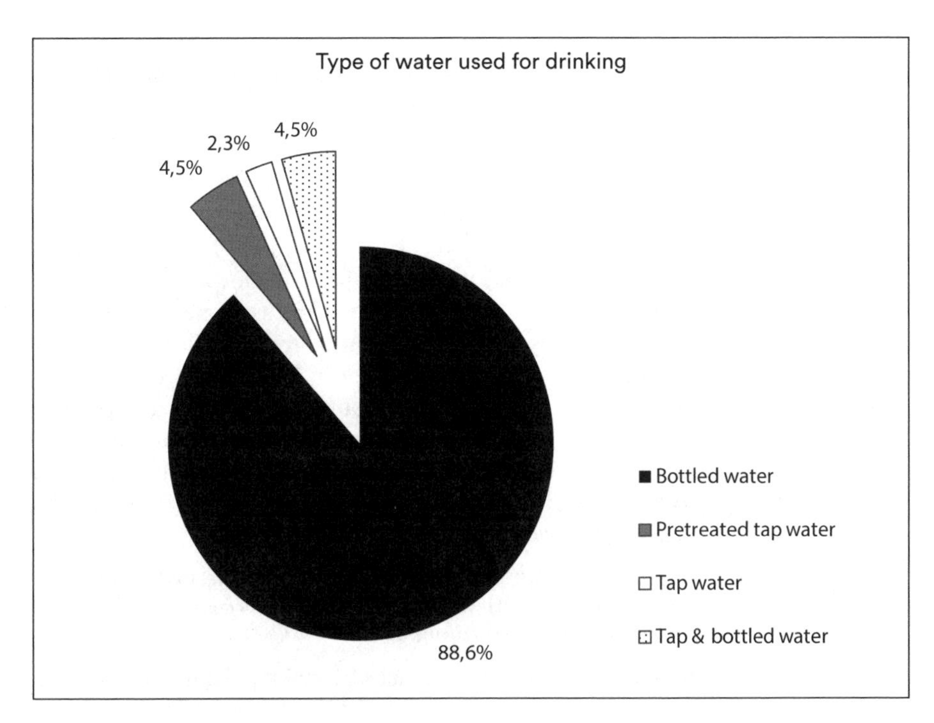

Figure 5.2: Type of water used for drinking in the interviewed households. Own elaboration based on 44 interviews

However, Oscar says tap water smells of chlorine, and his neighbor Catalina at times also perceives some limitations in terms of taste and appearance. Both households combine the consumption of untreated tap water with bottled water. In Oscar's household, around 17 liters of bottled water from a local purification plant are drunk per capita and month, at a total cost of 0.67€ per household member. Bottled water consumption in Catalina's household was lower, as she turned to it only on certain occasions:

> *We usually buy bottled water when we make a fruit drink or (...) in order to recover more quickly when we are sick.* (Catalina E18: 24, Cuauhtémoc)

On average, each member of Catalina's family consumes 11 liters of bottled water per month, resulting in expenditures of 1.91€ as they use the Bonafont brand and do not purchase in bulk but smaller bottles of one or two liters. Cheaper water from a local purification plant is not a viable alternative to tap water for her:

> *I really don't know if they purify the water or give me tap water (...). It's more secure with a brand, isn't it? (...) I have tried [...bottled water from a local plant] and to me, it tastes the same as the brand. It looks the same, it looks clear. But my doubts always remain.* (Catalina E18: 72–76, Cuauhtémoc)

While Catalina and Oscar drink both tap and bottled water, their neighbor Isabel is the only one in the entire sample who currently uses untreated tap water for all domestic purposes, including drinking and food preparation, and buys no bottled water at all. Recently separated from her partner, and caring for her three teenage children on her own, she leaves no doubt about her motives:

> *[We drink...] from the tap, because there is no [money] for bottled water. [... Normally] I buy from a purification plant just around the corner, but due to a lack of funds there is none now.* (Isabel E21: 241, Cuauhtémoc)

Saving on a product which reportedly would cost 14 Mexican peso (around 0.79€) per 20 liters might seem surprising, but there are other hints towards the family's precarious economic situation. Not only is Isabel's job as a cleaner the only household income, she also recently changed apartments for a smaller one in order to lower the monthly rent. Isabel describes tap water in Paulino Navarro as odorless, colorless and tasteless, and of a good overall quality. However at the same time, she stresses the importance of conditioning the body to endure the consumption of tap water without experiencing negative health impacts – an argument which can be linked back to both her precarious social status and her experiences with water supply limitations in past dwellings.

> *I always accustomed my children to bottled water (...), and if there is none, well, then from the tap. Their stomach has to adapt to both types of water – if your stomach gets used to only one water, it will suffer, with infections, with diarrhea.* (Isabel E21: 379)

Isabel's strategy of inuring her children's' stomachs by making them drink both tap and bottled water follows what can be called a deeply classed logic: The conditioning of the popular class body for anticipated future living conditions, as will be discussed in 5.2.3.

5.1.2 Pre-Treatment of Tap Water

Mexico City's tap water is considered undrinkable by an overwhelming majority of interviewees – at least as long as it is not pretreated in any way prior to domestic use. However, the perceived tap water quality itself is a potential limiting factor to this practice, and what is more, it is so in a highly spatialized way. While Iztapalapans tended to abstain from water pre-treatment as a source of drinking water, residents of Cuauhtémoc deemed it a more viable option (at least theoretically), as this chapter will show. All in all, there are a number of domestic techniques of pre-treating tap water in order to make it apt for human consumption, which can be subsumed in three groups: filtering, disinfection, and purification. The latter refers to the complete process of potabilization, and both filtering and disinfection form part of such a purification process. Filtering is a means of retaining suspended particles, whereas disinfection refers to the elimination of microorganisms in the water – mostly by applying heat or chemical agents such as chlorine (see Huerta Mendoza 2004: 43). These methods are low-tech, whereas disinfection by ultraviolet light or ozone is common on the industrial scale and has not (yet) spread widely into the domestic realm. According to a 2008 study, which found roughly 19% of households in the Federal District were practicing tap water pre-treatment, boiling continued to be the most common method, followed by filtering and chemical disinfection (see Jiménez Cisneros *et al.* 2011: 260). Yet no more than two of those interviewed for the present study relied on these techniques – at least when it came to water designated for drinking.

Disinfecting Tap Water

Not least upon the backdrop of questionable agricultural production methods such as irrigation of food crops with untreated waste water[8], food disinfectants based on chlorine, iodine, or ionized silver are commonly used in Mexico City's households in the preparation of meat, fruits, and vegetables. But at times, they are also employed to disinfect tap water for human consumption. Berta from the El Manto neighborhood in Iztapalapa is the only interviewee who employs this kind of tap water pre-treatment, and does not purchase any bottled water for drinking. This seems to be motivated mainly by health issues – as she is unable to lift heavy weights (such as a 20-liter container) as a result of a traffic accident – yet she also considers the quality of tap water in her neighborhood as maybe not drinkable but taste- and odorless, and hence as fit for pre-treatment:

> *I use the [.. disinfectant] and it works for me. I like the taste of this water better than [... untreated tap water] and bottled water (...). And moreover, [tap] water always runs clean here; it's not yellow or anything alike.* (Berta E29: 264–266, Iztapalapa)

Berta also stressed the lower cost of this method in comparison with bottled water. A 15-ml-container of Microdyn, a commonly used product, yields about 560 liters of disinfected water, and is sold for about 0.80€ on the local market. This product removes bacteria but not waterborne protozoa such as Giardia (see Procuraduría Federal del Consumidor April 2012: 68). A traditional means of disinfecting tap water is boiling – but it requires a "rolling boil" to effectively remove bacterial and protozoan contamination (see WHO 2011: 143). Some have argued that in particular an increase in tariffs of natural gas, which feeds most of Mexico City's stoves, has effectively abated this practice since the mid-1990s (see Castano 22.2.2012). Just as Guadalupe, many took to bottled water as a solution which also comes at a cost but requires less domestic efforts:

> *You spend the same, since you boil it and gas is also very expensive. So what else can you do? You buy bottled water and save some gas.* (Guadalupe E3: 128, Iztapalapa)

Even though it cannot be quantified here whether this argument is economically viable[9], a substitution of the practice of boiling tap water for the purchase of bottled water in order to reduce domestic efforts was a domestic strategy reported in a

8 Almost 900 km^2 of agricultural land in the Mezquital, Tula and Los Insurgentes valleys, where vegetables and cereals for Mexico City's population are produced, are irrigated this way (see Legorreta 2006: 56 ff. and Sosa Rodríguez 2010: 137).

9 From an empirical point of view, besides the price of natural gas, such a calculation would require detailed information on the efficiency of the employed domestic stoves and containers, in addition to the duration of boiling the respective household deems fit for sterilizing water.

number of cases. In effect, although several interviewees remembered boiling water in the past, none of them took to this practice anymore.

Filtering Tap Water

A number of filtering methods can be observed in Mexico City's homes – from sophisticated ceramic filters to the most make-shift solutions, such as tieing a cloth to the tap in order to retain any floating particles. Most filters are not stand-alone devices but are installed directly at one or several taps in the dwelling, frequently at the kitchen sink. However, most of these filters were used to improve the tap water quality for hygiene and cleaning purposes (see 5.2) and sometimes, food preparation, rather than to make it potable in the literal sense of the word. Mario, who rents a flat in the neighborhood of Cuauhtémoc in the borough of the same name, was the only interviewee in my sample who relied on filtered tap water as a drinking water source. While he was drinking tap water fed by a local spring during his childhood in the 1960s, living in the southern Federal District, the tap water provided at his current home in the city center made him consider pre-treatment:

> *Supposedly this filter percolates even fecal matter (...). You could say that you can actually drink it directly. [...When moving to this apartment] we bought bottled water at first. Then we said: "What for? We'll try the filter". (...) when we arrived [...we] tried [tap] water and it actually smelled bad, strongly of chlorine, and the taste... Everybody complained about [...its quality]. Also, whenever there is no water, it's almost like mud on the day it returns. (...) so you have to wait that it gets cleaner, or at least let it settle. (...) That's what the filter serves for. [... we use the filtered water] for cooking, washing fruits, and for drinking.* (Mario E42: 135 and 161–170, Cuauhtémoc)

Whereas for Mario, it was the local tap water quality which made him take to filtering, the very same aspect seems to inhibit tap water pre-treatment in a number of other neighborhoods, as will be discussed below.

Purification of Tap Water

Lately, some types of sophisticated micro-purification plants which resemble the set-up of a professional purification plant have been developed for the domestic realm. It can be argued that these devices most clearly represent an ongoing shift of responsibility for potability of water from the public into the private realm. Nurtured by and further deepening widespread mistrust in tap water quality, this kind of water purification is painted as a clever lifestyle choice of the urban elite. As an advertisement for domestic water treatment device available on the Mexican market puts it:

> "There's only one way to say it: unconditioned water has no place in your lifestyle. You can eliminate it from your home – and your life – with one of the most sophisticated water treatment systems available today." (General Electric: Pentair Residential Filtration 2014, accessed 27.11.2014)

Typically requiring an initial investment of about 250 € for the entire system[10], these devices come with up to five different purification steps. Featuring processes like sedimentation, filtration with activated carbon and ceramic filters, UV-radiation and reverse osmosis, these purification devices allegedly retain suspended particles and chlorine, inhibit bad smells and flavors, and eliminate bacteria and parasites. None of my interviewees owned one of these sophisticated devices – but some had heard about them and were considering whether to install one in the future, to substitute bottled water. Joining in with Mario from the neighborhood of Cuauhtémoc, his neighbor Alan was the only one who deemed local tap water quality as a good starting point for home purification and had recently taken to order such a device.

> *It's a bit expensive, but I think it's worth it because paying 35 pesos for a [20-liter jug] is too much squandering on the long run. (...) In the end you never know whether [bottled] water is pure.* (Alan E30: 113 and 129, Cuauhtémoc)

Generally speaking, there seem to be two main limiting conditions when it comes to tap water pre-treatment as a source of potable water: required investments, and more importantly, the local tap water quality to begin with. As already became clear with Berta and Mario, the perceived quality of tap water – often specific to one area or neighborhood – was a crucial precondition. In both cases, they received water which was not considered apt for direct consumption, yet still good enough to be made potable by domestic pre-treatment. But this precondition varies widely across the city, as Hilda, who had moved from the borough of Benito Juárez to Iztapalapa some years ago, pointed out:

> *In the [neighborhood of] Narvarte I was probably more confident to use some [disinfectant] drops and that's it. (...) Sometimes I bought bottled water. But it was easier to drink [tap] water in Narvarte. It didn't have that color, for instance, and didn't smell bad.* (Hilda E5: 130–134, Iztapalapa)

Particularly in those neighborhoods where tap water quality was perceived as low, interviewees doubted that a water filter (or any other method of domestic pre-treatment) would function properly in providing potable water. This is mostly an Iztapalapan issue, as the example of Dolores, a domestic worker living in the Chinampac de Juárez housing complex, shows. She would favor filtering – yet only considered it an adequate method to obtain potable water at her employer's house in the borough of Álvaro Obregón:

> *The water here is filthy, yellow (...). They have filtered water over there, they bought their filters (...) and I drink it. (...) They have much nicer [tap] water, much cleaner than we do.* (Dolores E8: 113 and 187–189, Iztapalapa)

Like Dolores, a total of thirteen interviewees from all of the studied neighborhoods in Iztapalapa considered the local tap water quality as so low that to them, filtering was not a viable way of obtaining potable water. Such a practice seemed most unimaginable in the neighborhood of Miravalle, where several interviewees first misunderstood the question on filtering as related to their storage tanks rather than

10 According to several online shops consulted in November 2014.

some kind of home purification. A filter would here prepare water for domestic use in general – but certainly not for direct human consumption.

> *[Tap water] comes super yellow; I guess that would clog it up very fast (…). We have never tried it but (…) at first, I put clean fabric on the tap to strain it (…). It turned black, black, black, very nasty. So I guess that the same would happen to a filter.* (Maria E2: 301, Miravalle)

The Sierra de Santa Catarina where Miravalle is located is notorious for its supply problems – tap water is not only rationed, running for several hours once or twice a week, but also of a very low quality. The water utility itself lists Miravalle as one of the 60 neighborhoods in the Federal District provided with water of limited quality due to elevated levels of iron and manganese – while claiming it is still "apt for consumption without health risks" (see SACM 14.04.2014). The severeness of these limitations is illustrated by the fact that domestic water bills in Miravalle and a number of other neighborhoods in the Sierra de Santa Catarina have been suspended for the same reason since 2008 (see Jefatura de Gobierno del Distrito Federal 05.04.2013: 3–6). Abstaining from tap water pre-treatment because of insufficient water quality is thus a strongly Iztapalapan issue[11] – thought it did not account for all neighborhoods in the same way, as the case of Berta shows. In contrast, there was only one interviewee from the borough of Cuauhtémoc who had given up her prior practice of filtering: Amelia, who reported that water quality in the Tabacalera neighborhood had deteriorated that much over the last five years that she now took to buying bottled water. However, even where tap water quality is considered as sufficient to make home purification a (theoretically) viable option, it is often turned into an economic question. The strain put on the household budget by the required initial investment is what seems to keep many, like Elena from Tlatelolco, from home-purification as a means of obtaining potable water.

> *[A filter] like that is actually one of our dreams. A friend has one that cost her 3000 pesos*[12]*, and it's supposed to be marvelous. It has stones and a light filter and I don't know what. (…) every six months (…) it costs 120 pesos to change the filters, that's not expensive. Yet what is expensive is the initial investment.* (Elena E43: 90–100, Cuauhtémoc)

As much as five interviewees from the Cuauhtémocan neighborhoods of San Rafael, Paulino Navarro, Tabacalera and Tlatelolco indicated that they might have installed a domestic water filter to produce potable water if it had been less expensive. In addition, some linked this to questions of homeownership. Apart from Mario, who filtered his drinking water, none of the other tenants considered the installation of a water filter. Like Berta and Fabiola, they would only be willing to invest in such an installation if they owned their dwelling.

> *If I had my own house – because I am a tenant – I would invest to buy a purification device for the tap. Because on top of that, it's expensive.* (Berta E29: 288, Iztapalapa)

11 Though there are hopes that the recent installation of several new purification plants by SACM in the area will improve these conditions (see SACM 2012: 73).

12 About 169€ by exchange rates of November 1, 2013.

> *As we are renting (…) we didn't want to install anything in the dwelling at all (…). Now we are about to buy an apartment (…). I think when we settle down for good, we'd have to do something to solve the drinking water question.* (Fabiola E37: 64 and 76, Cuauhtémoc)

Here is not the place to analyze the economic viability of different means of domestic water pre-treatment in comparison to bottled water (see 5.1.3). While local tap water quality is clearly a limiting factor, such a calculation seems to be highly dependent on several factors, not least the method of pre-treatment and local bottled water prices. In the case of Mario, the filter he uses in his Cuauhtémoc flat to produce potable water for his household of two comes at an initial cost of 450 Mexican peso (roughly 25€). As the filter cartridge needs to be changed twice a year, running expenditures are about 42 Mexican peso (2.37€) per month (not including the tap water bill). As will be discussed below, the monthly running expenditures per person for water used for drinking, cooking and washing fruits in Mario's household are more than half of what other residents from the same borough spend on bottled water for drinking only. Disinfection, the pre-treatment method practiced by Berta comes at an even lower cost per unit.

In summary, it seems as if in Mexico City, domestic water pre-treatment has been almost entirely replaced by bottled water consumption over the last years. Whether people consider pre-treatment a viable option strongly depends on the perceived water quality in their neighborhood. The location and the respective infrastructural conditions are hence relevant in limiting the options for drinking water sources down to bottled water for many – though there is also a small minority which takes to tap water consumption due to economic reasons, as mentioned in 5.1.1.

Figure 5.3: Local purification plants in the Ermita Zaragoza housing complex, Iztapalapa. Author's pictures, 2014

5.1.3 Bottled Water

Bottled water is the main source of drinking water in Mexico City and also amongst those interviewed for the present study – as much as 41 of all 44 interviewees reported buying bottled water for this purpose. Over the last decade or so, bottled water has gained a strong visibility in Mexico City's urban landscape as its increased consumption is paralleled – and sometimes also reinforced – by the massive proliferation of water vending points and local purification plants throughout the city. Mexico City's growing bottled water market is divided between the three major brands (Bonafont, Electropura and Ciel) and "unknown brands", as one interviewee called them (Raul E13: 92). The latter type is produced and bottled by small independent purification plants on the neighborhood level (see Fig. 5.3).

The bottled water market displays a strong spatial differentiation, as local purification plants are common in Iztapalapa but hard to find in the inner city where brands are predominant. Explorative walks and mappings conducted by the author in two of the studied neighborhoods of a similar size found a total of 14 purificadoras in Ermita Zaragoza, a large housing complex in Iztapalapa, but only three in the neighborhood of San Rafael, Cuauhtémoc, where shops and stores selling branded bottled water were more common by far (see Tab. 5.1).

Table 5.1: Number of shops selling branded and non-branded bottled water in San Rafael, Cuauhtémoc, and the Ermita Zaragoza housing complex, Iztapalapa. Own elaboration based on field survey in January 2014 (refers to refillable 20-liter-jugs only)

Neighborhood	Local purification plants	Stores selling branded bottled water	Total
Ermita Zaragoza, Iztapalapa	14	11	25
San Rafael, Cuauhtémoc	3	27	30

This pattern of brands providing bottled water for the city center, and local purification plants for the more peripheral Iztapalapa also emerges from the interview material. As much as 15 interviewed bottled water consumers from Iztapalapa drank water from local purification plants, whereas in Cuauhtémoc, all but six of those drinking bottled water resorted to brands instead (see Tab. 5.2).

Table 5.2: Number of interviewed households consuming bottled water (distinguished by type) in the boroughs of Iztapalapa and Cuauhtémoc. Own elaboration based on 41 interviews

Borough	Bottled water from local purification plants	Branded bottled water	Brands for drinking; local bottled water for other purposes	Total
Iztapalapa	15	3	2	20
Cuauhtémoc	6	15	0	21

In general, local, unbranded bottled water serves as a low-cost alternative to brands. In consequence, a clear distinction between the boroughs of Cuauhtémoc and Iztapalapa is not only observable in the types of bottled water on offer but also in price levels. Whereas this might to some degree reflect differences in domestic purchase power, it remains to be studied in how far this market is more demand- or supply-driven. Local brands for Iztapalapa, established ones for better-off neighborhoods in the city center: Such a differentiation between bottled water types on sale in specific parts of the city reflects what Di Nucci has termed an inferior and superior cycle of the (bottled) water economy (see Di Nucci 2011). Without doubt, even bottled water from a local purification plant comes at a much higher absolute price than tap water (supposing the latter were actually considered potable). The field study revealed that in 2013/14, prices for the home-delivery of a 20-liter jug ranged between 8 and 23 Mexican peso (0.45–1.30€)[13] for local, non-branded water, and 30 to 38 Mexican peso (1.69–2.14€) for water bottled by one of the major brands such as Bonafont. Bulk bottled water prices also showed some spatial differentiation throughout the city. Home delivery prices for branded bottled water in bulk seem to be largely stable throughout the city – minor differences could be explained by delivery costs. It is a whole different story with prices for local brands, which tend to stick to a very similar level within the same neighborhood, but varied significantly when compared on a larger scale. At the time of the mapping of local water selling points and purification plants in early 2014, prices for a 20-liter refill at a local purification plant ranged from 12 to 18 Mexican peso (0.68–0.90€) in San Rafael, and from 8 to 10 Mexican peso (0.45–0.56€) in Ermita Zaragoza. Reported bottled water prices from 23 interviews, covering all 12 studied neighborhoods further strengthen the impression of a profound spatial differentiation (see Tab. 5.3).

Despite the burden it puts on the household budget, bottled water is often considered more convenient than tap water pre-treatment, as apart from a bottled water dispenser, no special devices or additional running costs (e.g. for replacing filter cartridges or disinfection solution) are required to get potable water, and there is no waiting time as during pre-treatment. However, bottled water consumption does not necessarily come effortless. Brands and local bottled waters alike – for an additional fee of some pesos, Mexico City's water vendors and local purification plants typically offer home delivery of the bulky 20-liter jugs. Ordering bottled water to the home is a common practice in colonias populares and in housing complexes dominated by multi-storey dwellings. Whereas more than half of all interviewees relied on home delivery, the purchase of bottled water has become a modern-day manner of water fetching for others, who usually use simple hand carts to transport 20-liter jugs. In order to save on expenditures for home delivery, nine out of 41 bottled water consumers interviewed for the present study took to stores or purification plants themselves to purchase or refill their jugs.

13 All conversions from Mexican peso to Euro in this subchapter are based on currency exchange rates of November 1, 2013 (1 MXN = 0.0564€).

Table 5.3: Bottled water prices (for a refill of a 20-liter jug, home delivery) according to borough and neighborhood. (The usual deposit on refillable jugs is not included in this calculation.) Own elaboration based on 23 interviews

Borough	Neighborhood	Local bottled water, bulk home delivery, per 20 liters	Branded bottled water, bulk home delivery, per 20 liters
Iztapalapa	Miravalle	0.48€	n.a.
	El Manto	0.97€	1.75€
	Ermita Zaragoza	0.56€	n.a.
	Fuentes de Zaragoza	0.79€	n.a.
	Chinampac de Juárez	0.68€	1.80€
	Presidentes de México	0.70€	n.a.
Average Iztapalapa		0.70€	1.78€
Cuauhtémoc	Paulino Navarro	0.79€	n.a.
	Roma Norte	1.30€	n.a.
	Tabacalera	1.24€	1.80€
	San Rafael	Self-service: 0.79€	1.97€
	Cuauhtémoc	n.a.	2.14€
	Tlatelolco	n.a.	1.80€
Average Cuauhtémoc		1.11€	1.93€

At 52 liters of bottled water per capita and month, average levels of bottled water consumption amongst the interviewees exceed the findings of previous studies (such as Monroy/Montero 22.01.2014). The use of bottled water throughout the city is quite heterogeneous as consumption levels per capita reveal strong spatial differences, with Iztapalapans consuming twice as much as residents of Cuauhtémoc. The average per-capita consumption of bottled water of interviewees living in Iztapalapa was 72.4 liters per month, compared to 33.6 liters, in the borough of Cuauhtémoc. How can these elevated and varying levels of bottled water consumption be explained?

First, the steadiness of supply seems to make a difference, with higher per-capita bottled water consumption in areas experiencing non-permanent water supply. On average, all 30 interviewees who perceived water supply in their dwelling as permanent consumed roughly 49 liters of bottled water per capita and month and consumption rose to 60 liters amongst those supplied intermittently or subject to water rationing. This indicates that to some degree, bottled water consumption could serve as a means of substituting tap water and adapting to non-permanent water availability.

Second, and more importantly, bottled water is often used for a range of other purposes, too – and this practice is at least in part related to perceived and experienced tap water quality, which can vary locally, as we will see. Three different kinds of bottled water users can be identified from the interview material: Those who use it exclusively for drinking, those who both drink and cook with it, and those who employ it for multiple domestic purposes. These three user types hold almost equal shares: Of all bottled water-consuming interviewees, 14 used it for drinking and several other purposes, 12 for both drinking and cooking, and another 15 only for drinking.

Two aspects immediately catch the eye when comparing the amount and purpose of bottled water consumed in both boroughs (see Fig. 5.4 and Fig. 5.5): Iztapalapans generally use more bottled water, and they employ it not only for drinking but also for other domestic tasks. This indicates that in several of the studied neighborhoods, tap water is not even deemed fit for more technical purposes by its residents. As much as 17 of all 25 interviewees using bottled water at least for drinking and cooking lived in Iztapalapa, as well as all but one of the 14 interviewees using bottled water for a whole range of purposes – from drinking and food preparation to cleaning teeth. In contrast, bottled water was used exclusively for drinking by just two interviewees from that borough – a practice predominant amongst residents of the borough of Cuauhtémoc. This alone provides a strong indication that the quality of tap water is perceived as much better by residents of the borough of Cuauhtémoc than by those living in Iztapalapa. Apart from serving as a direct reference to a different socio-economic status of these two boroughs, the substitution of tap water by bottled water as a source of water not only for drinking but also for spectrum of other domestic purposes is likely to play a prominent role in the deepening of social inequalities through spatial disadvantages. This accounts particularly for Iztapalapa with its less well-off population and the given deficiencies in water infrastructure.

As they are strongly linked to questions of tap water supply both in terms of quality and quantity, the kind of water employed for drinking, and the use given to bottled water can often be tracked down even further, to the neighborhood level. Table 5.4 gives an overview on interviewees' potable water sources and predominant bottled water use for each of the twelve studied neighborhoods.

Table 5.4: Potable water sources and predominant bottled water purposes, by borough and neighborhood. Own elaboration based on 44 interviews

Borough	Neighborhood	Bottled water predominantly used for
Iztapalapa	Miravalle	Multiple uses
	Fuentes de Zaragoza	Multiple uses
	Ermita Zaragoza	Multiple uses
	Presidentes de México	Multiple uses
	El Manto	Drinking and cooking; Multiple
	Chinampac de Juárez	Drinking; Drinking and cooking
Cuauhtémoc	Tabacalera	Drinking and cooking
	San Rafael	Drinking; Drinking and cooking
	Paulino Navarro	Drinking
	Roma Norte	Drinking
	Cuauhtémoc	Drinking
	Tlatelolco	Drinking

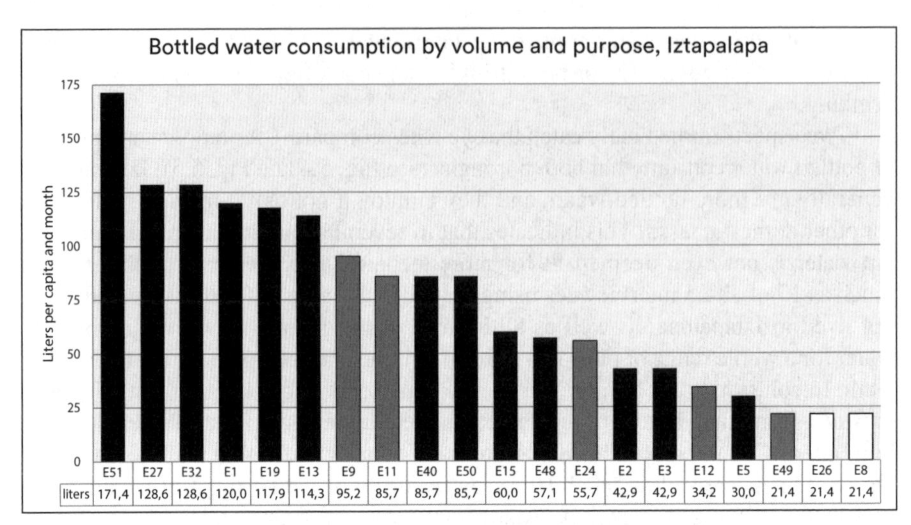

Figure 5.4: Average bottled water consumption in liters per capita and month; interviewees from Iztapalapa. Cases are distinguished by purpose of bottled water use: Black: Multiple uses; Grey: Drinking and cooking; White: Drinking only. Own elaboration

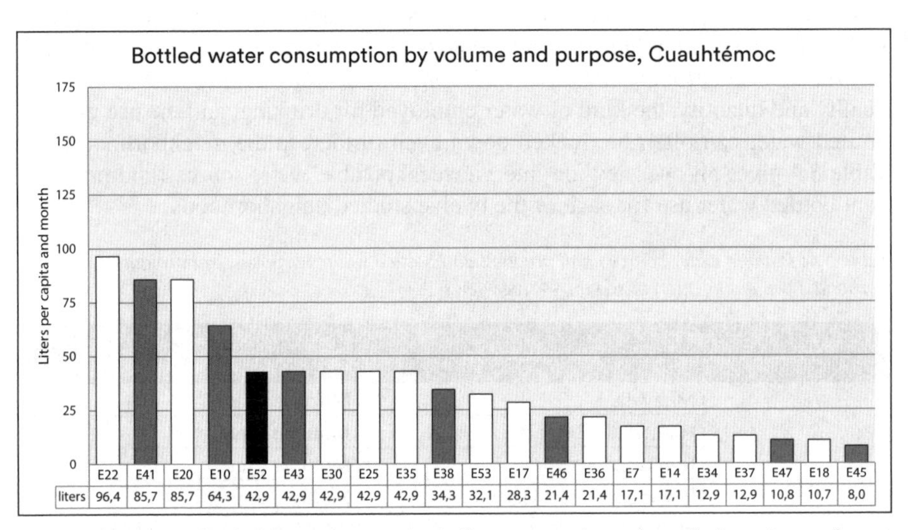

Figure 5.5: Average bottled water consumption in liters per capita and month; interviewees from Cuauhtémoc. Cases are distinguished by purpose of bottled water use: Black: Multiple uses; Grey: Drinking and cooking; White: Drinking only. Own elaboration

Bottled Water for Drinking Only

An exclusive use of bottled water for drinking was predominant in five of six neighborhoods studied in the borough of Cuauhtémoc, including the neighborhood of the same name, Roma Norte, San Rafael, Tlatelolco and Paulino Navarro. It is worth noting that all 15 interviewees who drank bottled water but did use tap water for all other domestic tasks perceived water supply in their neighborhoods as permanent. Remarkably, only two of these bottled water drinkers did not live in the borough of Cuauhtémoc. Both were from the Chinampac de Juarez housing complex in Iztapalapa, where water supply is permanent. In contrast to their immediate neighbors, both Silvia and Dolores voiced no concern over the use of tap water for cooking, washing fruits and vegetables and cleaning teeth. Limitations seemed to become less important in comparison with other neighborhoods they had lived in or where they knew about tap water quality.

> *As it doesn't stink like everywhere else, I can't complain about [tap water] quality.* (Silvia E26: 26, Iztapalapa)

Both agreed that tap water can be used for food preparation because it is boiled:

> *As it boils with the meal, supposedly that's when the microbes in the water die. But drinking from the tap? No. (...) I have never tried tap water.* (Dolores E8: 125–131, Iztapalapa)

In contrast, many bottled water drinkers from Cuauhtémoc expressed their doubts about an alleged limited water quality which did not entirely match what they experienced in their neighborhoods. For sure, some such as Pablo from San Rafael and Carlos from Roma Norte mistrusted tap water in general, fearing for their health:

> *I have never tried it, and I wouldn't dare to drink directly from the tap. That would be like saying: "Ha, I want to go to the hospital". (…) I think it's not clean at all; it's not ready for drinking*. (Carlos E35: 122–124, Cuauhtémoc)

Interestingly, drinking bottled water instead of tap water was linked to their respective housing conditions by two interviewees. Living in a flat in a turn-of-the-20th-century building in San Rafael, Antonia preferred bottled water as she felt that the building's internal installations, in particular some asbestos tanks, might affect tap water quality to a degree that does not allow for pre-treatment at home. For one of the residents of the Paulino Navarro neighborhood, a lack of cleanliness in the cistern shared by ten households in the small housing complex he lives in was the reason for drinking bottled water:

> *I don't drink from the tap. (…) Because (…) there are the tanks and all that. [...Our tap water smells moldy…] since the tanks aren't cleaned very frequently.* (Jorge E17: 141–143 and 215, Cuauhtémoc)

Hence for Jorge, it was the maintenance of domestic storage devices rather than local tap water quality per se that was thought to have an impact upon the quality of the water running from a domestic tap. This was supported by the fact that the three other interviewees from the Paulino Navarro neighborhood drank tap water without any pre-treatment, some combining it at times with bottled water for drinking. Where tap water was perceived as quite clean, questions of taste moved to the fore – such as those voiced by residents of the Tlatelolco housing complex:

It's risky [to drink tap water] because it is dirty. You don't see that at first glimpse (…). It does run clear but tastes a bit of chlorine, which is unpleasant. It's not tasteless. (Esther E7: 152–156, Cuauhtémoc)

In effect, a neutral taste becomes the very definition of good and potable water, as Martina from the same neighborhood – who had recently installed a water filter at her kitchen sink to obtain water for food preparation and save money on bottled water – explains:

It tastes of chlorine, therefore I don't like to cook with tap water either (…). The less taste the water has the better it is. (…) I relish E-pura [bottled] water (…). It doesn't taste of anything. Prepare your coffee with tap water, and prepare it with bottled water – that's a different thing. (Martina E53: 158–162, Cuauhtémoc)

The hegemonic belief that Mexico City's tap water should not be drunk at any cost is widely supported by people's perception of limitations in water quality in terms of color, smell and taste. Multiple use of bottled water is often the consequence. In contrast, a majority of those interviewees who use bottled water only for the purpose of drinking hardly perceived any limitations in water quality in their respective neighborhoods – Cuauhtémoc, San Rafael and Roma. Even though they did not go as far as drinking it, the debate over tap water potability became a more abstract issue to them.

Supposedly water from the tap is harmful – who knows if that's true or not, but it's a custom [not to drink it]. (Eva E22: 100, Cuauhtémoc)

Despite relying on bottled water for drinking, some even challenged what they suspected to be a "myth" of undrinkable tap water. Take Ana, a resident of San Rafael, who has an academic background in chemistry and prefers to drink bottled water yet was critical about the idea of non-potable tap water:

I don't drink from the tap because there is this myth – I tell you I have never verified it – that the water is not entirely potable. Also, the amount of bacteria it might contain, there are traces of metals and all that (…). You can get used to germs, the intestine gets used to it, but [not] the amount of metal (…). [A former flat mate] said: "But this is foolish – I'll drink tap water." She always [...did] and it seems as if nothing ever happened to her. (Ana E36: 74–76, Cuauhtémoc)

Ana was not alone with this critique; two other residents of the city center whom I interviewed also questioned the validity and origins of the established system of beliefs related to Mexico City's tap water quality. Crucially, this seemed to be related first and foremost to their own perceptions, which differed from the experiences with low water quality reported from Iztapalapa and elsewhere. Alan for instance, who also lives in the neighborhood of Cuauhtémoc, said he generally drinks bottled water, yet mistrusts bottled water quality at least as much as tap water.

It doesn't smell, it doesn't taste; in fact, I sometimes drink it when I have no bottled water. (…) I believe tap water has never done me any harm. (…) I do believe (…) that it's not that clean (…). But I think it's also a myth, created by the companies (…), the distributors of bottled water. (…) In the end, who guarantees that it is 'clean', in a manner of speaking? (Alan E30: 151–153, Cuauhtémoc)

Fabiola from the Roma Norte neighborhood voiced similar doubts based on her own experiences at home and abroad, and also spoke about a possible "myth":

[Tap water] has a strange taste (...). Therefore we prefer to buy purified water rather than boiling it (...). That's a problem, actually, because one should have access to potable water without having to pay for it all the time. (...) when you go to other countries, you can drink from the tap and it's the most normal thing. Here (...) everybody tells you that you can't drink it. (...) I don't know if that's myth or reality. My boyfriend for instance (...) drinks it. But (...) from early on in childhood we are told that you shouldn't drink tap water, ever. (Fabiola E37: 38–40, Cuauhtémoc)

Motivated by these reflections, and based on their own positive experiences with tap water quality at their homes, both Alan and Fabiola announced that they were planning to abandon bottled water consumption altogether and take up home purification. Chapter 8.3 will further discuss the debate around a widely perceived non-potability of Mexico City's tap water.

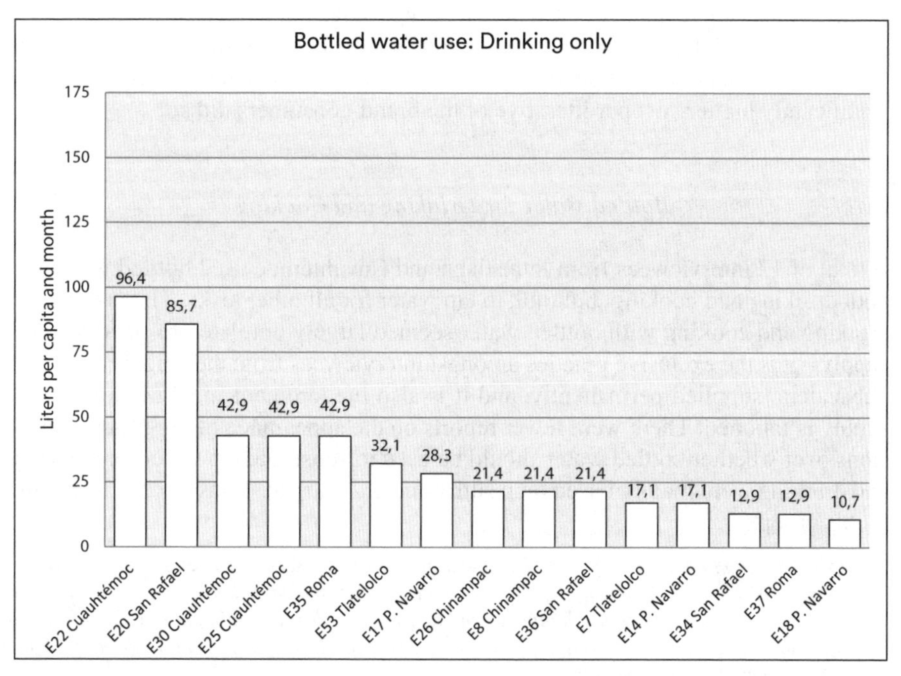

Figure 5.6: Consumption of bottled water per month and capita amongst the 15 interviewees using bottled water for drinking only. Own elaboration

As was to be expected, monthly bottled water consumption was significantly lower amongst those only drinking it when compared to the two other groups – 33.7 liters per capita on average. With the exception of Eva (E22) and Antonia (E20), both middle class residents of the borough of Cuauhtémoc, consumption levels of bottled water drinkers ranged between 10 and 43 liters per person and month. Surprisingly, consumption was highest in the well-off neighborhood of Cuauhtémoc (see Fig. 5.6), where quality limitations were scarcely reported by the interviewees. In San Rafael and Roma Norte, consumption varied strongly, and it was lowest in the neighborhood of Paulino Navarro, where it was mostly combined with tap water consumption. While interviewees' age and occupational status could also exert some impact on consumption levels, it seems as if these differences can mainly be traced back to the overall social status of the neighborhood, with the most bottled water drunk by residents of the better-off areas such as the neighborhood of Cuauhtémoc near Paseo de la Reforma. Where bottled water was exclusively used for drinking, average expenditures per household member and month remained at 2.83€ in spite of a much lower per capita consumption compared to the two other groups. In other words, those who only drank bottled water seemed to pay a higher price per unit. This could be explained by the fact that eight interviewees used bottled water by one of the well-known brands, whereas the remaining five resorted to local purification plants. This choice also seemed influenced by an absence of purification plants in the neighborhoods of Cuauhtémoc and Tlatelolco. Moreover, it was typical for residents of Cuauhtémoc who drank relatively low volumes of bottled water (11 to 21 liters per capita and month) to purchase branded water in smaller containers at local supermarkets and stores rather than 20-liter jugs. At a significantly higher cost per liter, five of the brand-consumers did so.

Bottled Water for Drinking and Cooking

A total of 12 interviewees from Iztapalapa and Cuauhtémoc used bottled water for both drinking and cooking, but took to tap water for all other tasks. The practice of drinking and cooking with bottled water seemed largely unrelated to steadiness of supply – it is the exclusive practice amongst interviewees from the neighborhood of Tabacalera, supplied permanently, and it is also predominant in El Manto, where supply is rationed. There were fewer reports on the appearance of tap water – decisions over whether bottled water should be used to substitute tap water for cooking and drinking were mainly based on health concerns and questions of taste for many in this group.

> *I was in Colombia and saw people drinking at the sink and it seemed something inconceivable to me. Who has grown up with the idea that water is dirty, that you shouldn't drink it, that you must be careful… (…) I almost fainted when I first saw a colleague [in Bogota] who opened the tap and started to drink. It would be unthinkable here in Mexico (…) – you would die.* (Elena E43: 38 and 380, Cuauhtémoc)

> Clara: *…We don't trust it – water just gets here too chlorinated, and as far as I know, chlorine can also cause health problems…*

> Miguel: *And the other issue [is…] the municipal network. It's a very old network with iron pipes and pipes of asbestos, so that is reflected for instance in the sediment which forms (…) both in the cisterns and roof-top tanks. (…) Water is probably quite potable to begin with, but it gets contaminated in the [distribution] network.* (Clara and Miguel E10: 251–257, Cuauhtémoc)

In addition, there was more evidence that the use of bottled water for anything other than drinking had substituted the pre-treatment of tap water as the local water quality worsened. Such was the case with Amelia, who resides in the Tabacalera neighborhood since 1972, and abstained from filtering in favor of bottled water when she perceived a deterioration of tap water quality during the past five years. In a similar logic, despite stressing that *"one should mistrust tap water"* (E11: 15), David from the El Manto neighborhood had recently perceived such improvements in tap water quality that he hoped to soon be able to give up bottled water consumption and obtain potable water from a domestic water filter – a practice his family had given up over a decade ago. On the other hand, Teresa from Ermita Zaragoza and Alma from Miravalle were exceptions in their respective Iztapalapan neighborhoods in using bottled water only for cooking and drinking but not for other purposes – despite a reported low tap water quality. In part, this seemed to be motivated economically – when the interview was conducted, 75-year old Teresa, who spend 0.63€ per capita and month on bottled water, had just returned from the local purification plant hauling a 20-liter jug of water on her hand cart in order to save on home delivery costs.

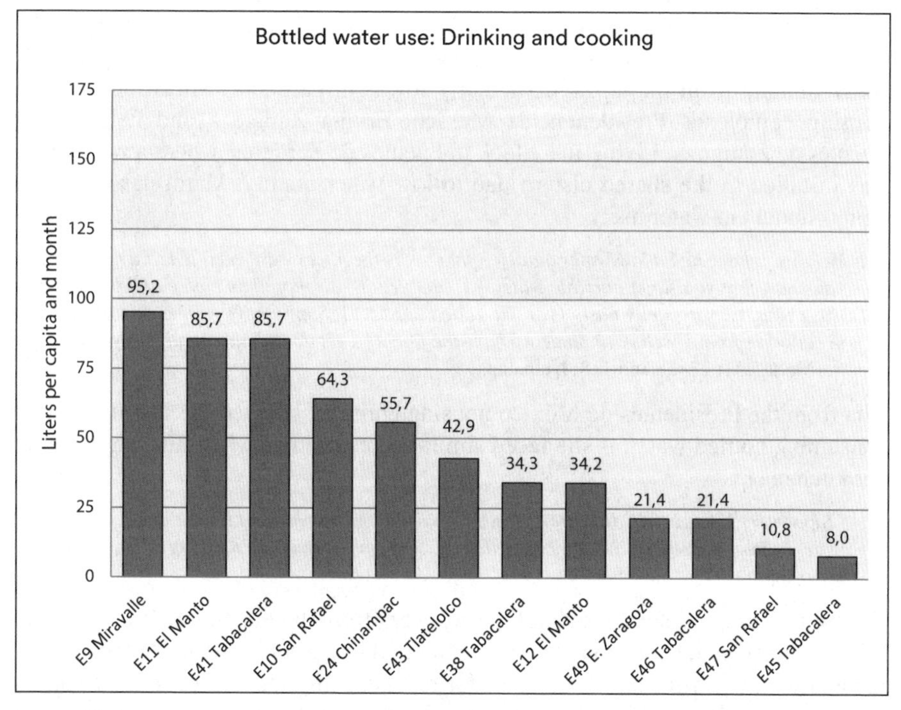

Figure 5.7: Consumption of bottled water per month and capita amongst the 12 interviewees using bottled water for drinking and cooking. Own elaboration

At 46.6 liters per capita and month, the average consumption of those using bottled water for drinking and cooking was lower than amongst multiple users. However, the consumption levels also showed some variability, ranging from 8 to 95 to liters per capita and month (see Fig. 5.7). This could not entirely be attributed to similar supply conditions in certain neighborhoods, as the case of David (E11) and Regina (E12) from El Manto highlights. Both perceived water supply in their dwelling as non-permanent, but each member of David's family consumed almost three times as much as Regina's. There were strong differences in consumption levels between neighbors particularly in El Manto, Tabacalera and San Rafael. The highest consumption was registered in the household of Alma (E9) – unsurprising given the multiple bottled water use with high consumption levels which was found to be the norm in her neighborhood. As for expenditures, households using bottled water for both drinking and preparing food spent 2.71€ per capita and month on this product on average. Six of them – those from Iztapalapa – obtained it from local purification plants; the other half purchased branded bottles.

Multiple Use of Bottled Water

Making use of bottled water not only for drinking but also for several other domestic tasks was a practice found almost exclusively in the borough of Iztapalapa. The practice is most common in those neighborhoods where the perceived quality of water running from the tap is particularly low – in particular Miravalle, and the housing complexes Presidentes de Mexico, Ermita Zaragoza and neighboring Fuentes de Zaragoza. Living in a block in Fuentes de Zaragoza where a water filter was installed in the shared cistern due to low water quality, Marta describes her doubts about tap water use:

> *We always have had a low water quality – that's why the filters were installed. If a filter breaks, you know that you'll get horrible water. (...) not yellow, it's more like coffee and often there's trash in it. (...) it smells nasty (...) like rotten eggs (...). Supposedly we get this low quality because it's from a well. (...) Since we have the filters (...) it doesn't taste bad, but you mistrust it.* (Marta E51: 28–38 and 118, Iztapalapa)

Rita from the Presidentes de México housing complex stressed the inevitability of consuming bottled water as she faced similar shortcomings when filtering water at her kitchen tap.

> *[Tap water] is extremely disgusting. I even have a filter at home, but I don't drink [the filtered water] directly because the filter is yellow – it doesn't stay clean for long.* (Rita E48: 167, Iztapalapa)

Led by similar experiences, several interviewees from the Fuentes de Zaragoza and Ermita Zaragoza housing complexes used bottled water not only for drinking, cooking and washing fruits and vegetables for raw consumption, but even for cleaning their teeth. Such a multiple bottled water use is also the predominant practice in Miravalle, where tap water quality is perceived as particularly low, and is also classified as such by the authorities.

> *Water is very filthy; (...) black, with a bad smell (...). No, I have never tried it. Neither would I dare to try it – I wouldn't even clean my teeth with this water.* (Gabriela E19: 160 and 232, Iztapalapa)

> *I get about three or four garrafones per week, because I use it (...) for everything, for chicken, coffee (...). I use tap water for things that will be boiled and strained, nopales for example. But for soup (...), everything that goes into food, it's all bottled water.* (Maria E2: 234–239, Iztapalapa)

A multiple bottled water user who was an exception from the usual practice in her neighborhood was Mercedes from Chinampac de Juarez. Other interviewees from this popular housing complex in Iztapalapa, which is supplied with tap water permanently, took to bottled water for drinking only, or for drinking and cooking. Yet Mercedes even objected to the use of tap water for dental hygiene, as she mistrusted the maintenance of shared storage facilities.

> *[I don't like to cook with tap water] for the same reason: I told you it gets from the cistern into the roof-top tank and from [there] to the house. Therefore, the water sometimes smells very nasty (...) or pigeons get up there and shit into the tanks. (...) It's bottled water [for cleaning teeth]* (Mercedes E27: 87–100, Iztapalapa)

Despite this link to domestic storage, relying on bottled water to fulfill multiple domestic tasks was common in neighborhoods with rationed and those with permanent water supply. In this context, perceived tap water quality seemed to matter much more than steadiness of supply – apart from the extraordinary case of Carmen. For the only multiple bottled water user from the borough of Cuauhtémoc, mistrust in tap water quality accumulated with problems stemming from an intermittent water supply in her neighborhood.

> *There is no maintenance of the pipes (...), maybe the purification is good but the pipes (....) are broken, so they get contaminated. Imagine there's the water pipeline and the sewage pipeline, and both are broken, that's a fundamental problem. (...) I guess that in order to avoid that (...) they put in a lot of chlorine. That's no solution. (...) I can't trust it – imagine I'm cleaning [my teeth] with who knows what kind of water. For showering, it's unavoidable – no way am I going to wash myself with bottled water. I only use it when there is no [tap] water – I have no alternative.* (Carmen E52: 138–140 and 158, Cuauhtémoc)

At only 0.4 m^3, the storage capacity per dwelling in the building's shared storage tank is relatively low, and Carmen often returns home from work just to find that water tanks have already been emptied by her neighbors. While all other interviewees not only from her neighborhood but also from various parts of the borough of Cuauhtémoc generally used bottled water for drinking and maybe cooking – and (very few) took to tap water consumption – Carmen was the only one to use it for a wide spectrum of domestic tasks. She was also the only multiple bottled water user in the entire sample to purchase it not in bulk but in smaller containers from various renowned brands at a local supermarket. This was common practice in the borough of Cuauhtémoc, mostly for reasons of convenience, and for the lack of local purification plants in the area. However, most others who did so mostly only drank bottled water, and had a much lower per capita consumption. As the water in smaller packages comes at an elevated cost, Carmen spent as much as 5.32€ on average per month on the 43 liters of bottled water she consumed.

All in all, multiple bottled water use is a strongly Iztapalapan practice, where steadiness of water supply seems to have a less relevant impact than (perceived) tap water quality. Multiple bottled water users in the present sample consumed 87.7 liters per capita and month on average, and all but three of them purchased their bottled water from local purification plants, where it comes at a lower cost than Bonafont, Ciel or Electropura. Even so, bottled water put a burden of 3.31€ per capita on the average multiple user's monthly household budget. Residents from the Fuentes de Zaragoza and Ermita Zaragoza housing complexes had the highest levels of bottled water consumption per capita amongst all multiple users, together with those from Miravalle (see Fig. 5.8). Consumption levels of multiple bottled water users ranged from 30 to 171 liters per capita and month. Yet remarkable variations in consumption levels could be observed amongst residents of the same neighborhood – note for instance the case of Maria (E1) with 120 liters per capita and month, twice as much as Diana (E15). Both are from the same neighborhood of Miravalle with its rationed supply and low water quality.

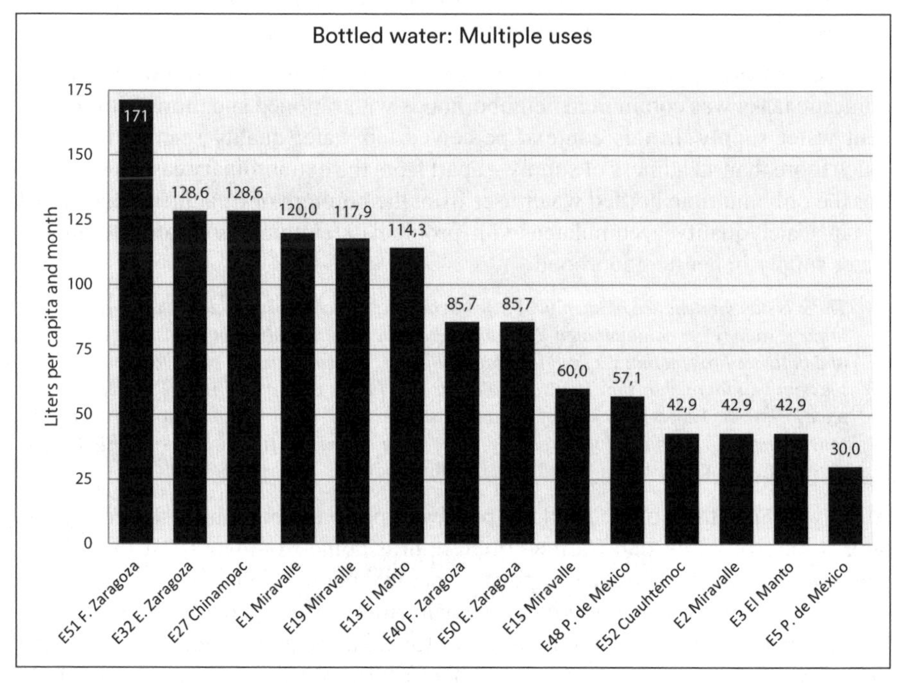

Figure 5.8: Consumption of bottled water per month and capita amongst the 14 multiple bottled-water users in the sample. Own elaboration

So what does bottled water consumption tell us about the mode of drinking water supply in Mexico City? Whether due to actual limitations of tap water quality, or a persistent myth of it being unfit for human consumption – in Mexico City, it is nowadays almost exclusively bottled water which is perceived as potable in the truest sense of the word, leading to increased domestic spending. On average, interviewees consuming bottled water[14] spend 2.97€ per capita and month to fulfill this basic need. Monthly expenditures amounted to 8.93€ for an average bottled-water-consuming household – corresponding to roughly 9% of the Mexican minimum

14 All calculations on expenditures are based on those 38 interviews providing information on how much the household paid for bottled water. No such data was available for three bottled water consumers, and another three interviewees did not consume bottled water at all.

wage[15]. It is surprising that while average consumption levels vary impressively between the three purpose groups, expenditures show only very moderate differences (see Tab. 5.5). This indicates that Iztapalapans, who tended to use bottled water for a whole range of purposes as they faced low tap water quality, paid less per liter (in absolute terms[16]) than their Cuauhtémocan counterparts who mostly purchased bottled water for drinking only. Less present and less used in the borough of Cuauhtémoc, the existence of local purification plants offering lower prices than established brands seems to be the main reason.

Table 5.5: Bottled water consumption by purpose – average volumes and expenditures per capita and month. Own elaboration based on 44 interviews

Purpose	Average consumption per capita and month*	Average expenditures per capita and month**	Practice predominant in
Drinking only	33.7 liters	2.83€	Cuauhtémoc
Drinking and cooking	46.6 liters	2.71€	–
Multiple uses	87.7 liters	3.31€	Iztapalapa

* Average consumption based on 41 interviews where volumes were reported.
** Average based on 38 interviews where expenditures for bottled water were reported.

Complementing indications from the existing literature (see for instance Constantino Toto *et al.* 2010: 250 f., CEDS 2010: 52–54, Inter-American Development Bank (IADB) February 2010, or Monroy/Montero 22.01.2014), the findings of the present study clearly show that in Mexico City, bottled water consumption is by no means neutral with respect to social status. First of all, the use of bottled water is ubiquitous in Mexico City across all social spectrums – rather than serving as a means of social distinction and conspicuous consumption – as is the case with certain mineral waters in Europe (see Strang 2004). But more importantly, bottled water is a clear marker of a low social status, with often under-supplied Iztapalapans employing elevated levels of this product not only for drinking but also for a larger spectrum of domestic purposes. In a highly spatialized manner, this is directly related to supply limitations (especially a perceived low tap water quality) in a considerable number of neighborhoods. It is hence mainly those living in these under-serviced neighborhoods who are integrated into an emergent bottled water economy, shouldering a financial burden which is likely to perpetuate social inequalities. In short, bottled water serves as a substitute for proper tap water supply

15 For 2013, the Mexican minimum wage for the Federal District was set at 65 Mexican peso (3.65€) per work day, adding up to around 95€ per month (see *Comisión Nacional de los Salarios Mínimos,* April 2015). Calculation based on 26 work days per month, and currency exchange rates of November 1, 2013.

16 The relative cost of domestic bottled water consumption with respect to people's household budgets, and possible implications hereof for the reproduction of social disparities remain to be studied elsewhere in the future.

in particular to many of Mexico City's poorer inhabitants and plays a role in reproducing their poverty.

Generally speaking, drinking water directly from the tap is an exception rather than the rule in Mexico City. The widespread mistrust is partly based on everyday experiences, which literally bring low water quality to the light, such visible floating particles in tap water. On the other hand, it seems to be nourished by a discourse which both reveals mistrust in public institutions (see Hamlin 2000 and Jiménez Cisneros *et al.* 2011: 81) and an aggressive marketing strategy by the beverage industry ("Drink two liters of Bonafont per day"). Particularly when it comes to drinking water – as opposed to water used for other purposes such as personal hygiene – the commodification of water in Mexico City is almost universal by now. To explain the practices and sources of drinking water as well as the consumed volumes, steadiness of supply in the respective neighborhood seemed to be of less relevance than the perceived quality of tap water. Whether multiple consumption or drinking only – it seems as if similar supply conditions and tap water quality (which could be expected to apply roughly for residents of one and the same neighborhood) were able to explain part of the practice of bottled water consumption.

5.2 HYGIENE AND CLEANING: TECHNICAL WATER AND THE BODY

In Iztapalapa, [tap water...] is dirty and may be used for mopping floors or things like that. But (...) using it directly as it is – no, [not] even if you boil it.
(Rosa E41: 208–210, Cuauhtémoc)

Not only with respect to drinking, specific water qualities are required to fill an entire range of needs for personal and domestic cleanliness. It becomes clear that a distinction is made between water for body-related purposes and those of a more technical nature. This chapter treats the examples of showering (as a body-related practice where water is used externally), doing laundry (as a more technical practice), and the domestic reuse of grey water. In particular the latter can only be fully explained by taking a hierarchization of required water qualities into account. Though a highly private matter, personal hygiene finds its most public expression in one of the bathhouses which can still be found in particular in some inner city neighborhoods of Mexico City, such as San Rafael and Paulino Navarro. Despite restrictions imposed during the period of colonization, the public bath has one of its main roots in the Aztec bathing culture[17]:

> "From the emperor down, residents of Tenochtitlan were encouraged to bathe daily. However, when the Spanish first encountered Native American populations in the sixteenth century, they were convinced that frequent bathing contributed to the massive smallpox deaths occurring

17 Another remainder of a strong pre-Hispanic bathing culture – which was apparently deformed but only partially overwritten by Catholicism in the course of colonization – is the joyful public water use during the *Sabado de Gloria* holiday. To the author's knowledge the widespread and now penalized tradition of publicly tossing water at each other on the street on the occasion of Joyous Saturday (preceding Easter Sunday) is not practiced anywhere else outside Mexico.

among indigenous peoples. (…) Europeans generally held that dousing the body in water opened the pores to evil forces (…) causing illness or death. (...) Moctezuma II, head of the empire when the Spanish first arrived in 1519, was known to bathe up to four times a day, and he encouraged daily bathing among the rest of the population. While the homes of the Mexica nobility and middle classes were constructed with private bathing quarters, bathhouses were established throughout the imperial capital for the poorer classes." (Bailey Glasco 2010: 91).

In particular the Aztec *temazcal*, a ritual steam bath with medicinal herbs in a specially constructed sweat lodge, can still be found in the periphery of Mexico City. It is praised for its health effects and recommended particularly for the post-natal period.

As a modernized form, bathhouses seem to serve for relaxation and health rituals to some degree, but were mainly introduced as predecessors for domestic bathrooms. Modern Mexico City's urbanization came with a proliferation of bathhouses during some time. Similar as in Western Europe, they covered the population's need for personal hygiene at a time when most residential buildings had no private bathrooms (for Mexico, see Ayala Alonso 2010). Their number seems to have diminished over the last decades – baths still to be found mostly stem from the late 19th century to the 1960s. They were prominent features in particular in those dense inner-city neighborhoods where tenants in *vecindades* often shared a single toilet and tap alongside some sinks for doing laundry in the yard. Clara and Miguel recall their childhood in such dwellings in the city center:

Clara: *My [older] sisters had a day off on Saturdays [...and] they took us to the baths. (…) We are seven women; (…) we went all into one of the rooms and took a bath (…). The heat, the steam – it's a nearness between sisters in the end.*

Miguel: *We used to live in vecindades, and (…) there was no bathroom (… and) much less a shower. (…) There were only toilets, so I believe that's why people had to go to public baths.* (Clara and Miguel E10: 674–679, Cuauhtémoc)

Apart from substituting a shower at home, baths were used for relaxation after heavy physical work, as in the case of Oscar, a former baggage porter.

I am the only one [in the family] who went to the steam [bath...]. My work was a bit heavy so I showered daily [... and] I gathered some tips to go to the steam. (...) When I was eleven years old, over there in Veracruz [... my uncle] took my brother and me to the steam once a week. That was a custom he and his brother had. (Oscar E14: 176–198, Cuauhtémoc)

Oscar and others also recalled that going to the public steam bath was an all-male ritual. In effect, public baths tend to have separate sections for women and men, and taking a steam bath seems to remain a highly gender sensitive issue. In some families the visit to the public bath was a traditional habit amongst relatives, but other than the temazcal, steam baths were often connoted with the male, popular-class body. A strong popular class issue and a family tradition in the past, attending bathhouses seems to have become obsolete over the last years. All in all, more than half of all interviewees (or one of their household members) had used such baths in the past, often sharing it as a childhood memory. Today, only a small minority continue to frequent them. Catalina from the Paulino Navarro neighborhood for instance reported that her husband and teenage son went to a local bathhouse every once in a while. Considering it unhygienic, she would never enter such a bath herself, and moreover was fearful of the social context her son might be exposed to in that particular setting. Moreover, bathhouses were attended for personal hygiene by some better-off inner-city dwellers as a mere emergency measure during supply suspensions or intermittence.

Private bathrooms are common in most dwellings today – however, around 17% of inhabited dwellings in the Federal District did not dispose of a shower in 2010 (see INEGI Censo de Población y Vivienda 2010). Even if the public bath serving as a bathroom to the popular class had lost most of its relevance over the last decade or so, the private bathroom still seems to hold some relevance as a status symbol in middle class housing. The actual use of bath tubs seems rather insignificant[18] – but newly constructed condominiums and apartment blocks in Mexico City are nowadays frequently promoted with a private bathroom or toilet assigned to each bedroom.

This directly leads to the question how water use is related to infrastructural conditions. The relation between practices of personal hygiene and the material spatial setting is analyzed for questions of supply steadiness and perceived tap water quality. Steadiness of water supply is not the only aspect potentially influencing domestic practices of hygiene and cleaning. Facing a limited availability of water considered potable, the water quality standards Mexico City's residents consider apt for practizing personal hygiene seem to be somehow lower than what is required for drinking and food preparation. For instance, a vast majority of interviewees used tap water for cleaning teeth – despite strong doubts over its potable quality. Eight of 44 interviewees opted against the use of tap water for this purpose and took to bottled water instead. They mostly were from neighborhoods with reportedly low water quality[19]: Miravalle, Ermita Zaragoza and Frentes de Zaragoza in Iztapalapa.

5.2.1 Showering

With respect to showering, tap water quality did generally not seem to impose any limits on its usability – despite some reports about skin irritations. Rather, water needs to be available to begin with – something taken for granted as long as water runs from the domestic tap without failure. Just how essential water is for personal hygiene comes as a shocking insight once water supply fails – take the first-person narrator in Vicente Leñero's autobiographic novel *La gota de agua*[20]:

18 In contrast to showers, bath tubs are a rare appearance in Mexico City's dwellings of the middle and popular class – found in no more than six of the 44 interviewed households, all of which were older buildings in the Cuauhtémoc and Tabacalera neighborhoods. Even there, most tubs seemed to be unused since a long time.

19 Rita from the Cuauhtémoc neighborhood is a rare exception, as she used bottled water to clean her teeth not so much because of tap water quality but because of the effects of intermittence and a lack of domestic storage.

20 Vicente Leñero (1933-2014) was a Mexican writer, playwright and journalist who most prominently published in *Proceso*. Originally trained as a civil engineer at UNAM, his 1963 novel *Los albañiles* (The bricklayers) gained him critical acclaim. *La gota de agua*, published in 1983, is a novel with strong autobiographic features, providing a contemporary record of

"There is no water. To these bad news, I woke up definitively grungy on Sunday, January 31. I thought I might be unable to open my eyes because my eyelids would be plastered with sleep, as hard as resistol. I felt prematurely filthy, sweaty, smelling like a goat, bearded. Stiff hair, crumpled face, black fingernails, the soul completely converted into a rubbish bag which I would have to carry around during the entire morning, afternoon and night of this unhappy Sunday.

– Don't exaggerate – said Estela when she heard me complaining."(Leñero 1983: 7)

As a basic need, personal hygiene is not only about health but also comfort and self-esteem. A lack of water undermines the autonomy to clean oneself and the dwelling when deemed necessary, disrupting rhythms of domestic everyday practices. Facing internal water rationing and spontaneous supply disruptions in the San Rafael apartment block he lives in, Pablo also spoke of shame and inconvenience.

... my girlfriend comes to visit, just imagine being without water for the toilet! That's a big problem, (...) it's uncomfortable or embarrassing. (...) Or you get up quite early to be ready and all, and when you open the tap, there is no water. So you know you have to be presentable (...) it destroys your entire routine. (...) Or you have (...) an appointment and you have no water – that's something you don't expect to happen. (Pablo E34: 236–240, Cuauhtémoc)

Living in the same neighborhood, a temporary lack of water in Antonia's dwelling during the first large-scale supply interruption she ever experienced generated a similar sense of hardship.

I had never experienced a lack of water – and back then there was not even water for the toilet, so it was very traumatic (...), all the dirty dishes, and to bath yourself in any possible way. We even used bottled water to clean a bit. (Antonia E20: 256, Cuauhtémoc)

When taking into account how those who are used to being supplied permanently experience temporary supply disruptions, the question emerges whether this also applies to intermittence. Daily showering is common amongst residents of Mexico City. 31 of 44 interviewees reportedly took showers every day – some even twice, citing hard physical labor or representability at work as reasons. Take the case of Jorge, an employee in a car workshop to whom daily showering was essential.

Imagine [being] without water! (...) I am used to take a shower daily, and I sweat (...) a lot, so imagine if I didn't wash myself, I (...) would feel bad. If I didn't shower, I'd be scratching my head because of the dirt the other day, no less. That's why [water...] is actually very important for me. (Jorge E17: 131, Cuauhtémoc)

domestic water use in the Federal District from an everyday perspective. Pre-dating the 1985 earthquake, when many inner-city residents experienced limitations in water supply for the first time, it elaborates on the experiences of a middle-class family in the neighborhood of San Pedro de los Pinos, borough of Benito Juárez, during a suspension of water supply in early 1982. Meandering between detailed, journalistic documentation (including sketches of installation plans) and seemingly non-fictional observations by the first-person narrator, it covers a period of six weeks between January and March 1982, describing the family's efforts to secure water availability and their growing distress given an unprecedented suspension of supply. Through the ideas embraced by the first-person narrator, the novel provides an example of the wide-spread technological optimism of the time.

It is noticeable that only eight of those 31 taking showers daily lived in dwellings where water supply was intermittent or rationed. But then again, the group of 13 interviewees who tended to shower every second or third day did not share similar supply frequencies – some lived in areas with permanent supply, some did not. Instead, what unites this group is that all but one of them took to domestic grey water reuse (see below) – as did 22 of those taking showers day by day. In effect, that hints toward a more general water-saving attitude rather than an influence of supply patterns. A remarkable logic lay behind the way water was assigned to members of the family in some of the non-daily showering households. Even as laundry was done only once a week in her household – on the only day water was running from the tap in the neighborhood of Miravalle – Alma's grandchildren showered daily and were given new school uniforms every day without any hesitation.

> *I take a shower every third day, but as they [Alma's daughters and grandchildren] go to school and to work, they shower daily.* (Alma E9: 252, Iztapalapa)

63-year old Dolores, also from Iztapalapa, made a similar difference between herself and her grandchildren. Despite a permanent water supply, Dolores is taking showers every third day only in order to save both water and gas for the water heater. Such a preferential water use and distinction between family members according to their leaving the domestic realm is not uncommon amongst popular class interviewees from Iztapalapa. Raul pointed out how his brother, who works as a technician repairing domestic water heaters, has to pay particular attention to his appearance due to his clientele:

> *In my family, my brother for instance showers daily. (…) because of his kind of job he is a lot on the street and goes into homes where people have money, so that's why he showers daily.* (Raul E13: 130–133, Iztapalapa)

A difference is hence made between a logic of cleanliness within the domestic realm, and another which applies to workplace and school – a place shared with other people. Interestingly, Raul works in public space himself, selling vegetables at a small stall in front of his mother's house in the El Manto neighborhood. The logic of cleanliness hence seems to vary – in a sphere where people of differing social status meet, standards seem to differ from those in one's own neighborhood or context.

Apart from bottled water consumption, the present study did not cover volumes of water used in everyday practices. Yet it has to be stressed that a similar frequency of showering does not automatically imply that a similar amount of water is used for this task. Showering with buckets is a way of saving water but also a necessity in those nine of the interviewed households which do not dispose of proper shower installations but simple drains. As the use of washcloths is uncommon, residents of these dwellings used to take their daily bath by occupying plastic buckets, scooping water with a smaller recipient. Called *"bañarse a jicarazos"*, this is almost exclusively a popular class practice by Iztapalapans, and implies that less water is used on an everyday basis – usually no more than 20 liters (one bucket) per person and bath. That is less than 10% of what recent estimates by the Federal District's water utility assign to showering given an unmindful water use – 264 liters per person and day (see Ávila 11.03.2013).

5.2.2 Doing Laundry

In addition to supply disruptions, varying water qualities pose another limit to domestic tap water use, as 5.1 has indicated. In some neighborhoods, tap water quality is perceived as too low for purposes directly involving the body – such as cleaning teeth, and bathing babies. Whereas water for more sensible purposes (and not just drinking) is sometimes substituted with bottled water, pre-treatment is also applied – and required – for other purposes, in particular for doing laundry. Though rather shy during the rest of the interview, Eugenia expressed anger upon remembering how she had to handle water prior to using it for laundry when living in a housing complex in the vicinity of Ermita Zaragoza and Fuentes de Zaragoza in Iztapalapa:

> *My white clothes actually got stained and I got very angry. (...) It left horrible stains [... which] could not be removed (...). I was really fed up for sure, furious about that. That's what annoyed me most [in Iztapalapa]. (...) Later on, I got into the mode and put up three rags as filters where water pours into the washing machine. Or I used water I had stored (...) – you could see some sludge down there, coffee-colored, nasty.* (Eugenia E46: 176–182)

Rather than striving for potability, measures like these are taken to make tap water apt for doing laundry and conducting other domestic cleaning tasks. Filters are installed directly at the water tap – typically in the kitchen. Another method is letting tap water rest in huge containers for some time to allow for sedimentation of floating particles. A similar effect is often obtained unintentionally in domestic cisterns where a layer of sediment tends to form after a certain period (see 5.3). And in the Iztapalpapan housing complex of Fuentes de Zaragoza, additional filters were installed directly in each building block's cistern in reaction to a particularly low water quality reported by residents and water supplier alike (see SACM 14.04.2014). Lucia highlighted the need for this measure in order to allow for basic use:

> *Before we installed [the filter, water] was dirty – it even stained our clothes.* (Lucia E40: 44, Iztapalapa)

As with Eugenia, this is probably an effect of elevated levels of iron and manganese reportedly present in the groundwater of this area. Water quality hence is a determining factor for the practice of doing laundry – and steadiness of supply has an even stronger impact: Households which receive water intermittently and do not dispose of sufficient domestic storage capacity adopted their laundry rhythm to the supply schedule. In the small ground-floor building without any cistern where she resides with four other tenant families, Guadalupe does laundry in the early morning before leaving the house for her job as a domestic worker:

> *Here, we usually have water in the morning, but (...) at 10, at 11, at 12, at one or two in the afternoon, it stops and will not resume until the other day. (...) Therefore, all that is done here with respect to laundry, that's in the morning. In the afternoon you can't do anything. (...) Usually I get up at six in the morning to do my laundry.* (Guadalupe E3: 80–84, Iztapalapa)

This practice is even more time-consuming if laundry is done manually, often resulting in several hours of work. 17 interviewees did at least wash some clothes by hand, and seven of them did so exclusively, as they did not dispose of a washing machine. (For details on this immediate use of water for doing laundry, see 5.3.1).

5.2.3 From Body-related to Technical: The Cascade of Domestic Water Use

Finally, water employed for body-related hygiene practices such as showering or washing clothes, where a higher quality is requested, is afterwards often collected and reused for other purposes. Essentially, this practice of domestic water reuse is based on a functional distinction between water types for different purposes – in other words, the transition from body-related water into technical water. As many as 35 of 44 interviewees took to reuse in one way or the other. It is noticeable that all those who did not undertake any reuse were living in the borough of Cuauhtémoc and enjoyed a permanent water supply. Yet surprisingly, the practice of reuse seemed to be widely independent from the steadiness of water supply in the current dwelling. Although all but one of the interviewed households who perceived water supply in their current dwelling as non-permanent were practicing reuse, another 11 interviewees also did – despite enjoying a permanent water supply. A concern for saving water in the interest of oneself and other residents of Mexico City – as voiced by many interviewees – provides the backdrop, while past experiences also seem to exert an influence (see 7.2).

Figure 5.9: Collecting grey water for reuse, Tabacalera neighborhood, Cuauhtémoc. Author's picture, 2014

Recycling domestic grey water is a simple, low-tech domestic measure common in Mexico City, where grey water is usually collected and stored in the dwelling itself.

Hence it becomes immediately visible upon entering the bathroom or yard whether a household practices reuse: Plastic buckets and containers of various sizes are set up for collecting water running off the shower or washing machine. The common types of washing machines are semiautomatic so that water in- and outflow can be managed manually. Figure 5.9 shows the tubs in the courtyard of Amelia's house in the Tabacalera neighborhood where she collects grey water from doing laundry, then reuses it for washing darker-colored clothes or mopping floors. Such multiple use of laundry water can involve two or even more steps, as Paula explains:

> *I separate clothes for the washing machine: first the light-colored, then the medium-colored, and at last those that bleed or are black. (…) I wash all three with the same water, and afterwards I use it for cleaning the dog's kennel.* (Paula E32: 186–189, Iztapalapa)

Besides fetching laundry water, the most common method to obtain water for reuse is to collect the run-off from the shower. Some stand in a small tub while showering, or keep a bucket in the shower cabin in order to retrieve some of the employed water. And the gas-fed water heaters installed in most homes in Mexico City require some preheating, providing another way to obtain water for reuse:

> *When I take a shower I have a bucket to collect the cold water. I use it for the toilet, or, if no soap got in, for my plants.* (Ana E36: 48, Cuauhtémoc)

Apart from flushing toilets, interviewees used grey water for a number of other purposes, including cleaning floors, yards and sidewalks, and watering plants. Typical reuse chains are showering – flushing toilets, showering – cleaning floors, or washing clothes – cleaning floors – cleaning sidewalks. In this sense, water reuse is mostly a way of down-cycling rather than literal recycling. Typically, water is used more than once in what can be called a hierarchical cascade of water use. Partly inspired by the World Health Organization's *Hierarchy of water requirements in emergency situations* (see WHO Water Engineering Development Centre 2011: 9.1), this scheme is derived from the empirical material and illustrates how water types are linked to specific purposes. By not only employing different kinds of water for different purposes but also by reusing water, the output of one practice is being turned in to the input of the next – in short, a cascade scheme (see Fig. 5.10). Such a relation is particularly noticeable in a context such as Mexico City, where drinking water is not only generally set apart from tap water but further distinctions between water types are also made.

The distinction between body-related and technical water use is a crucial one in this context. It seems as if in Mexico City today, the domestic tap is mainly a means of providing water for more technical purposes, while water for the highest ranked body-related practices (such as drinking) is generally substituted by bottled water. Yet according to the interview material, even bottled water consumption is often subject to a fine-grained distinction between branded water (for purposes of drinking) and the one obtained from local purification plants, which is used for food preparation or personal hygiene. Some Iztapalapan interviewees took to a sophisticated strategy, relying on as much as three different kinds of water for domestic use.

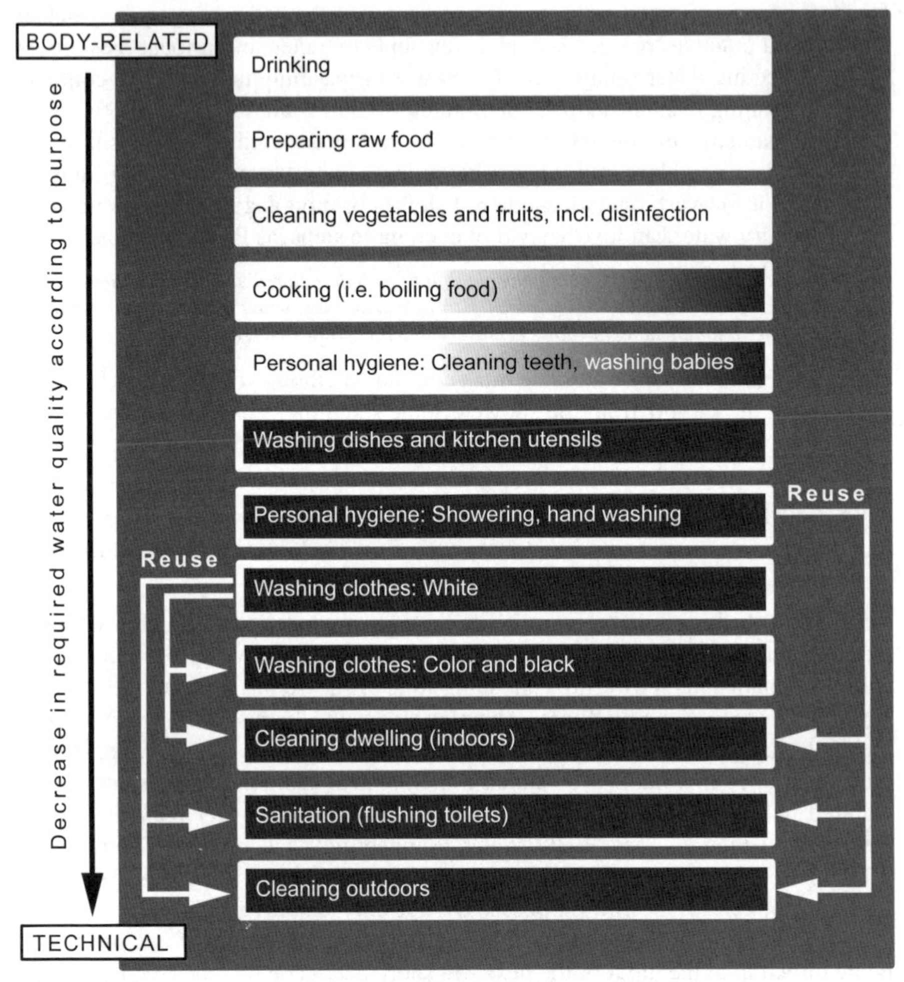

Figure 5.10: The cascade of domestic water use in Mexico City. White indicates a predominance of bottled water use, grey indicates tap water use. Own elaboration based on interview material

For instance, tap water was used only for practices such as showering, flushing toilets, doing laundry and cleaning the house in Felipe's family in Iztapalapa, who in addition managed two different types of bottled water:

> *[We use] water from the local purification plant (…) when preparing food (…), it boils and there are not so many problems. For drinking, well we feel that (..) a brand is (…) of better quality.* (Felipe E24: 100, Iztapalapa)

It is striking how different water types and their (perceived) quality were assigned precisely to certain purposes. All in all, three interviewees from housing complexes in Iztapalapa used only branded bottled water for drinking, water from a local purification plant for cooking, food preparation, and eventually for dental hygiene, and tap water for no more than the remaining domestic tasks. Such a fine-grained differentiation in the realm of the cascade of water use which directly involves the human body seems to be motivated by a combination of economic aspects – local bottled water coming at a lower absolute cost than established big brands – and with respect to the (perceived) quality of these waters.

When it comes to practices of hygiene and cleaning, generally speaking there is an impressive homogeneity in both practices and the employed water types – yet the steadiness of supply type or perceived water quality did make a difference at times. For the more technical purpose of washing clothes, tap water quality imposed a need for pre-treatment in some areas. Simultaneously, intermittent supply schemes dictate laundry routines – or require domestic storage to allow recuperating the autonomy of domestic rhythms (see 5.3). By and large, the practice of domestic water reuse seemed to be independent from the steadiness of water supply in the current dwelling, and more about past experiences (see 7.2). With respect to personal hygiene, a preferential water use for certain family members could be identified in some popular class households – independent of whether they were supplied permanently or intermittently. Those undergoing educational training or conducting wage labor were assigned greater importance than other family members when it came to domestic water use. This highlights the classed character of some water-using practices.

In addition, the specific water type employed for a practice can also make and mark a (social) difference. Not only is the multiple use of bottled water a typical popular class practice, it is also complemented by attempts of adapting the body to the given water supply conditions, in particular in terms of quality. A strategy of conditioning the body to a certain water type – as discussed in 5.1.1 by the example of offering children untreated tap water for drinking in order to desensitize their digestive system – is also reported with respect to personal hygiene. Living in an Iztapalapan neighborhood which is intermittently supplied with water of an often yellowish color and with suspended particles, Maria was reminded of the importance of increasing young children's' resilience rather than spoiling them:

> *When my children were babies (…) I sometimes would have liked to wash them with bottled water, but people told me: No, if you start bathing them this way (…) you will only get them used to it (…). It is better if from the very beginning, they adapt to the kind of water we have.* (Maria E2: 94, Iztapalapa)

These practices aiming at adaptating the body to a (perceived) water quality are as much about current infrastructural conditions as they are about anticipating under which conditions one will live in the future. As popular class residents, these interviewees display a clear notion that their future generation's living conditions are expected to be very much the same as today, including a non-permanent supply with non-potable tap water. The reproduction of social status turns here into an issue negotiated via the body itself.

The externalization of domestic services presents another attribute of class: What tends to distinguish the middle class (or at least the better-off amongst them) from the popular class is the employment of a domestic worker, who takes over the responsibilities of cleaning, washing dishes, doing laundry and often also food preparation. This paid domestic work is usually conducted by women from the popular class, who eventually return home only to find their own domestic tasks waiting for them. Concierges fulfill a similar role when securing permanent water availability through storage and water purchase from private vendors (see 5.3), whereas washing clothes is sometimes outsourced to commercial laundries. All in all, both social status and gender seemed to define who conducts a water-related practice rather than the practice per se. As women continue to be "the managers of the process of social reproduction" (Bennett 1995: 80), most domestic tasks involving water use – from water purchase and food preparation to doing laundry and domestic cleaning – continue to be strongly regarded as female duties (see for instance Massolo 1992).

5.3 STORING WATER: SYNCHRONIZING RHYTHMS OF SUPPLY AND USE

As we have the cistern, we can use [water] whenever we like. (Margarita E1: 319, Iztapalapa)

Just like other basic urban infrastructures and services, a permanent water supply is taken for granted as long as it functions smoothly and access is guaranteed. Yet where this system fails to function properly and permanently, meeting domestic needs becomes more complex, requiring additional efforts. As Margarita's statement shows, storing water is what allows people to regain certain autonomy in domestic water use by reducing the dependency on intermittent water supply intervals. Disposing of a domestic cistern is what seems to grant the freedom of using water according to one's own needs and own rhythm. In Mexico City, colliding patterns of supply and domestic water use have yielded a number of domestic coping strategies, as this chapter will show. Where water is not provided permanently and with sufficient pressure – in spite of dwellings being hooked up to the water supply network – responsibilities for permanent and sufficient water availability are shifted to the domestic realm. One particular device symbolizes and materializes this shift most clearly: The roof-top tank, or *tinaco*. Over more than five decades, the urban landscape of the metropolis has been influenced by the reliability of the water supply scheme – and the population's lacking trust in it, which resulted in the installation of these tanks all over the city. Leñero, for instance, depicts them as a common urban feature in his 1983 autobiographic novel:

> "Then we walked by San Pedro[21]. I took a look at the rooftops. Cubic tanks, oval tanks, spherical tanks, *tinacos* of all shapes and sizes crowning the rooftops of the neighborhood and the entire city. I had never noticed the obvious: There were as many tanks as houses, as dwellings, as TV antennas. Photographers used to shoot these forests of antennas in order to decry the phenomenon of collective alienation in newspapers and magazines, but I had never heard about any photographer interested in tracing the aerial landscape invalidated by these concrete monsters: dents of all construction, anti-aesthetical appliances for storing the vital liquid (...). There they are, always, everywhere, like crouching chickens, blemishing the little serial houses

21 San Pedro de los Pinos, a neighborhood in the Benito Juárez borough, Federal District.

of the *colonias proletarias,* disrupting the architectural harmony of a middle-class *fraccionamiento*. Only in the [upper-class] residential areas, some developers saw to it to hide the horrible tanks in boxes of concrete, as if hiding a flaw (...). The rich don't like their tinacos to be seen. For the poor, on the other hand, they are a symbol of progress" (Leñero 1983: 95).

Storing water in private roof-top tanks had apparently become a regular practice in the inner-city Federal District as early as the 1980s. Today, such domestic water storage facilities are ubiquitous all over Mexico City. In 2010, 78% of all inhabited dwellings in the Federal District featured a roof-top storage tank, and 54% were equipped with a cistern (see INEGI 2010). This clearly demonstrates a perceived need for prevention against water shortages amongst the population and those who construct housing. On the one hand, domestic water storage has become a crucial strategy as water continues to be intermittently provided in a significant number of neighborhoods throughout the metropolis. According to official figures, roughly 18% of all dwellings in the Federal District – some 430,000 dwellings in total – did not have access to permanent water supply in 2010, despite of in-house water tap installations (see INEGI 2010). In addition, and in spite of official rationing schemes such as the *Programa de Tandeo* set up by the water utility, the rhythm of water supply is often subject to spontaneous variations (see SACM 2013), for details 5.1.4). On the other hand, the prevalence of storage devices in the Federal District's dwellings suggests a much wider scale of domestic water storage, including neighborhoods provided with water permanently – at least according to the official census data.

The present study explores this issue by making reference to the interviewees' own perception of supply rather than relying only on the officially reported water supply conditions. This approach is based on the assumption that everyday practices of water-use are likely to be influenced by actual water availability at the domestic tap and less by abstract supply schedules which might or might not be perceivable to the domestic water user. For the 44 households studied in depth, there seem to be enough reasons for keeping water in stock. A vast majority – 36 of all studied households – lived in dwellings equipped with some kind of domestic storage device, typically an underground cistern. This accounts both for dwellings with steady and those with unsteady supply. As chapter 4.2 has indicated, water was officially rationed in five of all six studied neighborhoods in Iztapalapa, and reported as permanent in all six neighborhoods in Cuauhtémoc. However, the interview material shows that housing complexes Chinampac de Juárez in Iztapalapa and Tlatelolco in Cuauhtémoc, as well as the Roma neighborhood in the same borough were the only locations were all interviewees reported a permanent supply of a stable pressure at their dwelling. Interviewees perceived temporary variations in water pressure[22] in half of all 12 studied neighborhoods, and water supply was generally perceived as non-permanent in another three: El Manto and Miravalle in Iztapalapa (both subject to SACM's rationing scheme), and the inter-

22 Low water pressure decreases the through-flow volume, thus influencing the time required to fill cisterns. Where pressure is insufficient to fill roof-top tanks, the installation of an electric pump at the cistern is required to build up sufficient pressure for water provision within the building.

mittently supplied neighborhood of Cuauhtémoc in the borough of the same name (see Tab. 4.10).

In the present context, domestic storage is understood as a way of securing water availability and synchronizing non-permanent patterns of water supply with often quite divergent domestic rhythms of water use. As Mexico City's tap water is usually not used for direct human consumption[23] (see 5.1.1), the present chapter discusses the storing of water for domestic use of a more 'technical' nature, from showering and doing laundry to domestic cleaning (see 5.2.3).

Where water is not provided permanently by the water utility, or where supply is feared to fail, even temporarily, securing permanent water availability turns into an individualized task. This involves a number of strategies which will be discussed in the present chapter. In the absence of proper storage facilities, water use is adapted directly to supply patterns (see 5.3.1). In order to regain a certain autonomy of domestic rhythms of water use, cisterns and other domestic storage devices are installed and require a management of supply, which can be organized individually for each dwelling, or for a group of dwellings (see 5.3.2). Flat dwellers in particular usually share water from a collective cistern, a practice which brings about questions of cooperation and competition (see 5.3.3). And finally, there is the controversial issue of cleaning domestic cisterns (5.3.4).

5.3.1 Immediate Use of Water

The relevance of domestic storage in facing scheduled supply interruptions is highlighted above all by a strategy employed in absence of proper domestic storage (or given very limited domestic storage capacities) in combination with intermittent supply: the immediate use of water, or *"we run when water starts running"* (Alma E9: 299) as Alma and her daughter phrased it. They were in fact busy continuing with their weekly laundry at two run-down washstands in the yard of their house in Iztapalapa even during the interview, as water was provided at that very moment. Doing laundry and fulfilling other water-intense practices in combination with stockpiling water in a variety of containers once the rationed water provision starts: A similar strategy was applied by another four interviewees all residing in the Iztapalapan neighborhoods of El Manto and Miravalle, such as Gabriela:

> *I do my laundry every week, whenever water comes. (...) I fill the containers when water starts running, and go! Wash everything.* (Gabriela E19: 23 and 86, Iztapalapa)

As domestic practices requiring elevated amounts of water such as doing laundry and cleaning the house are done once water starts running from the tap, the domestic rhythm of water use becomes heteronomous: It is the water supply frequency which structures the daily routine of domestic work for these women. One of Alma's

23 With bottled water as the main substitute, securing drinking water availability does not seem to require any particular kind of domestic storage amongst the interviewed: It is usually purchased on a weekly basis rather than stockpiled at home.

and Gabriela's neighbors recalls how she reduced her dependence on water supply patterns by installing an automated storage system.

> *I just switch on the pump to have water in the roof-top tank and in the ducts. When we didn't have [the pump] we took advantage of the day when water ran in order to avoid having to extract water from the cistern. We supplied our sink with a little hose connected to the tap, and did laundry and all that.* (Maria E2: 142, Iztapalapa)

Indeed it seems as if the mode of urbanization is what makes the difference here: An immediate use of water is specific to residents of colonias populares subject to water rationing, particularly those living in self-built dwellings without proper cisterns. Both in El Manto and Miravalle, the collective water infrastructure (mostly installed ex-post), and the self-built character of dwellings, consolidated as they may be, are shared issues. In this context, it is remarkable that proper domestic water storage installations seem to both symbolize and materialize the consolidation and formalization of the dwelling and neighborhood, as Berta recalls from her childhood.

> *My father delayed the construction [of a cistern] for about five, six years. Besides there being no water (..) pipelines, there were no economic resources (...) and they thought we wouldn't get the land title because a lot of it was ejidal land. That's why my father said: "I can't make expenditures if they will take it from me". Because land titles were not yet legalized at that time. Time passed [...and] when they saw that water was being installed, drainage, (...) and all that, my father said: "Now we'll stay." That's when he began to set up the house properly, because the roof was metal sheets when we arrived.* (Berta E29: 450, Iztapalapa)

Moreover, cisterns have turned into a symbol of autonomy in domestic water use in neighborhoods supplied intermittently, as this chapter's introductory quote by Margarita has shown. Yet amidst supply limitations, a cistern alone is not what guarantees permanent water availability.

5.3.2 Being at Stand-By: Filling the Buffer

Where water supply is subject to interruptions, or suspected to fail, a majority of dwellings are now equipped with some kind of storage installations. Stockpiling, i.e. storing water in smaller, mobile devices within the dwelling at all times, is sometimes used as an alternative to domestic cisterns in areas with intermittent water supply. However, even in areas where water supply is permanent and domestic storage facilities available, some residents keep an extra water reserve in their dwelling to prevent any kind of unpredictable interruption of water supply. This reflects mistrust in the reliability of official water supply schemes or prior experiences with intermittent supply in former places of residence (see 7.1). In addition, such individual stockpiling is generally performed in preparation for Mexico City's infamous *cortes de agua* – temporal supply interruptions which can last several days (see 4.1.3). As an emergency measure, it is applied by those with and without access to domestic storage installations. Take Clara, who lives in a building with two cisterns and a number of roof-top tanks:

> *Normally, people prepare by keeping buckets or barrels in the bathroom, mainly for the toilet. (...) Sometimes we are notified via television (...) that there will be no supply on Friday, Saturday and Sunday.* (Clara E10: 141–143, Cuauhtémoc)

Serving as a last resort, the stockpiled water is used for the most immediate needs such as flushing toilets, washing dishes and personal hygiene. Yet apart from this exceptional measure, most interviewees regularly had access to water stored in a domestic cistern: 36 of all 44 interviewed households disposed of one. The most common domestic storage set-up in Mexico City comprises an individual or shared subterranean cistern connected to the public water supply network and an electric pump to lift the water from the cistern to one or several 700 to 1,100-liter plastic tank(s) *(tinacos)* on the roof or an upper floor of the building. This ensures a steady water pressure appropriate for distribution and use within the building while at the same time providing some additional storage capacity. Other than simpler water storage devices such as barrels and tubs to be filled manually, domestic cisterns are restocked automatically: Most dispose of a floater to interrupt water inflow as soon as the cistern is full. More advanced systems include an additional floater in the connected roof-top tank which will automatically start the electric pump once water levels fall below a certain level. However, even the most sophisticated storage system requires maintenance and management, and more so in areas with intermittent supply. The task of enabling an autonomous domestic rhythm of water use independent of the actual intervals with which water is provided through the public network is either performed by the water users themselves, or delegated to a third party.

Delegating Domestic Water Management

In apartment buildings and housing complexes, the responsibility for collective water storage (see 5.3.3) usually lies not with the residents. Either a delegated neighbor or an employee is responsible for the maintenance of the water storage system, and manages the storage and internal distribution of water within the building (if not automated) as well as the purchase of water from tankers if need be. In middle-class condominium towers, such as those studied in the neighborhoods of San Rafael and Cuauhtémoc, the responsibility for storage and water availability is usually professionalized and formalized as a concierge employed for the entire building. Tasks of these employees include securing a sufficient water pressure by pumping water from the cistern to roof-top tanks, but also limiting internal supply if need be. The intermittent supply experienced by inhabitants of some buildings resulted from an internal rationing scheme applied by the building's administration rather than non-permanent supply to begin with. Given an insufficient in-flow of water at a low pressure from the public network, such internal rationing is for instance applied in the apartment block in San Rafael where Pablo lives.

> *Not much water enters, so the cistern takes a long time to fill up and that caused problems with the pump (...). [It] began to wear out, so currently it is only turned on in the morning and at*

> *night. (...) The situation is halfway critical now, (...) if you open the tap during the day, there is no water.* (Pablo E34: 50–54, Cuauhtémoc)

Limiting internal water supply to certain hours is often done to minimize the wear-off of electric pumps and avoid the extra cost of purchasing additional water from private vendors. The latter was regularly applied in an upper-scale apartment building with 22 flats in the Cuauhtémoc neighborhood, where it was the porter rather than the interviewed residents who pointed out supply limitations in the neighborhood and explained the coping strategy adopted in this building:

> *As concierge it's my job to order tanker trucks every time when the water level in the cistern is low. Sometimes 13,000 to 16,000 liters are consumed in just one afternoon. Facing this kind of demand we have to order a water reserve. (...) In general, we order a 22,000 liter-tanker truck three or four times a week.* (Concierge in apartment building of Eva E22, Cuauhtémoc)

While coming at a cost, this strategy allows residents to experience the convenience of permanent water availability in their own flat, similar to a neighborhood actually supplied permanently. Indeed, a guarantee for such permanent water availability is something that comes included with the apartment's purchase contract for Claudia, who recently bought a newly constructed flat in the neighborhood of San Rafael.

> *[Water supply] is good, it's permanent. And we rely on the administration – [...they] give you a guarantee that if water supply stops, if the cisterns are depleted, they will order tanker trucks.* (Claudia E47: 4, Cuauhtémoc)

Such strategies allow the individual resident to use water according to his or her own needs and entirely independent from actual supply intervals and water pressure provided by the public water network. Consequently, the residents' awareness of rhythms and characteristics of water supply as well as storage efforts is often quite limited. While such carefreeness or 'ignorance' might be seen as part of a universal human right to water, it is only made possible by privatizing these tasks in all those areas where water supply is non-permanent or subject to limitations to begin with. Better-off middle-class households, particularly flat dwellers, tended to delegate the task of internal water management to employees, as the case of Claudia illustrates. And in the case of smaller residential buildings with up to 25 units and less affluent inhabitants, administrative tasks were usually fulfilled by one of the neighbors. This job of administrator or *encargado de cisterna* can be of a more or less formalized nature, sometimes assigned by a communal assembly of all residents. Miguel, a long-term resident of San Rafael, recalls how he was put in charge of internal water management in the apartment building he lives in as water supply deteriorated over the years:

> *Earlier on, (...) there was a lot of water, and the pump worked day and night (...). People were (...) economically well off, so nobody complained about the expensive water bills, the electricity bills. (...) But as water became scarce, the pumps were strained a lot as they had to turn off when [water] stopped running and then came back on (...). [Our neighbors] started to complain about elevated electricity and water bills. So [... they made sure that] the pumps are used only during the day, and that someone is responsible to switch them off at night.* (Miguel E10: 210–217, Cuauhtémoc)

Being at Stand-By

In contrast to the convenience of third parties managing domestic water availability, storing water requires more time and effort when the storage system is not automated amidst unreliable or insufficient supply patterns. When living temporarily in an underserviced colonia popular in Iztapalapa in 1989, Dolores observed a particular way of water fetching:

> *They just had a little tap on the corner, for everybody who lived in that block. (...) the [neighbor] who noticed that water started running – at twelve, one in the night – informed all the others on the block. [... and] everybody lined up to fetch water (...). They worked all night gathering water, and instead of going to rest in the morning, they got ready and went to work. (...) Those few times I got up to help my mother [to fetch water at night], my feet buckled as I fell asleep when I was on the bus in the morning, standing there holding on to the handle bar.* (Dolores E8: 385–387, Iztapalapa)

Her experience is remarkable perhaps not so much for the lack of domestic water taps in the neighborhood – typical during early phases of popular urbanization (see 4.2) – but mainly for the reported impact on residents' daily routines. Water from the public tap is stored by these women as a way of coping with intermittent supply frequencies. However, this strategy does not seem to make them entirely independent of unpredictable supply patterns. By imposing a certain timing of storage activities, the frequency of water supply collides with daily routines, depriving people of relaxation and sleep. Public taps are now much less of a common sight on Mexico City's street corners[24] – but the issue of colliding patterns of supply and use of water has remained. It is the unpredictability of water supply which imposes such a *"stand-by"* modus, as Maria put it (E2: 232, Iztapalapa): securing water availability for domestic needs requires not only domestic storage facilities such as cisterns and tanks, but also a certain alertness by household members. Again, it is Leñero's 1983 novel which provides a good impression of how the wait for water might feel (and sound) like:

> "During the droughts of '79, of '80, of '81, the roof-top tanks (...) were empty by mid-morning, and water did not resume to ascend to them during the entire afternoon. But when night fell, at about twelve or one in the morning, a characteristic thunder in the ducts, the noise of hammer strokes, categorically announced the resumption of supply. Sometimes I woke up hearing the stream as it filled the toilet's water tank and sometimes I couldn't fall asleep until I heard it. (...) That is to say, my insomnia had the length of the waiting: my God, when will the pressure increase? When will the water come?" (Leñero 1983: 13)

Waiting for water supply to resume early in the morning on every third day in her Iztapalapan neighborhood, Berta was in a very similar monitoring mode:

> *The black roof-top tanks are above (...) my bedroom, and you can hear when water starts running. (...) At dawn I keep listening, I wake up, that's how I am at daybreak. (...) I already know when the tank is full or is a bit more than half-full. Then I get up, open the water tap and begin to set water aside.* (Berta E29: 78, Iztapalapa)

24 Today, some 60,500 residents of the Federal District still depend entirely on this unimproved source of water, of which 17% live in the borough of Iztapalapa, and a vanishing 0.6% in Cuauhtémoc (see INEGI 2010).

Apart from paying attention to acoustical indicators announcing the resumption of water supply, this preparedness also includes a frequent revision of taps – or, as Guadalupe called it, *"checking the tap to see at which hour"* (Guadalupe E3: 359, Iztapalapa). This allows starting to store once water begins to run and ideally exploit the limited provision period to its full extent. Eight of 44 interviewees[25] reported to be at stand-by amidst intermittent supply. It is typically the oldest woman in the household (and/or the female head of household) who is responsible for storage via this strategy. Living in a self-built dwelling without a cistern in an Iztapalapan neighborhood which receives water only once a week, Alma summarized it as follows:

> *Those who have a cistern don't have to be alert. But I don't have one, so I need to fill my tubs. (…) I get up; I almost don't sleep in order to fill them.* (Alma E9: 78–80, Iztapalapa)

Waiting for water to start running, or for cisterns, tanks and barrels to fill up, it is the (often unpredictable) water supply scheme which shapes the daily routine of these women, often disrupting other activities. As the examples of Dolores, Berta and Alma reveal, schedules imposed by supply patterns can deprive people of hours of rest, potentially even affecting their productive capacity. This again sheds light on the relation between infrastructural conditions and the reproduction of social inequality – all the more so as only women with a relatively low social status took to the practice of being at stand-by.

Turning Water Supply into a Locational Factor

By turning infrastructural conditions into a locational factor, the middle class adds a different kind of approach to this portfolio. Having recently moved to the neighborhood of San Rafael, Pablo is a good example: despite taking his clothes to a laundry, employing a domestic worker, and generally spending most of his time in his office rather than home, he felt strongly affected by the internal rationing scheme in his apartment building. Therefore, he is considering selling the apartment he bought just two years earlier and for which he is still repaying the mortgage:

> *The truth is that with those water problems, it is not worth staying here. You consider (…) transferring the apartment. (…) the city is too complex to have these difficulties in your own home.* (Pablo E34: 298 and 304, Cuauhtémoc)

Similarly, three other middle class interviewees cited local water supply conditions as a factor they take into consideration during house hunting. Facing not only much less of a choice on Mexico City's stratified housing market but also often much stronger limitations in water supply in their current dwelling, none of the less well-off interviewees voiced any considerations of this kind. On the contrary, and partic-

25 In addition, a less rigid stand-by strategy is also applied by flat dwellers such as Oscar from Cuauhtémoc and Regina from Iztapalapa: sharing storage installations with a few neighbors, they ended up with the task of refilling roof-top tanks amidst their neighbors' passivity or negligence (see 5.3.3).

ularly in colonias populares, it seemed as if initial hardship in terms of lacking urban infrastructures was understood by many as an almost 'natural' obstacle in the process of obtaining homeownership, only to be overcome after some undefined period of time. An unequal access to proper housing conditions, including water, hence tends to be tolerated if not naturalized by those with a lower social status. Middle class residents, in case they are subject to insufficient water supply, express hopes to opt out of these conditions by making use of their broader access to housing markets, moving elsewhere.

5.3.3 Sharing Stored Water

One aspect makes an essential difference when it comes to domestic water storage: whether storage devices are used exclusively by one household, or shared collectively. In contrast to those living in single-family houses, flat dwellers are usually connected to a shared water storage facility – independent of whether they reside in one of the huge housing complexes with several hundred dwellings, or in apartment buildings with a couple of flats only. These central cisterns serve either an entire building alone, or a number of adjoining buildings, and their overall capacity is often much higher than those of single-family dwellings. In housing complexes, collective cisterns may be shared amongst hundreds of households – one Iztapalapan housing complex was equipped with several huge cisterns of 300 m^3, each feeding some 120 apartments.

Numbers were usually much lower for apartment buildings, where a cistern was typically shared by between four and 25 dwellings. New apartment buildings in boroughs such as Cuauhtémoc are now mostly being constructed with huge underground cisterns. Two recently constructed condominiums in San Rafael, for instance, featured around 45 apartments in two adjoining blocks connected to a cistern of considerable size – 125 m^3 in one case, and 285 m^3 in the other. Older residential buildings, in particular those dating back to the 1960s and earlier, typically feature much smaller storage facilities. Usually, a cistern with low capacity connects to roof-top water tanks, or the latter are directly hooked up to the central water tap. In the past, electric pumps as well as huge cisterns seemed unnecessary, as tap water was provided with sufficient pressure to fill roof-top tanks directly (see for instance Leñero 1983). Adapting these buildings to a deteriorating water supply situation through the installation of larger storage tanks faces serious obstacles. Being a structural component of the building, the ex-post construction of underground cisterns covering several cubic meters can be a tricky issue. As Carmen, an inhabitant of a several decades old apartment building near Avenida Reforma, explained:

> *The problem is that we don't have enough space to install a large cistern. (...) the [building's] foundation might get damaged (...) and we also can't have a large plastic tank as it would block the entrance (...) so if there's an earthquake or something, how will people pass?* (Carmen E52: 92–94)

She faces water shortages, as water supply in her neighborhood has turned intermittent during the last decade, and at 0.4 m^3, the installed storage capacity per dwelling is low.

Generally speaking, the empirical data suggests that installed storage capacities per dwelling unit[26] vary considerably between shared and un-shared cisterns. The typically higher total volume of collectively shared storage tanks does not translate into a higher storage capacity per household when compared to individually used devices. Cisterns in single- or double-family dwellings in the sample had an average storage capacity of 7.4 m³ per household – more than twice the average volume available to flat dwellers in housing complexes and apartment buildings (see Tab.5.6).

Table 5.6: Domestic water storage: Average installed capacity per dwelling, distinguished by borough and dwelling type. Own elaboration based on 31 interviews

Dwelling type	Average share of installed storage capacity per dwelling	N° of dwellings connected to cistern	N° of cases
Flats	2.8 m³	2–480	26
One-or two-family houses	7.4 m³	1–2	5
Overall average	3.6 m³	-	31

Where storage installations are used collectively by neighbors, related tasks and costs are typically shared. This includes for instance the maintenance of cisterns and pumps, the replacement of older asbestos storage tanks[27], and the periodic cleaning of cisterns and roof-top tanks. The latter task is either outsourced and the costs shared, or conducted as collective labor missions in some buildings and housing complexes with well-organized inhabitants (see 5.3.4). If additional water is purchased from water tankers, or if the building's water management is delegated to a concierge, residents are requested to share these costs, too. Nevertheless, the sharing of bills and pending tasks is sometimes also the source of discontent amongst neighbors unwilling to cooperate. And amidst the neighbors' negligence of collective storage tasks, it is sometimes one of the other residents who ends up being responsible for internal water management in smaller apartment buildings where up to five flats are connected to the same cistern. As water supply is permanent but not of sufficient pressure to reach roof-top-tanks in the building Oscar lives in, water needs to be pumped to the roof by controlling the electric pump manually.

> *The negligence of the other tenants is that they don't switch on the pump when they wash their clothes. I am the only one who puts it on. That's why I go downstairs in the mornings (...), I switch on the pump and I'm at stand-by so water won't spill. (Oscar E14: 34,* Cuauhtémoc)

Limited to restocking roof-top tanks as an automated pumping system is not available, this practice only partly resembles the much more drastic stand-by strategy

26 Information on approximate installed storage capacities of domestic cisterns (see Tab. 5.6) were obtained for 31 of all 36 cases disposing of such devices. This provides a rough approximation of the entire storage capacity available to each household, as roof-top tanks – potentially providing some additional storage space – are not included in this calculation.

27 Reflected in common producers' names such as *Asbestolit*, water tanks were usually made of a mixture of concrete and asbestos during the 1970s and 80s (see for instance Leñero 1983).

applied by those not disposing of an automated storage system amidst intermittent supply (see 5.3.2).

When asked whether she keeps an additional water stock in her own flat to prevent shortages, Catalina gave an unexpected answer:

> *We don't set aside containers and more containers and barrels – precisely because we have other neighbors who also need water. Therefore, it's unfair if you stockpile plenty of containers and they are left without water, isn't it?* (Catalina E18: 210, Cuauhtémoc)

A tenant of a run-down inner-city vecindad, she shares a small cistern with three other neighbors. Her awareness of other people's water needs highlights a crucial aspect: those who share storage facilities are hence not only dependent on the frequency of water supply from the public network but also their neighbors' patterns of everyday water use. Accordingly, it was a similar sense for using water economically during supply interruptions that secured water availability for Antonia, a tenant in an old apartment block in San Rafael.

> *It's not that we would need a water tanker to fill our roof-top tanks, because our neighbors are also quite conscious about it. (...) We share the roof-top tank with them and they, too, don't wash any clothes [during supply suspensions] and also don't waste water.* (Antonia E20: 56–58, Cuauhtémoc)

Sharing storage facilities and particularly sharing stored water hence seems to bring about a certain co-dependency amongst neighbors over each other's rhythms of water use. Internal distribution and temporal patterns of water use therefore turn into a focal point of discontent and neighborly conflict amongst those hooked up to the same cistern. Berta wrapped it up as follows:

> *I leave for work and my daughter for school. Then those neighbors who stay [at home] clean dishes, use the water (...) and when I come back I don't have water.* (Berta E29: 88, Iztapalapa)

Mainly households such as the ones of Berta, Ana and Hilda, where women were the main income-earners and no one stayed at home during the day reported difficulties in obtaining enough water to fulfill their domestic needs.

> *The average family in my building are (...) five persons or more. (...) water rises to the roof-top tanks and the families start to fill or launder or do things. That implies an inequality: some can do everything they need to do (...), and those who come home later will not have the same pressure anymore.* (Hilda E5: 228 and 336, Iztapalapa)

> *I live alone in this flat, so in theory I shouldn't have problems with water. If I had a tank of my own (...) I would never use it all. But (...) there is a tank which supplies the entire building, and (...) entire families reside in several apartments. [... In some] there are maybe five or six persons. Hence I believe that water is not distributed equally. So I come home at night and even though I have not used water during the entire day (...) there is no water for me to take a shower.* (Ana E36: 8, Cuauhtémoc)

Despite having their access to water in theory secured by a shared cistern, de-facto water availability was limited at times due to neighbors' asynchronous patterns of water use. In addition to neighbors' varying consumption levels as an result of household sizes and composition, it remains an open question whether this is also an effect of a generally lower storage capacity in shared cisterns compared to those exclusively used by one family and/or a supply with insufficient water pressure, resulting in a slower refill of cisterns.

While some flat dwellers' water use was limited indirectly as an effect of divergent temporal patterns of water use between neighbors sharing a cistern, others had their water use directly regulated by landlords. This applied specifically to tenants in owner-occupied buildings such as Isabel.

It was (...) getting up early, stockpiling water, because the landlord there didn't like us to take water after ten in the morning. (Isabel E21: 151–153, Cuauhtémoc)

Sharing the same building or renting out single rooms, landlords reportedly defined rules for water use, such as scheduling tenants' laundry hours or obligating them to stockpile water, in several other cases.

Contestations between neighbors over water use patterns are also what explains the barrels and tubs piled up in the yard of Guadalupe's small and simple one-bedroom flat in a vecindad in Iztapalapa. Here it is not stored but running water which has to be shared, as the building containing three flats has no cistern. A low water pressure during the morning hours, when water is supplied in this neighborhood influences on Guadalupe's ability to use water simultaneously with the other residents.

I usually get up at six in the morning to do laundry. When I notice that my neighbor is washing, I wait, because we are drawing water off each other. When they open [the tap] near the entrance (...) that doesn't allow [water] getting here. (Guadalupe E3: 84)

Not much seems to have changed since neighbors shared taps rather than storage tanks in inner-city vecindades. In these colonial buildings refitted with multiple flats and rooms rented out to the popular class, it was common to share water taps located in the courtyard together with other facilities such as bathrooms, toilets and laundry stands. Water was then internally redistributed with hoses and buckets. In particular in neighborhoods with intermittent supply, competition over access to water from shared taps caused conflictive situations similar to the ones related to sharing stored water, as Marta recalls from the Doctores neighborhood in the early 1980s:

[Water] came only during the night, and it didn't reach the flats. There was only one general tap where we connected our hoses. There was a lot of trouble over the hoses – it's my turn, no, I had put it up... It was a clamor. (Marta E51: 164, Iztapalapa)

The negotiation over access to water from shared taps can even result in the direct restriction of water use by neighbors, as Teresa recalls from a vecindad in the northern part of the Federal District, where laundry stands were shared with several other tenants:

We had a problem with the laundry stands, and worse, over water. (...) sometimes, they didn't let you do your laundry. (Teresa E49: 298–302, Iztapalapa)

It is remarkable how the challenge of sharing limited water resources stored collectively led to cooperative actions marked by mutual respect and solidarity amongst neighbors in some cases, whereas ignorance or a conflictive competiveness ruled in others. It remains an open question to which extent these contradictory manners of dealing with shared resources and current supply conditions can be traced back to common past experiences – for instance the collective organization of neighbors in housing projects under the roof of the *movimiento urbano popular,* or the shared struggle and collective work typically applied during the consolidation process of irregular settlements. Whether a limited access to stored water (or shared taps) is imposed directly or results from asynchronous patterns of water use amongst neighbors who share storage devices – stockpiling water in the own dwelling is an attempt to regain some independence. This provides a last resort in case the shared cistern runs dry. Take Mario and his partner who share a cistern with eleven other flats in intermittently supplied Cuauhtémoc and work in a different part of the city during the day:

> *As there is always a chance that we come home and there is no water left, we keep some plastic jugs in the utility room.* (Mario E42: 22, Cuauhtémoc)

However, there are often spatial limitations to stockpiling water within the own dwelling, in particular in some of the smaller flats in housing complexes. In areas with permanent and reliable water supply, an insufficient storage capacity of shared facilities or incompatible rhythms of water use amongst neighbors are only revealed during the periodic supply interruptions scheduled by the water utility. Dolores from a popular housing complex recalls a supply suspension when the huge shared cistern had to be restocked by municipal water tankers, and neighbors started consuming all the water immediately:

> *As everybody takes advantage to do their laundry, to shower (...), and as they are at home [during the day] they stockpile water in order to have some when it runs out. And those who are not [at home]? We don't get any.* (Dolores E8: 105, Iztapalapa)

As an individual strategy, stockpiling hence serves to make flat dwellers less dependent on neighbors' water use rhythms. Ironically, such reactions to competition with neighbors over shared water simultaneously tend to amplify these problems by privatizing a bigger share of the overall available amount of water.

5.3.4 Cleaning Storage Devices

Apart from efforts stemming from the management of storage facilities and the redistribution of water within the building, domestic installations such as cisterns, tanks, pumps or filters require maintenance. One of the major tasks is the cleaning of storage devices, which has emerged together with the proliferation of domestic storage in Mexico City over the last decades. The disinfection of storage tanks for instance was a service already being offered via the Federal District's phone book by the early 1980s (see Leñero 1983: 94). Yet how would someone who is not an expert in the analysis of water quality know when the time has come to enter the domestic tank and start scrubbing? In an approach similar to that for discerning whether tap water is apt for drinking (see 5.1), residents refer to the appearance, smell and taste of water at the domestic tap as indicators that cleaning of storage devices is due.

> *When the roof-top tanks are dirty (...), [tap water] runs with a golden color (...), or smells strongly of iron. That's when [our neighbors] tell us: (...) Let's clean them already.* (Silvia E26: 159–165, Iztapalapa)

As a more extreme example, a cistern in the Presidentes de México housing complex in Iztapalapa featured tiled walls covered by a layer of sediment of a brownish, slimy consistence during refill. The interviewee, responsible for administrative tasks in her block, displayed it as a proof of low water quality, stressing that the cistern was cleaned just three months earlier. Though this is probably an extreme case, most of the interviewed – residents of Cuauhtémoc and Iztapalapa alike – reported the accumulation of some kind of sediment in storage tanks with time. Maria, who lives in a neighborhood in Iztapalapa where yellowish tap water with floating particles is provided once a week, describes how their domestic cistern made of concrete is cleaned:

> *My husband and sons clean the cistern (...). One enters and the other one is outside, receiving (...) the buckets to pour away the dirty water. In the last turn (...) it's like chocolate. Really, when we wash it off it's viscous. Using soap water and chlorine (...), my husband scrubs the walls (...). [The cistern] gets soiled in quite a nasty way because of all the polluted water that flows in all the time. (... it) stays there for several days, it all settles down, and the water comes out clearer, but the sediment remains.* (Maria E2: 216, Iztapalapa)

Both authorities and tank manufacturers recommend sanitizing domestic cisterns and plastic storage tanks every three to six months. In general, interviewees in the present study were well within that margin, reportedly sanitizing their cisterns about three times per year on average. However, striking differences emerge. Cleaning frequencies seemed to be independent of steadiness of supply in a neighborhood, but with roughly four times a year, cisterns were cleaned twice as often in Iztapalapa than in Cuauhtémoc. These differences appear to be linked to two aspects: the neighborhood – i.e. the (perceived) local water quality – and the dwelling type, i.e. whether cisterns are shared or not. On average, interviewees reportedly sanitized individually used cisterns or roof-top tanks as often as 7.6 times per year, and shared cisterns in apartment blocks and housing complexes 1.5 times per year (see Tab. 5.7).

*Table 5.7: Domestic water storage: Average frequency of cleaning cisterns and tanks, distinguished by borough and dwelling type. Own elaboration based on 32 interviews**

		Cistern is cleaned ...times per year	N° of cases
Dwelling type	Flats	1.5	22
	Single-family houses	7.6	10
Borough	Cuauhtémoc	1.9	13
	Iztapalapa	4.2	18
Overall average		3.2	32

* Based on those 32 of all 44 interviews where a cistern or roof-top tank was installed and information about cleaning frequencies was provided by the interviewee.

Those sharing cisterns were more inclined to hire someone for cleaning storage facilities, while those using own tanks tended to clean them not only more often but also by themselves. A more direct perception of water supply conditions (including water quality) and of the condition of the domestic cistern in single-family dwellings might be the reason. For shared cisterns, cleaning was a collective task fulfilled by residents themselves only in some exceptional cases. Dolores for instance spoke about these collective missions, which were common in the first years after the housing complex she lives in was built by UPREZ, a leftist grassroots organization, in the early 1990s. Eventually, this practice was replaced by the outsourcing of cleaning tasks, which is common amongst those sharing cisterns. Moreover, shared cisterns had not been cleaned whatsoever during the last three or four years in some studied apartment buildings in San Rafael, Paulino Navarro and Cuauhtémoc. A lack of cooperation from neighbors – be it in sharing costs or taking part in the actual cleaning work – was often claimed as a reason. But most remarkably, domestic storage tanks were generally cleaned less frequently in those neighborhoods where interviewees tended to employ tap water either for all kinds of domestic use, abstaining entirely from bottled water use, or using bottled water exclusively for drinking (see 5.1). Not only was tap water quality in the neighborhoods of Paulino Navarro, Cuauhtémoc, Roma Norte, San Rafael and Tlatelolco perceived as higher – its residents also reported that cisterns were cleaned less often, about once a year or less. In contrast, cisterns tended to be sanitized between six and eight times a year in neighborhoods such as Miravalle and Ermita Zaragoza, where residents took to a multiple use of bottled water given a perceived low quality of local tap water. Some interviewees from these neighborhoods took to sanitizing their cisterns every month. In short, cisterns were generally cleaned more often in areas with low water quality, implying that the latter is an issue of the supplied water to begin with. This is at odds with claims by the water utility and government officials alike that a lack of maintenance in domestic cisterns is what mainly causes limited water quality at the domestic tap to begin with (see for instance SACM 2012: 149). According to the present findings, it can be assumed that a need for cleaning is strongly linked to the (perceived and experienced) quality of supplied tap water in a given neighborhood, contradicting the idea that domestic users themselves are the first to blame for low water quality. Similar to the practice of drinking water, storing is hence a practice where local water quality in the neighborhood matters.

To conclude, domestic water storage can be seen as a reaction to rhythms of water supply which are asynchronous with those of domestic water use. Serving as a preventive measure, it provides a buffer to overcome periods of insufficient supply and supply interruptions of both expected and unexpected nature. Storing water is a means of synchronizing the rhythms of water use and supply: by what could be called "everyday infrastructural planning" (Allen 07.11.2014), autonomous rhythms of water use are facilitated and negotiated in the light of supply limitations. However, the Federal District's public water utility is not alone in imposing heteronomous rhythms of water use, for instance by rationing water supply. Neighbors' divergent rhythms of water use also play a key role when stored water is shared via a collective cistern. Storage strategies and the perception of water supply tend to vary according to the type of housing, and in consequence, the type of storage device. In general, domestic storage devices were present in most households, whether supplied permanently or not. However, steadiness of water supply was of the essence for storage practices. Amidst intermittence and low water pressure, storage facilities were either automated or else the management of storage turned into a major task disrupting and orchestrating daily routines. As particularly the practice of 'being at stand-by' reveals, a heteronomy in own water use rhythms and schedules imposed by supply patterns can deprive people of hours of rest, potentially even affecting their productive capacity. This is all the more relevant as storage strategies appear as a classed issue to begin with: in the present sample, the better-off usually had a permanent water supply, and if not, the task of domestic water storage was delegated to housekeepers and concierges. Popular class women, in turn, were most likely directly involved in securing the availability of water for domestic use when water supply was unsteady. Amidst the lack of an automated storage system, they took to immediate water use and the strategy of being at stand-by. In consequence, water supply patterns increasingly dominate daily routines, limiting the time available for and autonomy over activities such as education, wage-earning and care work, leisure time and rest. A patriarchal logic is hence reproduced and reinforced as women – carrying a double responsibility for both care and wage-earning work – are affected by extra efforts and limitations arising from insufficient water availability (see also Bennett 1995). It is precisely the attempt to regain some autonomy over domestic water use (which itself forms an essential part of care work) by building a buffer of water which turns in to a limiting factor for these women. In other words, "water can become one's master rather than one's slave" (Swyngedouw 2004: 57). In this sense, it can be said that social inequality along lines such as class and gender are spatially mediated via water supply infrastructure, in particular intermittent supply frequencies.

5.4 IMAGINING URBAN WATER

> Sonia: *Those in charge [of water purification] probably rip off the chlorine, the fluorine, the activated carbon for their homes – and they say: "Yeah, people over here [in Iztapalapa] endure that. They are already used to it, they have developed antibodies, and nothing happens to them".*
>
> Alba: *(…) even during the last elections they said: "Oh well, people in the East are already used to be badly off."* (FG III: 87–88, Iztapalapa)

Beyond water running from a tap in the own dwelling, the water supply system as a whole tends to remain a highly abstract concept with limited relevance to everyday experience. Nevertheless, the overarching logic of urban water supply is imagined in one way or the other. These narratives often refer to multiple, interrelated questions of space and power, thus placing them within a territorial framework – as the conversation between Sonia and Alba during a focus group discussion in an Iztapalapan housing complex prone to unclean tap water highlights. Such urban visions – or what will be referred to as imagined landscapes of water supply in the present chapter – pin down the way in which the (inter)relation between the individual and its surrounding world is experienced, imagined and lived on a daily basis. With respect to water supply disparities – a matter of common knowledge and a frequent topic in the local media, in addition to own experiences – how are logic(s) of supply handled? Following this question, the present chapter will explore to what extent these imagined landscapes are related to current infrastructural conditions in the form of steadiness of water supply.

Referring to collectively shared ideas around the logic of urban water supply, these imagined landscapes form part of what García Canclini and colleagues have defined as *imaginarios urbanos* (see García Canclini 1997, Lindón 2007). Evolving over time in a process of sedimentation, these imaginarios are at once spatial and operational – in the sense that they are able to exert an influence on spatial everyday practices (of which they simultaneously form an integral part) (see García Canclini 2013: 39 f.). In the present work, such symbolic landscapes of water supply are explored via material obtained from both focus group discussions and individual interviews. The empirical material derived from focus group discussions is of particular interest here as it provides insight to public discourses while also grounding them through concrete experiences shared by the discussants. Hence, this material is able to bridge the gap between material and representational space by linking it via what Lefebvre (1991) calls spaces of representation. By informing practices, these imagined landscapes of supply are not mere preconditions or material containers within which actions are somehow dictated by supply frequencies. Rather, they constantly evolve from an interrelation and interaction between everyday practices and planning decisions materialized in pipelines and dams, the latter (re)enforcing a strong path-dependency. What people literally 'make' of these supply conditions is hence of considerable relevance for everyday practices of water use (see 2).

When asked about the logic of intermittence and water supply disparities in Mexico City, four principal perspectives on the imagined landscapes of water supply evolve from the interview material. Supply disparities are either sketched as a symbol of an overpopulated city close collapsing (5.4.1), of incomprehensive urban planning (5.4.2), as a straight-out expression of social inequality (5.4.3), or – perhaps most peculiar – as an educational method employed by the paternalistic state (5.4.4). While the imagined landscapes of water supply of most interviewees are ultimately made up of a combination of these collectively shared metaphors, they are separated here to clarify the picture.

5.4.1 *Ciudad Colapso*

Starting with probably the most nihilist of metaphors, *ciudad colapso* (Breakdown City) provides a good example of an imaginary which was initially triggered by water supply disruptions (or other infrastructural shortcomings) yet gained a much wider relevance and autonomy as a symbol of urban (or state) failure in general. Pablo, a journalist living in the city center, brought it to the point:

> *I don't call the Federal District Federal District – I call it Breakdown City, as it's a city which will eventually collapse. (…) Due to the huge number of people the day will come when the city will fail in one way or the other.* (Pablo E34: 318-321, Cuauhtémoc)

As much as 18 of all 44 interviewees drew a similar picture by stating that an alleged overpopulation of the city is the key factor which not only was depleting resources and yielding ecological destruction, but might ultimately also lead to a collapse of the city. Apocalyptic scenarios of this kind also feature in fictional and non-fictional literature such as Aridjis (1994) and Simon (1997) (see also Larochelle 2013). Tinted by a strong anti-urban undercurrent, *ciudad colapso* transpires a feeling of fearful fascination with an imagined (uncontrolled) urban growth. Evoking Malthusian lines of thought, a scarcity of resources is imagined as a direct result of uncontrolled population growth, and in consequence, urban growth (for a critique of this concept, see Harvey 1996: 137 ff.). Being subject to very divergent living and water supply conditions, Pablo, Nadia and several others nevertheless were unanimous when imagining the logics of supply:

> *The water problem is due to Mexico City's growth. So much population and as far as I understand, many people from the provinces came to Mexico City – that's the reason why it grew and I believe that's the problem.* (FG III Nadia: 72, Iztapalapa)

> *All the problems of Mexico City result from us being millions (...). The entire urbanized area hits 20, 21, 22 million people. So there is no solution, no program, no infrastructure which can handle such a demand. The principal problem of Mexico City and the entire country is the population's concentration in a very small part of the territory.* (Pablo E34: 108, Cuauhtémoc)

Herself experiencing good water supply and praising the benefits of her own neighborhood, Martina from the Tlatelolco housing complex went further and painted a picture dominated by straight-out classism, with the less well-off to blame for both demographical and urban growth, and in consequence, for supply disparities.

> *The foundation of the entire problem is an overcrowding (...) of the city. If we speak of people in general, we are a Third World country. (...) The most marginal people who have the least opportunities to study are the ones having the most children. (...) If you ask me, those kids should be taught at school: You shouldn't have a family at a very young age. As they can't offer anything to their family, they'd have to control their sexuality.* (Martina E53: 48–50, Cuauhtémoc)

Whether blaming specific groups or not, in this overly simplified logic, disparities in water supply are thought to result from Mexico City's population growing in an uncontrolled manner. This representational space is reined by a reification of socio-spatial differences as it steers clear of posing questions on power relations or any other underlying logic of urban development. The enormous, expanding city is featured as a symbol and result of a societal (or natural) sickness, to be treated by urban experts as 'doctors'. Such a naturalization of urban conditions effectively replaces a critique of urban conditions (as a product of a capitalist system) by a pathologization of urban space (see Lefebvre 1991: 99).

5.4.2 The Untamed Urban Leviathan

In contrast to fears on overpopulation and disastrous 'natural' urban development – but often also drafted as a strong weapon (or cure) against it – the conflicted landscape of supply is turned into a symbol of failed or insufficient management of urban growth. Almost two thirds of all interviewees shared a perspective according to which Mexico City's limitations in water supply were a result of severe shortcomings in expert-led urban and infrastructural planning – either exclusively, or in combination with other imaginarios. As a baseline, this metaphor harbors the hope that strategic urban planning, properly implemented, might tame the "urban leviathan" (Davis 1994).

> *There has been a tremendous invasion of dwellings here, due to the corruption which still tolerates that housing is being built even as water services don't yield more than a certain capacity, right? So there is this problem (…) that it grows badly planned.* (Esther E7: 138, Cuauhtémoc)

Such imagined landscapes of supply are often based on references to specific processes of urban development, such as inner-city redensification and the negotiated space produced by popular urbanization. Some inhabitants of inner-city neighborhoods such as San Rafael and Cuauhtémoc actively linked their imaginarios on own experiences, blaming ongoing processes of upgrading and redensification for increasing limitations in local water supply. Noting a deterioration of supply since around 2010, Miguel, a resident of the San Rafael neighborhood since the mid-1980s, provided the following explanation:

> *[Here in San Rafael] the pipelines are very old, (…) and another factor is that (…) a lot of apartment buildings are erected in the entire area, also causing lower water pressure.* (Miguel E10: 169, Cuauhtémoc)

Intermittence and spontaneous supply disruptions are hence thought to result from adaptation of urban infrastructures not catching up with the pace of urban development. Trained as an engineer, Carmen was startled by what she saw as ignorance and lack of foresight by local authorities in the light of a local construction boom. She angrily blamed it for frequent and spontaneous supply disruptions in her flat near Paseo de la Reforma in the borough of Cuauhtémoc, home to recently-built headquarters and prime office space alike.

> *Many many buildings have been built on Reforma, so water consumption is high. (…) Formerly, when there weren't so many buildings, we had no problems [with supply]. It's fine that they allow the big enterprises to build big buildings, but the infrastructure should also be created. (…) It causes huge problems and more buildings are being constructed all the time. I don't understand how you can do something without planning. (…) that's what everything is rooted in, everything: planning. […it's] illogical – if there are water problems, I don't understand why they keep approving more construction in Mexico City.* (Carmen E52: 62 and 82–96, Cuauhtémoc)

Moreover, the logic of water supply is structured along a dichotomy rooted in urban regulation for many: formal and informal modes of urbanization. Whereas this might be expected from a highly negotiated space[28] (see for instance Becker *et al.* 2008), it is remarkable that the imaginario still most prominently features the informality of popular urbanization in Mexico City, rendering other negotiations in the process of spatial production[29] rather invisible. Urban informality is doubtless one

28 For a thorough introduction to the context and modes of popular and irregular urbanization of Mexico City, see Azuela de la Cueva 1989, Gilbert/Varley (1991), Davis (1994) and Ward (1998), amongst others.

29 Sticking to the example of water supply, a number of middle and upper-class projects and neighborhoods - for instance on the slopes of Ajusco to the South and Santa Fe/Cuajimalpa to

of the most prominent features in the academic debate around 'the Latin American city' or, more general, of cities in the developing world (instead of many, see Burdett/Sudjic 2007). To what extent the term informality provides an adequate approach to processes of urbanization remains an open question in the field of urban studies (see the critique by Roy/AlSayyad 2004). However, a distinction between so-called 'regular' and 'irregular' modes of urbanization continues to form part of *imaginarios urbanos* and everyday perception in Mexico City, as the interview material reveals. Indeed, the image of an urban leviathan in urgent need of regulation seems to be nurtured by this distinction for many. A minority sketches a landscape of supply wherein popular and informal urbanization is perceived in the first place as a result of a need for housing not fulfilled by the regular housing market. In this logic, limitations in water supply and other urban services are a secondary effect of stratified housing policies forcing the urban poor to resort to self-help and a production of space with limited resources. The ex-post implementation of proper urban infrastructure is both required due to and complicated by the individualist character of this mode of urbanization. Berta had herself experienced this during her childhood in Iztapalapa and again the neighborhood on the slopes of Cerro de la Estrella where she lived at the time of the interview:

> *Building neighborhoods such as this one is tolerated: due to their need, people built it in any possible way. And most likely it's not well planned for putting a good water network in place (..) because it's a rocky or volcanic area. (…) There is nobody who'd tell them [how to do] a good urbanization.* (Berta E29: 72, Iztapalapa)

> *There are irregular settlements in the periphery, and you know perfectly well that this is another huge problem, as there is no planning, (…) there are no water taps, no power lines and nothing. But that's being installed afterwards, if the government tolerates it (…). There are conflicting interests: [the citizens] need to live, they need urban services, but the government doesn't give it to them (…). It's a problem [between] government and society.* (Rita E48: 82, Iztapalapa)

Reversing this logic, others paint bad living conditions and a lack of urban services almost as a deliberate individual choice, thereby giving the metaphor of the urban leviathan a neoliberal twist and shifting the responsibility for any disparities in supply to those affected. In short, people choosing to live in informal settlements are thereby opting for water supply disruptions, according to this imaginario.

> *Sometimes it's irregular settlements, the infamous paracaidistas*[30]*. People see a large plot and start to arrive, put up their cardboards. Logically, the government didn't take these people into consideration [to begin with…], therefore they lack water.* (Silvia E26: 63, Iztapalapa)

> *People even want to build in places where they shouldn't. Areas (…) close to rivers and places like that, where it's forbidden. (…). People go there and squeeze in. And thereafter (…) they are victims… No, we don't have water; we don't have electricity (…). But these areas are not for*

the West of the city - have been developed against planning and zoning laws, without proper water and waste water infrastructure. Many of these rely on water supply by tankers. The mass production of 'formal' industrialized housing by private developers in the utmost periphery of the city since the turn of the 20th century often operates in ignorance of or exempted from local regulations, too, increasing the pressure on local water resources (see Legorreta 2006: 139 ff.).

30 Verbatim: sky divers, this term refers to those directly occupying land for purposes of dwelling rather than purchasing it on the informal land market.

dwelling (...). That's why I tell you, we live in Mexico – that's the expression. "Oh, we'll just get in here and do something." The rules are not abided by as it should be. (...) people arrive, settle down: "Let's put up some paperboards and start living here". It seems all very easy to them, very simple. (Martina E53: 46, Cuauhtémoc)

At times, this imaginario strives to establish a strong contrast between the negotiated space ascribed to informal urbanization and the well-conceived and -implemented space of the modernist housing complex, such as the one Martina lives in:

Tlatelolco is a blessing, so to speak. Because Tlatelolco was built in '62 (...). Some French engineers, architects (...) won the competition and (...) the entire housing complex is very well-conceived. (...) accordingly, the entire water distribution was done very well. (Martina E53: 56)

Such a yearning for a proper engineering of the city's (infrastructural) needs falls in line with the modernist water policies employed by CONAGUA and other governmental institutions ever since the post-revolutionary phase (see Wester 2009). Based on a similar embrace of technology and planning as the ultimate solution to water supply problems, an expansion of water extraction by developing new sources as the ultimate answer to supply disparities continues to form the base of the hegemonic political discourse on Mexico City's water supply to this day (see 4.1.1). It seems as if not much has changed since the installation of the Cutzamala system was praised as a guarantee for a future access to water for all, as expressed in Leñero's novel in the early 1980s:

"I spoke about the problems of the metropolis: traffic, smog, water scarcity. All of them were most severe, yet all had a solution. With respect to traffic, you could already sense the relief granted by the ejes viales[31] and soon, the Metro network, extended to the remotest corner of the city, will render the use of cars almost dispensable (...). With respect to the scarcity of drinking water, an old classmate, engineer Hectór del Mazo assured me (...) that there was (..) an answer to this problem. The answer was in sight: Cutzamala. Once water from the Cutzamala arrives to the metropolis, all its inhabitants will swim in abundance." (Leñero 1983: 131–32)

In the face of common and widespread supply limitations, the myriad coping strategies applied by Mexico City's inhabitants – such as measures of domestic water storage described earlier – harbor a similar optimism towards technological solutions, individualizing it through self-help. Back to collective infrastructures, the metaphor of the untamed metropolis also shines through whenever disparities in supply are ascribed to topographical features hindering an adequate water distribution.

Maybe there is no infrastructure or water pressure (...) to supply all those hills – as the valley has now been plagued and all the hills are full up to the very top with people who live there. Maybe [there is] a lack of infrastructure (...) for the most elevated parts of the city to be supplied. But apart from that I believe it's a question of giving priority to certain areas, no less, and actually leaving others in oblivion. (Antonia E20: 110, Cuauhtémoc)

31 *Ejes Viales:* A vast system of major connecting roads implemented in the Federal District during the 1970s to facilitate the flow of motorized individual traffic. Expanding existing roads and cutting new connections through existing neighborhoods, a significant number of buildings were demolished in the inner-city boroughs to create this grid-like system, often leading to the displacement of the inhabitants towards the urban periphery.

The urban leviathan hence keeps expanding in an uncontrolled manner, with hydraulic infrastructure unable to keep pace, or – in a kind of inverted technological optimism – seemingly incapable of overcoming topographic challenges. Somehow, the underserved are thought to have simply settled down in the wrong location – on the hills rather than the city's vast plains. This imagined logic of supply corresponds closely with the official discourse on water supply disparities, wherein topographical and technical challenges are quoted together with references to unregulated urban growth, but possible links to sociospatial differences largely remain beyond consideration (see for instance GDF/SACM December 2004: 80 and CONAGUA 2012: 25). Again, natural limits are invoked – though in Antonia's imagination, supply disparities were also linked to biases in the administration and implementation of urban water supply. Some of those directly affected don't seem to entirely trust the official explanations, either. Alma, who has running water only once a week in her house in a colonia popular overlooking the valley from the slopes of the Tetlalmanche volcano in Iztapalapa, expressed her doubts:

> *In the center they have daily [supply], but we don't. (...) They put a pretext: first, because of the altitude. That we are on the hill and that they can't pump it, supposedly. So (...) the pumps don't work, they wear off (...). At least that's what they tell us.* (Alma E9: 98–101, Iztapalapa)

As the reference to the city center suggests, there seem to be other fault lines structuring the imagined landscape of supply which need to be unearthed.

5.4.3 Preferential Treatment

When encouraged to talk about the logic behind inner-urban disparities in water supply, several issues arose during one of the focus group discussions held in an Iztapalapan church yet what prevailed was a sense of indignation.

> Esme: *If it's Iztapalapa, I reckon the suspension of water is probably not so much for cleaning pipes – [...it's] simply the area. Because at Los Pinos*[32] *and all that, the borough of Miguel Hidalgo, Coyoacán, water is never suspended (...). It's always boroughs like Iztapalapa, which leads the list, Cuajimalpa, to mention some. I figure it's the social class, first and foremost.*
>
> Interviewer: *How does that influence supply?*
>
> Esme: *Precisely in that we don't have supply. As far as I know, they never lack water in those boroughs. (...)*
>
> Susana: *I think it's due to politics. (...)*
>
> Ivan: *When we go and see who lives there and who lives here. Because the very rich (...) say: First we'll torment the most tormented.*
>
> Susana: *Our borough is very poor. But money does enter – it does. If you set yourself up here in front [on the street] to sell pepitas*[33], *they charge you for that space, too. So apart from being poor – the borough probably hasn't the same income as others – but apart from that, we have*

32 Official residency of the Mexican president, often used as a synonym for the Mexican government in general.

33 Roasted seeds of pumpkins, sunflowers and the like are a popular and cheap snack sold at street

a borough that is (…) very treacherous. That's the truth. [...gives an example] In our borough, [those governing us] always do with us as they like. (…) And that's always the problem, maybe because we are people with a bit less of culture or I don't know what, or we are not united. (…)

Esme: *When [the current administration] came to power in the borough, it was one of the boroughs which was given a very broad budget to do everything. So there actually is from where [to take] – but it's not used where it's needed.* (FG II: 78–98, Iztapalapa)

So people felt that water supply was an expression of local politics and a corrupt political culture but also, and perhaps more profoundly, a question of distribution and preferential treatment, wherein clear-cut differences were made between different parts of the metropolis. Far from representing solely an Iztapalapan point of view, a similar perspective was shared by Berta, Ana and Pablo in spite of remarkable differences in their supply conditions, locations within the city, and social status.

Preferences by the boroughs, probably political preferences (…). Those who have more economic power, purchase power need to be well, whereas those who possibly can't give a present to you as a borough don't. (Berta E29: 194, Iztapalapa)

The lack of supply in some neighborhoods (…) probably has to do with social status; they never lack water here in the upper-middle class neighborhoods. In contrast, I feel that (…) colonias populares are less taken care of. Or maybe it's related to property taxes – when I pay a higher property tax, I probably have more opportunities to get water. (Ana E36: 38, Cuauhtémoc)

Those who are able to pay have water. Economically spoken, they have a specific weight in society, they have a guaranteed service. (…) be it legal or illegal or (…) with certain favoritism, I think that applies, too. You can't, so to speak, leave Reforma 222 without water[34], because of all the businesses, all the commerce, all the money that circulates there. So I believe that there are actually preferences with respect to water supply. (Pablo E34: 118, Cuauhtémoc)

Their landscape of supply resembles an expression of social status, where access to water supply is negotiable over economic capital. The idea that unequal power relations materialize and thus become 'readable' in urban space, and not least in water supply, is the most prominent, shared by a vast majority of the interviewed (32 of 44). As social inequality is thought to be inscribed in infrastructural networks, limitations in water supply turn to symbols of neglect and social exclusion. After showing me the water filters installed in her apartment building's shared cistern in Iztapalapa to improve water quality to a level suitable at least for basic domestic needs such as laundering, Lucia voiced little doubt on what she saw as a preferential water provision to some parts of the city.

They have abandoned us (…). There should be no preferences, we should all be equal. (Lucia E40: 94–96, Iztapalapa)

Dolores, from a different housing complex in Iztapalapa, was convinced that lower tax payments were what gave her neighborhood a low priority in water supply, in

stalls. At the time of the interview, a new tax on all street vending activities had recently been implemented in the borough of Iztapalapa.

34 One of the numerous new skyscrapers on Paseo de la Reforma linking the borough of Cuauhtémoc and Miguel Hidalgo, designed by the Mexican architect González de León and completed in 2008 (see http://www.reforma222.com).

contrast to better-off parts of the city. Social hierarchization was symbolized by the type of water running from her tap:

> *You should understand that we are the forgotten, and only filthy, yellow water gets here.* (Dolores E8: 113, Iztapalapa)

A preferential distribution and a hierarchization of the city based on a center-versus-periphery dichotomy formed an integral part of this imagined landscape.

> *Here, if you wouldn't have a cistern you'd have no water, and in the center they don't even have cisterns and there's always water.* (Diana E15: 64–66, Iztapalapa)

> *You pay taxes and pipelines break. And what about us? Since both are paying taxes, also give priority to the people of the periphery and not just those in the center! So the problem is that there is no strategic planning to respond to the needs of all neighborhoods.* (Rita E48: 100, Iztapalapa)

Irrespective of their actual place of residence or supply conditions – there was little doubt for many that the location within the city made all the difference when it came to infrastructural conditions. Inhabitants of neighborhoods close to the Historical center, such as Paulino Navarro and Tabacalera, pointed out what they felt was a special treatment of the city center. Sitting in Oscar's small and frugal apartment in a not quite up-market side street of Paulino Navarro, he stressed how privileged he felt living so close to the Zócalo, the city's main square:

> *There is no lack of water here because, fortunately (…) we might say that we are in the center. (…) I imagine that a priority is given to all the [central] neighborhoods. Iztapalapa and other neighborhoods are very remote so I guess the water department doesn't provide supply for them.* (Oscar E14: 38, Cuauhtémoc)

The city center is hence understood as an area of privileges, independent of its residents' social status or overall dwelling conditions. Such prioritization symbolizes both its political and economic relevance, as mechanic Jorge explained while standing in a car workshop in one of the streets of Paulino Navarro:

> *Supposedly the center has everything. (...) here we have the whatshisname, the president, the National Palace. A lot of people come to visit the center. So it's logical: most attention needs to be paid where there's the most important (things), right?* (Jorge E17: 117, Cuauhtémoc)

A resident of the city center since her birth, 70-year old Rosa from the Tabacalera neighborhood also highlighted the economic weight of this part of the city:

> *The bottom line is that it's the center, so even if [Tabacalera] is a colonia, it's a very central one. So there is a lot of hotels, a lot of restaurants, a lot of sites where water is mandatory. So for sure they need to have supply, because otherwise they'd (…) protest. And in one way or the other everything that is a business is convenient for the authorities. They are the ones (…) that give the most money to the government due to taxes and all.* (Rosa E41: 206, Cuauhtémoc)

The metaphor of the city as a mirror and vehicle of social inequalities – it seems as if in this imaginario, some areas are thought to be almost automatically given priorities in supply over others. In addition to the city center, the imagined landscape of supply is often tied to a set of specific places in Mexico City which serve as imagined extremes along a spectrum of supply conditions.

> *Areas such as Polanco, (...) Lomas Verdes or Satélite, all these – I have never heard that they'd have water scarcity (...). So yes, that's where you see the inequality between people.* (Silvia E26: 65, Iztapalapa)

These references bear clear socioeconomic connotations and are repeated over and over when explaining the logic of differential water supply, thus reinforcing a certain kind of imagined landscape. For local residents and people living elsewhere in the metropolis alike, it is the borough of Iztapalapa which represents one of those urban areas given the lowest preference in the hegemonic logic of water supply, as contributions from two focus group discussions in Iztapalapa point out:

> *Here in Iztapalapa (..) we suffer most from bad water quality. It is one of the most desolate boroughs in that sense. So you will get good information. Yes, because Iztapalapa is the worst borough.* (FG III Manuela: 5, Iztapalapa)

Hence a vision of just distribution, voiced in another Iztapalapan focus group discussion, took this borough as a starting point:

> *[We need] a program for the supply of all boroughs (...). Because (...) they never lack water in important boroughs with a high social standard. Thus if they suspend the water, okay – one day there is no water for you. Something that is well-balanced, a just program for the Federal District.* (FG II Esme: 257, Iztapalapa)

In its role as a symbol of the neglected, Iztapalapa is set in contrast to the city center, as we have seen, or with traditional, well-known upper-class residential areas such as Lomas de Chapultepec and Coyoacán, which are handled as references for areas enjoying a preferential treatment in water supply.

> *Is a question of (...) social justice that it doesn't suffice everybody. (...) it's a right that is not guaranteed to all inhabitants of this city. (...) it's the poorest families who live in the periphery who have to disburse the most money with respect to water. (...). It's always that kind of neighborhood – allegedly with lot of surplus value, like Roma, Juárez, the Pedregal in Coyoacán – and oddly enough, there is less water in the poorest areas.* (Alan E30: 65–71, Cuauhtémoc)

However, this fixed idea of an almost self-evident infrastructural privilege of certain areas such as the city center apparently suffered heavily when it lost its material base during the first water supply disruptions faced by its residents. Antonia revealed how her idea of the neighborhood's status harshly clashed with her actual experience in centrally located San Rafael.

> *It seems to me that water has always lacked in Iztapalapa (...) and there were even heavy street fights, clashes between neighbors, due to water scarcity. But [... here] you'd have never imagined this. Because these are more or less good areas, centrally located, and I thought that not many water supply suspensions would occur.* (Antonia E20: 80, Cuauhtémoc)

Once there is a deception or rupture of the implicit image – here being the reliable, well-supplied and almost naturally privileged city center – it probably becomes most evident that representational space also has an operational character, exerting an influence on everyday practices in the sense of leaving residents unprepared for any spontaneous supply disruption. Why take precautions for a lack of water when you don't expect it to stop running from your tap, ever?

Though those who were subject to supply limitations often used these experiences as a reference when laying out their imagined landscape of supply, linking social inequalities and supply disparities was not an exclusive idea of those actually affected by these conditions. A majority of those enjoying permanent water supply also shared this collective imaginario wherein social status was decisive for the quality of domestic water supply in many (if not all) parts of Mexico City. In this imaginario, residents of certain areas were assigned less power over influencing infrastructural and overall living conditions, rendering them voiceless and invisible, whereas others exert a considerable influence and their needs are both voiced and heard. Within the Federal District, the studied boroughs of Iztapalapa and Cuauhtémoc serve as strong symbols of periphery and center in this sense. Residents from both boroughs – subject to a spectrum of different supply conditions as well as from different social backgrounds – unanimously shared this representational space, wherein *"the uppermost people have the most water"*, as a resident of one of the many undersupplied colonias populares in Iztapalapa put it (Margarita E1: 718). Organized along lines of social inequality, the landscape of supply is hence felt to be an expression and reflection of social disparities and contradictions. Yet it is not the only supply-related metaphor making a direct reference to questions of power.

5.4.4 Punishment: Disciplination via Water Supply

The idea of space simultaneously containing and implying social relationships – and, above all, inhibiting actions (see Lefebvre 1991: 142) – is perhaps most clearly articulated in the metaphor of water supply conditions as a sort of punishment, where a regulating if not paternalist character was assigned to urban water infrastructure and its management. The idea of the state as a planning agent and competent engineer managing water supply (see 5.4.2) turns even more powerful here: it assumes an authoritarian role and teaches 'ignorant' city dwellers how to use resources in a proper manner. This small group (5 of 44 interviewees) seems to be aligned by the feeling that domestic water taps and the flow of tap water are somehow adequate means of education.

> *Also because water is running short (…). As we ain't educated to take care of [water], they teach us that we have to take care. (…) that's why they ration it.* (Ada E1: 246–248, Iztapalapa)

> *They have noticed that (…) it's better to supply water by night (…). Maybe it's a means of controlling [water use], as (..) there is high water consumption in the morning (…). So when water is available, you take less care (…) whereas when you don't have running water constantly and you know that you won't have water for one or two days, you'll try to rationalize it.* (Carlos E35: 64, Cuauhtémoc)

Supply frequencies are hence imagined as a way of governmental control over domestic water consumption, as tools of disciplination, and in a surprising twist, are even welcomed as necessary regulation by some. This opinion did not differ perceptively between Ada and Carlos, supplied on divergent rationing schemes. Reading supply disruptions as a form of governmental control charged with unequal power relations, others expressed feelings of humiliation and unjust treatment – like David, from a neighborhood subject to limitations in water supply over several decades:

> *I ask myself: Why do they punish us so much? (...) It's all under control of the borough, you see. (...). People waste a lot, so I'm not so much against the rationing, but it should be [provided] for a longer period (...). The goal is less wastage (...), that's good, but the punishment is not very even.* (David E11: 73–77, Iztapalapa)

Others, however, turn this idea of the state as an essential (if somewhat unfair and biased) educator into its opposite, voicing a strong critique of a political culture which is perceived as rife with shortcomings, inaction and corruption by many. In this logic, water supply failures are a symbol of a government culture suspected to be untrustworthy in general, as the earlier excerpt from an Iztapalapan focus group discussion (see 5.4.3) and the following debate on responsibilities for water quality during a different focus group in Iztapalapa exemplify. As a starting point, participants were given a statement by Ramon Aguirre, boss of the Federal District's water utility SACM, in which he indicates that rather than the supplier, domestic users not cleaning their cisterns might be to blame for low tap water quality.

> Manuela: *He is an official, so what's he going to say (...). Like all officials and a majority of politicians, he'll wash his hands (...).*
>
> Linda: *Because he doesn't live in Iztapalapa.*
>
> Manuela: *Good point.*
>
> Alba: *And maybe it could be part of the problem (...) that it's more than a year that the building's cistern and the roof-top tanks have not been cleaned. They are not cleaned because one half doesn't want to pay, the other doesn't feel like it, because of laziness and all that (...). So how can it get a bit cleaner if I don't take care of my own pipes and my own storage? (...)*
>
> Linda: *That would be a minimal part of the official's comment, minimal. But the greater part [of responsibility] lies with the officials (...).*
>
> Alba: *Well, it's not the official's fault.*
>
> Linda: *Why? Because he looks away – he knows about the problems. We might say: "Oh, the poor thing, he doesn't know." Yes, indeed he knows! There might be the one or other ignorant amongst the officials who doesn't know what he's working on or talking about, but in general, they know.*
>
> Manuela: *The one who doesn't know what he's opening his hand for and earning a lot of money is acting the fool. (...) How much do they charge for doing nothing? (...) Yes, unfortunately, Mexico is number one in corruption at the level of officials.*
>
> Sonia: *Gold medal for Mexico.*
>
> Manuela: *It comes from the top.*
>
> Alba: *And how many of us have offered the police: "I'll give you some dough and you let me go"? We encourage it.*
>
> Manuela: *No, I don't.*
>
> Linda: *You don't, but people do.*(FG III: 241–259, Iztapalapa)

In addition to such a fatalistic view on institutional actors and everyday corruption enclosing everyone, repeated by several interviewees from both boroughs and literature alike (see for instance Tejera Gaona 2003), a political instrumentalization of supply disruptions is just one of the myriad examples of clientelism deeply entrenched in Mexico. One resident linked temporary supply suspensions in the Sierra de Santa Catarina area of Iztapalapa to attempts to widen influence by local political leaders.

> *I feel that's maybe politics, because they tell us: "Let's go, there will be a meeting" and (…) afterwards they provide us with [piped] water (…), or sometimes they told us that they'd send water tankers.* (Gabriela E19: 53–55, Iztapalapa)

Features of the imagined landscape of supply are carved out by what is seen as a particular manner of disciplination by a paternalistic state and its agencies, which are being closely linked to logics of corruption and clientelism. Water supply is hereby just one of the sounding boards of the first, seven-decade lasting hegemonic era of PRI government, which enjoys a continued and profound impact upon the Mexican society and what remains of its democratic institutions (see for instance Davis 1994, Tuckman 2012). In other words, whether thought to encourage water-saving or foster political instrumentalization, a domestic water tap running dry is sometimes sensed by Mexico City's residents to be an authoritarian way of exerting influence on their everyday practices. In the context of such imaginarios, water supply networks actually turn into a medium of social relations.

All in all, Mexico City's water supply rhythms and water qualities can be understood as forming and informing implicit images of the urban as a lived symbol of either preferential treatment based on social inequality and unequal power relations, of a wildly expanding entity in urgent need of control by planning, or of population growth driving the city close to collapse. Some residents of the metropolis even attributed water supply disruptions a disciplinary character, aimed at altering domestic water use and decreasing water consumption. To what extent are such imaginarios related to (or influenced by) own experiences with current conditions of domestic water supply? Clearly, the water supply limitations people experienced in their dwellings at the time of the interview did not seem to provide a strong background for these imaginarios. Rather, there was a wide and heterogeneous spectrum of combinations of the four metaphors shared by those who did and those who did not experience a limited water supply in their present dwelling. Current water supply conditions hence seemed to make no notable difference when it comes to imagined landscapes of supply. In other words, they seemed to be less driven by personal experience in terms of housing and supply conditions than by a wider *imaginario* of the city as a whole, an urban logic. As emphasized by references to other urban or social ruptures and fault lines – not just the aspect of water supply – that accounts for both the failing 'Breakdown City' and visions for a just distribution of water, which touch more profound questions of access to and distribution of urban services. This absence of a direct link between the kind of imaginario(s) people share, and their own everyday experience goes in line with the literature: indeed, imaginarios may evolve and continue to exist without any concrete reference to ongoing urban processes (see Lindón 2007: 10). And it may also indicate that other influences are at play when it comes to the imagined landscapes of water supply. While it remains an open and potentially fruitful task for future research to analyze in which way these imagined landscapes of supply refer to urban images drafted in media, arts and public discourses alike, the following chapters will stick to the micro level and explore the links between these metaphors and people's past experiences with water supply.

To conclude, people's own water supply conditions appeared to make no difference when it comes to imaginations of the urban. Nevertheless, the divergent urban landscape of water supply in Mexico City was felt by many to be an expression and reflection of social contradictions – whether they experienced supply limitations themselves or not. And by "command[ing] bodies, prescribing or proscribing gestures, routes and distances" (Lefebvre 1991: 143), widely experienced supply disparities indeed seemed to exert an influence on everyday practices and social status in a number of other ways – whether supply patterns and storage tasks disrupt daily routines, discontent arises over sharing stored water amongst neighbors, or water needs to be pre-treated to make it apt for doing laundry; whether a preferential water use was assigned amidst a lack of water to family members undergoing educational training or conducting wage labor, or drinking water needs to be purchased in bottles. To which extent such coping strategies are perpetuated continuously rather than showing a volatile character once supply conditions improve will be discussed in the following chapters by analyzing them from a historical perspective, linking them back to previous experiences with water supply limitations.

6. HABITAT BIOGRAPHIES: THE BECOMING OF HABITUS FROM A SPATIAL PERSPECTIVE

After having explored practices of domestic water use in Mexico City and their relation to current housing and water supply conditions, the present chapter turns to a historical perspective, looking into the question: *To which extent does the past sociospatial setting (and the experiences people made in that very setting) influence current practices of water use?* For a historical analysis of the individual and collective processes of socialization in a given sociospatial setting, the empirical method of habitat biographies – visual representations of a person's dwelling history – was developed by the author (for details on the method, see 3.2.3). With the help of this tool, a person's past sociospatial setting on the micro level of the home is analyzed as the place where water-related experiences were made. This contributes to a deeper understanding of a subject's current practices of water use. Aiming for a further spatialization of Bourdieu's concept of habitus which "makes the habitat" (Bourdieu 1991: 32) according to one of his most-quoted phrases, the question arises to what extend past habitats in turn exert an influence on the formation of habitus. The habitus-habitat-practice model (see Fig.2.2 in chapter 2) forms the backbone of such an analysis of the historic formation of habitus in a specific sociospatial setting and its repercussions on current everyday practices while avoiding the trap of spatial determinism. In other words, are a subject's past experiences with water supply relevant for a deeper understanding of current practices of water use? Do people learn and incorporate past experiences e.g. with water shortages and act in a different way afterwards, or are current living conditions the main factor influencing everyday practices such as storing water at home or imagining the logic of urban water supply? Linking habitat biographies with current practices will shed a light on these questions. Based on empirical material from 44 individual interviews conducted with residents of two boroughs of the Federal District, each habitat biography documents the following aspects for each of the dwellings an interviewee has lived in:

- the housing type (ownership and type of dwelling),
- the type and steadiness of water supply,
- the availability of storage devices,
- the duration of stay, and
- the location.

Memories of and experiences with water supply, processes of consolidation or deterioration of infrastructural and dwelling conditions, as well as comparisons between different places of residence are illustrated through direct quotes from interviewees for each dwelling they lived in. To increase comparability, a rough periodization was introduced ex-post to these biographies. It marks two specific moments in the interviewee's life course: first, when settling down on her or his own after leaving their childhood home (labelled as *first own household)*, and second, when becoming a home-owner, if applicable (labelled as *first own dwelling)*. The tool of a graphic habitat biography is employed to identify collective memories and shared experiences with water supply and water use in the past. Collective memories such as the 1985 earthquake's impact on Mexico City's residents trust in tap water quality as well as the traditional use of public bath houses were already introduced in chapter 5. Here, the habitat biography tool is used for the construction of empirically saturated types of habitat biographies. While infrastructural conditions, housing type, home ownership and place of origin could all be employed as categories for the construction of these types, the present study put a particular focus on water supply conditions. Four different types of habitat biographies were identified with respect to different experiences with water supply conditions and are introduced in this chapter. In order to identify these types of habitat biographies, the interviewees' dwelling histories were analyzed for perceived limitations of water supply. The term infrastructural conditions serves to outline whether the interviewee has experienced limitations in domestic water supply during any period of their life. This can refer to limitations in water quality, the available quantity, or the frequency and steadiness of supply. The definition of what qualifies as 'limited' is based on each interviewee's own perception rather than the author's judgment. Including such a subject-based perspective instead of an assumedly objective perspective is crucial here, using Denzin's life history approach (see Denzin 1989: 183 f.). Those periods of each subject's habitat biography when limitations in water supply conditions were experienced and reported by the interviewees are highlighted in the respective habitat-biography scheme by a particular background pattern (see Fig. 6.1).

Four basic types of habitat biographies can be distinguished when it comes to past and present limitations in water supply amongst the interviewees from Mexico City. There are those who

- currently live in a dwelling with water supply limitations and also experienced similar conditions in former dwellings,
- experienced supply limitations in former dwellings only,
- came to experience limitations in the current dwelling only, and
- those who never perceived any limitations of water supply in any of their dwellings.

All four types are introduced in detail below. In what is probably a reflection of the general situation in Mexico City, a vast majority of interviewees had experienced water supply limitations in one or more of their dwellings, while those who had always enjoyed an unlimited access to tap water without any limitations formed only a small minority.

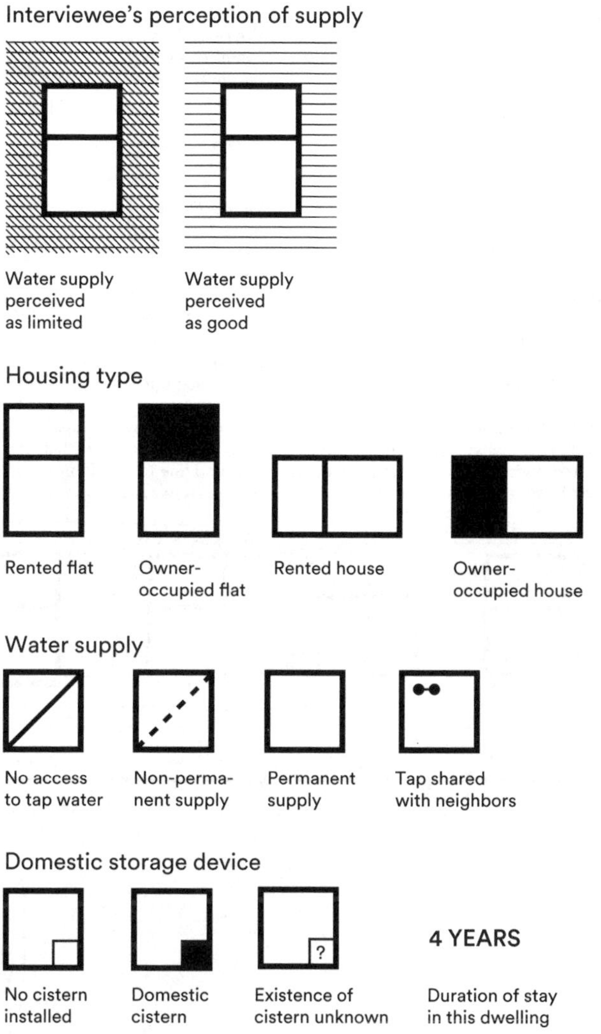

Interviewee's perception of supply

Water supply perceived as limited

Water supply perceived as good

Housing type

Rented flat

Owner-occupied flat

Rented house

Owner-occupied house

Water supply

No access to tap water

Non-permanent supply

Permanent supply

Tap shared with neighbors

Domestic storage device

No cistern installed

Domestic cistern

Existence of cistern unknown

4 YEARS

Duration of stay in this dwelling

Figure 6.1: Legend for habitat biography figures. Own elaboration

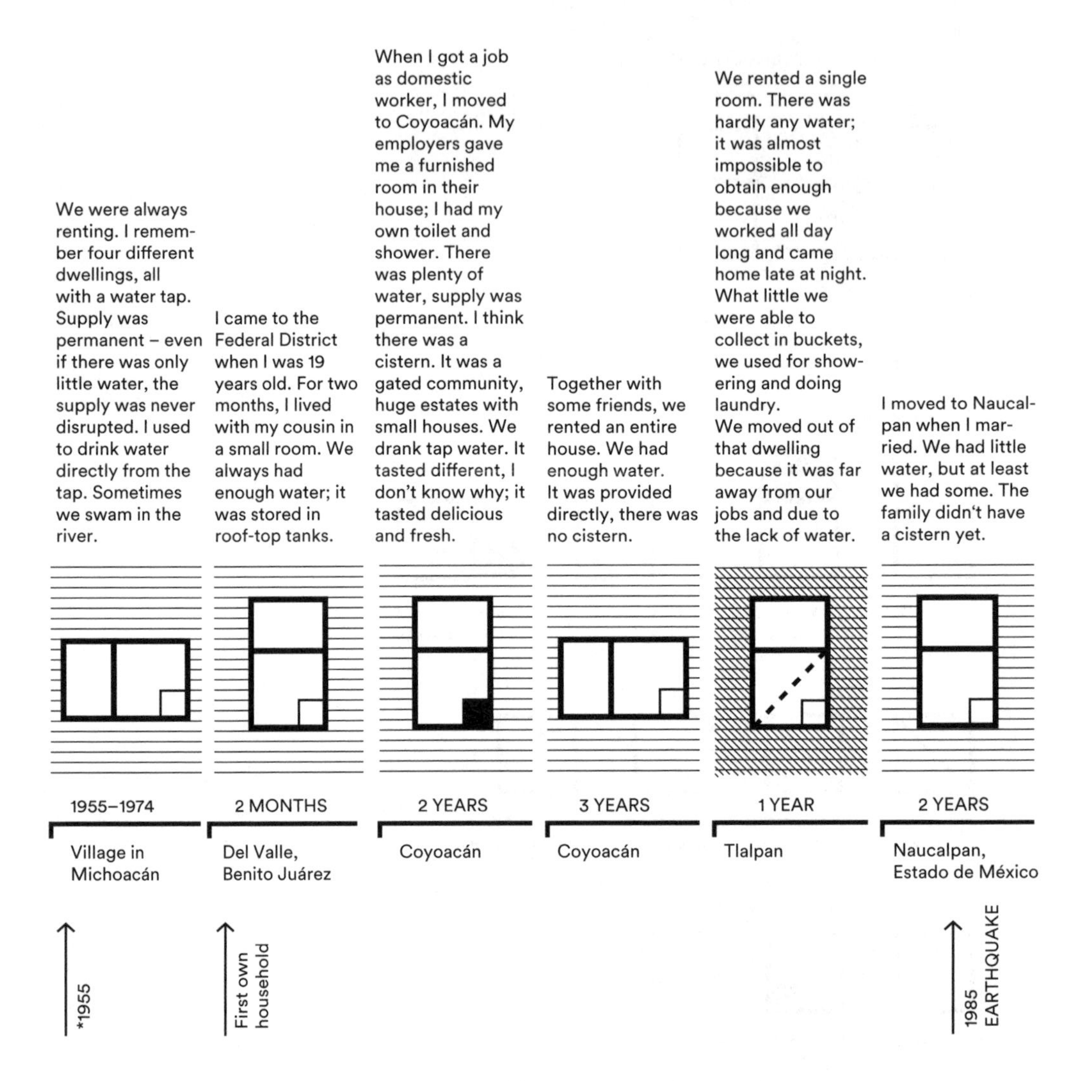

Figure 6.2: Alternating water supply conditions with limitations in both past and present dwellings. Habitat biography of Guadalupe (E3), domestic worker, Iztapalapa. Own elaboration

There was enough water, but we couldn't use much of it as we had trouble with the landlords. They sought to restrict our water use, but I told them: We need to do our laundry. I didn't stay long in this flat.

I worked as a house keeper in a condominum in Coyoacán. They gave me a furnished room where I lived with my kid. Water supply was perma-nent.

Tap water had a nasty smell and taste so we started to buy bottled water. We always had enough water even though there was only a make-shift water tank and a single faucet. My laun-dry day was on Sundays only, and we shared a bath-room with other neighbors. I was very happy in this flat because the landlady knew and respected me.

We lived in a vecin-dad without a cistern, sharing the only bathroom with everybody. There were many arguments and conflicts over water. Our neigh-bours used a lot of water and didn't allow us to gather enough of it. Therefore, I bought a barrel which I could fill with water from the courtyard tap. In the evening, we had to wait until water started running, as supply was intermittent.

Usually, our tap water smells. It's not potable. Water is supplied only during the morning, which is why everything needs to be done then. As we don't have any cistern or roof-top tanks, we have to wait until water starts running in order to be able to do anything. I usually get up at 6 a.m. to do my laundry. Afterwards, I fill my barrels in the yard and leave for work.

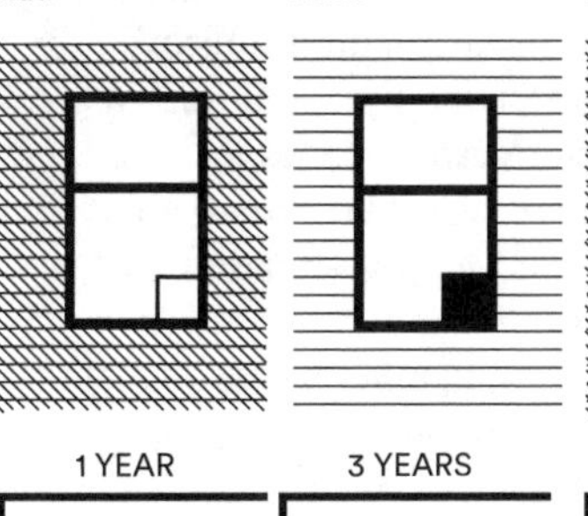

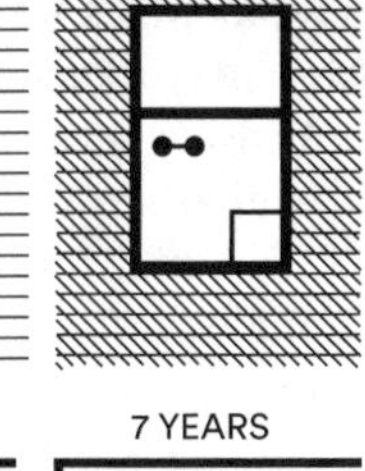

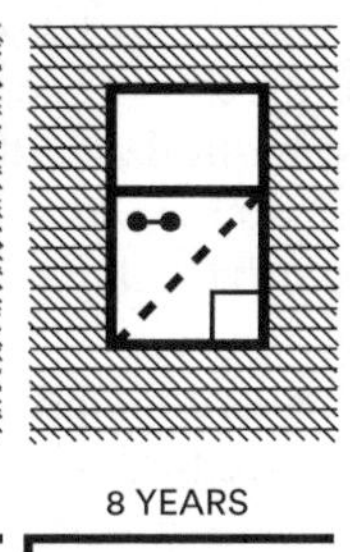

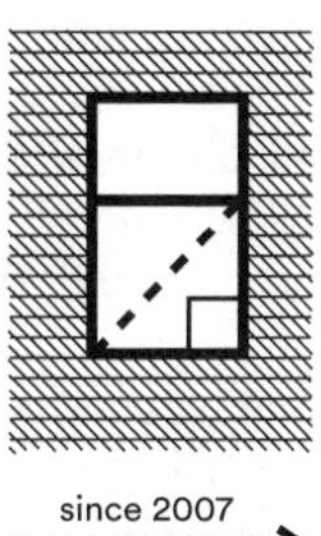

1 YEAR — Colonia popular, Iztapalapa

3 YEARS — Coyoacán

7 YEARS — Colonia popular, Iztapalapa

8 YEARS — Colonia popular, Iztapalapa

since 2007 — El Manto, Iztapalapa

Starts buying bottled water

Ongoing Limitations in Water Supply

The by far biggest group – half of all 44 interviewees – had not only experienced supply limitations in the past but still faced a lack of water in their home at the time of the interview. It is significant that a vast majority of them lived in the borough of Iztapalapa, amounting to 17 interviewees in total. As a whole, this group was socially heterogeneous, with educational levels ranging from one person who never attended school to others holding university degrees. Occupations and material housing conditions amongst the group were also diverse. Nevertheless, a tendency towards a lower social status was indicated by the fact that 14 of all 23 interviewees forming part of this group either had the most basic *Seguro Popular* health insurance, or none at all (see 2.1.2 for details on the stratified Mexican health insurance system). For those sharing this kind of habitat biography, periods of good supply over time usually alternated with limitations – depending on the neighborhood or the type of building that people moved to. Being employed as a domestic worker over the last four decades, Guadalupe, for instance, had changed dwellings quite often[1] and experienced a range of water supply conditions, often varying significantly between different homes (see Fig. 6.2). The path of infrastructural improvements is not necessarily a progressive one as sometimes water supply conditions in a neighborhood improved after initial hardship only to deteriorate decades later, as happened in the case of David (see Fig. 6.3). But there are also two remarkable cases, Maria and Paula from Iztapalapa, in which the water supply situation has never changed significantly, remaining limited in all dwellings they ever lived in – housing complexes and colonias populares alike (see Fig. 6.4). In other words, both women never lived in any dwelling with permanent and sufficient water supply during their entire lives.

Apart from the Iztapalapan majority, those six cases reporting not only past but also current limitations in water supply from the borough of Cuauhtémoc mostly resided in neighborhoods such as San Rafael and Paulino Navarro, where other interviewees also reported decreasing water pressure. If not for the current deterioration of water supply conditions in these neighborhoods, these interviewees would form part of the habitat biography type coined by past limitations only (see below).

1 It is common for domestic workers in Mexico City to receive part of their salary (typically below the minimum wage and without any social insurance) in the form of board and lodging in the employer's residence. As a consequence, women holding these live-in positions are forced to move when changing jobs.

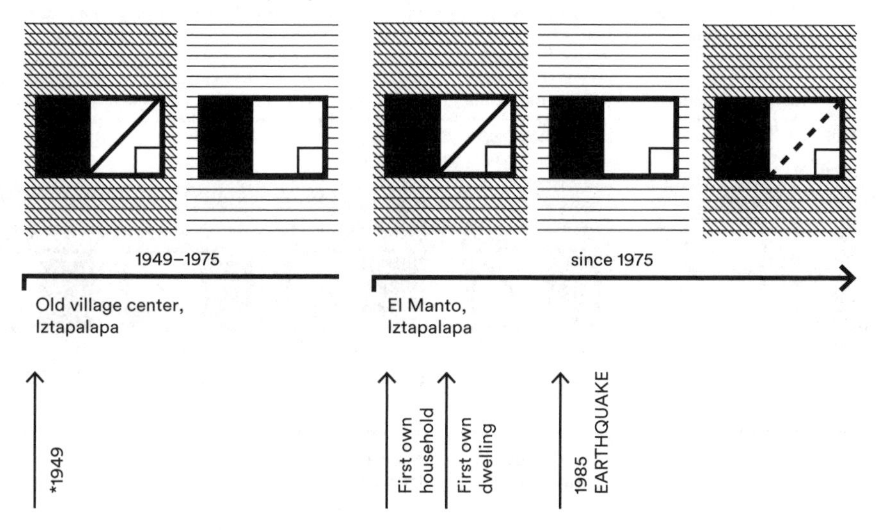

Figure 6.3: 'Twenty years on, history repeats itself': Improvements in water supply followed by its deterioration. Habitat biography of David (E11), retired engineer, Iztapalapa. Own elaboration

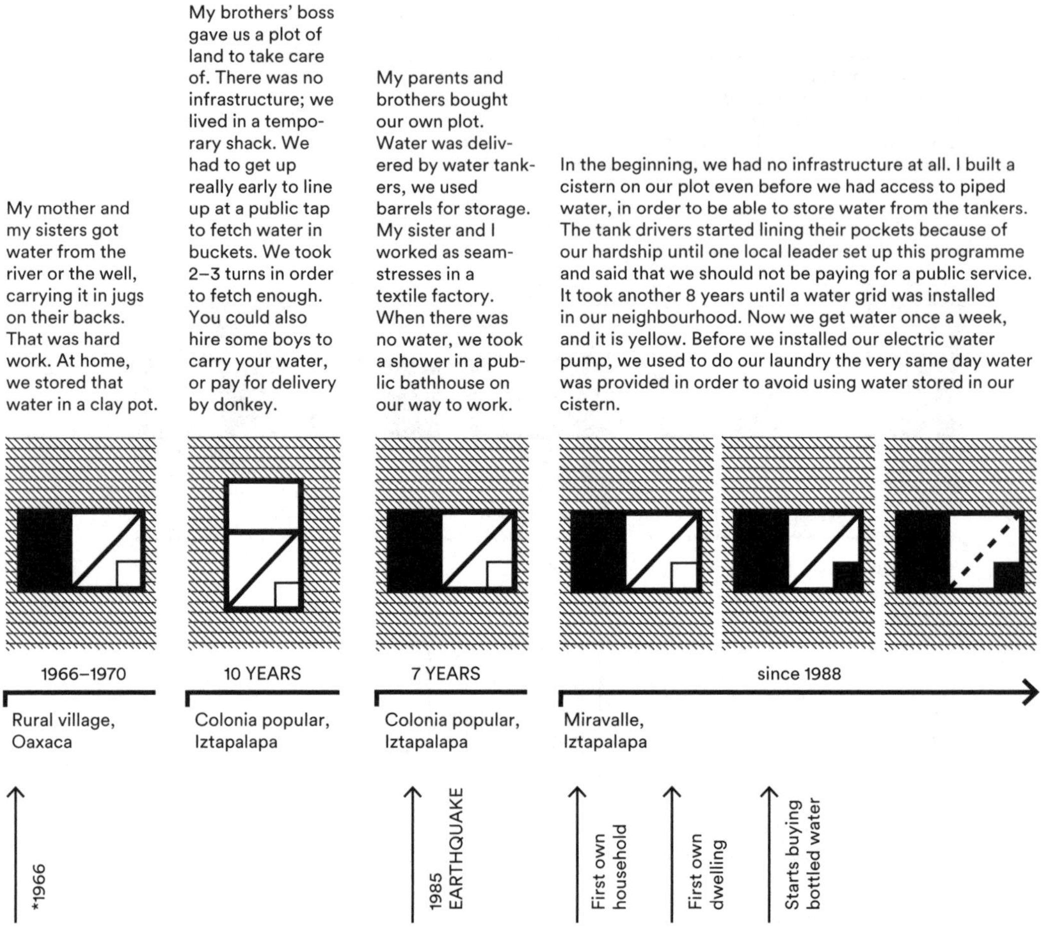

My mother and my sisters got water from the river or the well, carrying it in jugs on their backs. That was hard work. At home, we stored that water in a clay pot.

My brothers' boss gave us a plot of land to take care of. There was no infrastructure; we lived in a temporary shack. We had to get up really early to line up at a public tap to fetch water in buckets. We took 2–3 turns in order to fetch enough. You could also hire some boys to carry your water, or pay for delivery by donkey.

My parents and brothers bought our own plot. Water was delivered by water tankers, we used barrels for storage. My sister and I worked as seamstresses in a textile factory. When there was no water, we took a shower in a public bathhouse on our way to work.

In the beginning, we had no infrastructure at all. I built a cistern on our plot even before we had access to piped water, in order to be able to store water from the tankers. The tank drivers started lining their pockets because of our hardship until one local leader set up this programme and said that we should not be paying for a public service. It took another 8 years until a water grid was installed in our neighbourhood. Now we get water once a week, and it is yellow. Before we installed our electric water pump, we used to do our laundry the very same day water was provided in order to avoid using water stored in our cistern.

1966–1970
Rural village, Oaxaca

10 YEARS
Colonia popular, Iztapalapa

7 YEARS
Colonia popular, Iztapalapa

since 1988
Miravalle, Iztapalapa

*1966

1985 EARTHQUAKE

First own household

First own dwelling

Starts buying bottled water

Figure 6.4: Water supply limited at all times. Habitat biography of Maria (E2), shop owner and housewife, Iztapalapa. Own elaboration

Past Limitations in Water Supply Only

For a group of 11 interviewees, current water supply conditions were perceived as good and without limitation – but a look at their habitat biographies reveals that they were subject to water supply limitations in the past. People sharing this kind of habitat biography were flat dwellers and tended to live in the borough of Cuauhtémoc – though with one remarkable exception: all four interviewed residents of the Chinampac de Juárez housing complex in Iztapalapa. For them, moving to the tiny flats of this neighborhood in 1990 meant an end to a more or less lengthy experience of water supply limitations in a number of previous dwellings. The current material housing conditions, and the location and 'bad reputation' of the area in terms of security[2] provide a strong indication that these interviewees form part of the popular class though there are some slight variations in their educational level and some have access to IMSS health insurance, working as hairdressers, in offices or as domestic employees. However, the bigger part of the group formed by this habitat biography, seven cases in total, lived in several neighborhoods of the Cuauhtémoc borough. As indicated by their educational levels, types of health insurances and occupations – ranging from housewives and sales persons to chemical engineers – they are of diverse social backgrounds, forming part of the (lower) middle class. All in all, past experiences with water supply limitations were the only aspect that seemed to link the entire group – and most did not report to have lacked water when living in the city center, but rather elsewhere in Mexico City or beyond. To which extend are these past experiences of relevance for current practices? The example of Silvia from the Chinampac de Juárez housing complex shows how some in this group refer to past hardships when asked to judge current water supply conditions (see Fig. 6.6). Alan, in turn – having endured intermittent water supply in his parents' house in Ecatepec and afterwards as a tenant in the city center – now lives close to the Cuauhtémoc neighborhood in a flat with permanent water supply which he describes as a privilege (see Fig. 6.5).

2 Both the interviewed residents of Chinampac de Juárez themselves, and others who became aware that the author was conducting interviews with residents of this neighborhood, issued repeated warnings over the risky nature of this venture. One local interviewee, for instance, had repeatedly been robbed while returning home from work in the evening by public transport.

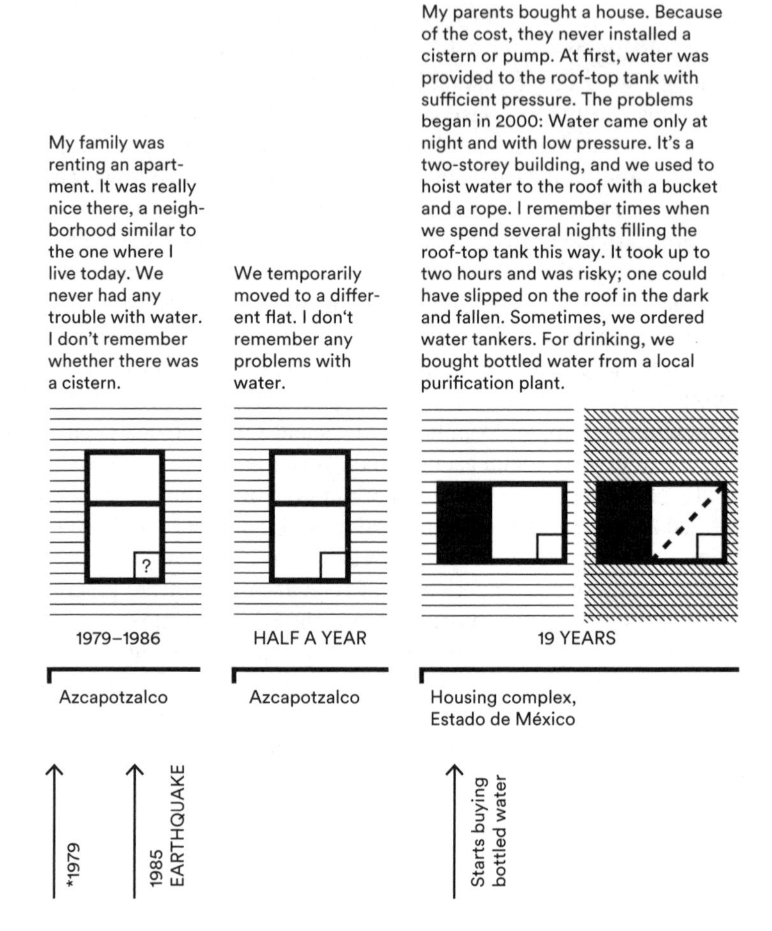

Figure 6.5: 'We are priviledged here'. Habitat biography of Alan (E30), writer, Cuauhtémoc. Own elaboration

We suffered due to the landlord's negligence. I shared a roof-top tank with my neighbor; the pump was controlled from her flat. When I ran out of water and she wasn't at home, I couldn't do anything. One of our neighbors even stole water from our roof-top tank with a hose. When we had no water I took showers in one of the public steam baths nearby.

Here, as in some other parts of Mexico City, we are really privileged in terms of water. In other areas, such as my mother's house in Ecatepec, there is less water. I have been very lucky indeed over here; I haven't had any problems, except once when they cleaned the roof-top tanks. It was two days without water, but I was fine as I had stockpiled some water in buckets.

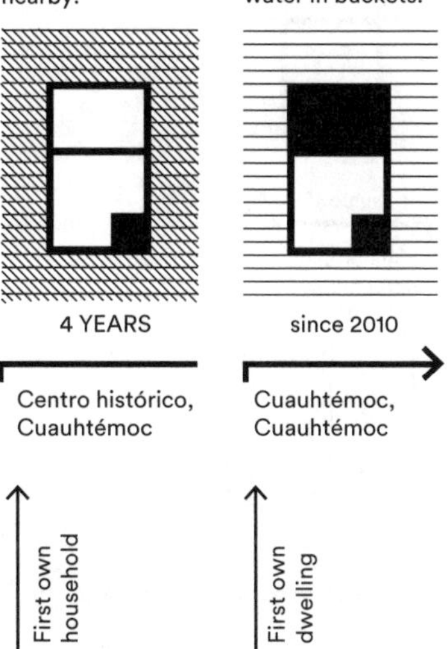

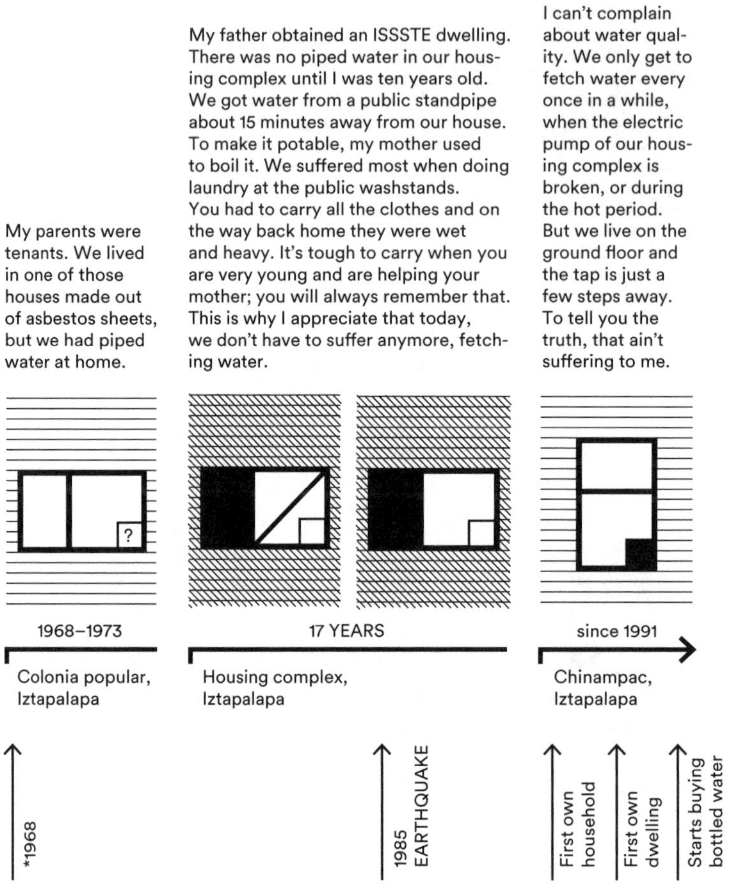

My parents were tenants. We lived in one of those houses made out of asbestos sheets, but we had piped water at home.

My father obtained an ISSSTE dwelling. There was no piped water in our housing complex until I was ten years old. We got water from a public standpipe about 15 minutes away from our house. To make it potable, my mother used to boil it. We suffered most when doing laundry at the public washstands. You had to carry all the clothes and on the way back home they were wet and heavy. It's tough to carry when you are very young and are helping your mother; you will always remember that. This is why I appreciate that today, we don't have to suffer anymore, fetching water.

I can't complain about water quality. We only get to fetch water every once in a while, when the electric pump of our housing complex is broken, or during the hot period. But we live on the ground floor and the tap is just a few steps away. To tell you the truth, that ain't suffering to me.

?

1968–1973

17 YEARS

since 1991

Colonia popular, Iztapalapa

Housing complex, Iztapalapa

Chinampac, Iztapalapa

*1968

1985 EARTHQUAKE

First own household

First own dwelling

Starts buying bottled water

Figure 6.6: 'This ain't suffering'. Habitat biography of Silvia (E26), housewife, Iztapalapa. Own elaboration

Current Limitations in Water Supply Only

Having always had good water supply conditions in the past, a minority of four interviewees experienced water supply limitations only in the current dwelling, as supply in their neighborhood deteriorated gradually with time. This was an issue exclusive to residents of the city center, more precisely, the borough of Cuauhtémoc with the neighborhoods of Paulino Navarro, Cuauhtémoc and San Rafael. These are the same areas where other interviewees also reported a deterioration of water supply conditions in terms of water pressure and intermittence during the last five to ten years. Carmen, who bought a refurbished apartment in the neighborhood of Cuauhtémoc in 2005, is one example of where water supply has generally deteriorated over the years and is intermittent now amidst a lack of storage space in the building (see Fig. 6.7). Her case is similar to the one of Antonia, who receives a permanent supply in her San Rafael apartment but felt severely affected by temporal water supply suspensions (see Fig. 6.8). Both are middle class residents who moved to the borough of Cuauhtémoc during the last decade or so and were rather upset to find deteriorating water supply conditions in these centrally located neighborhoods where they expected supply to be permanent and stable. One working as a tradeswoman and the other as a professor, both hold university degrees and spent part of their professional life abroad. In contrast, the other two interviewees sharing this kind of habitat biography only coincide with the former by living in flats in the same borough but have a lower social status. Catalina (E18) and Jorge (E17), both long-term residents of the less well-off neighborhood of Paulino Navarro, noted a gradual decrease in water pressure in their dwellings over the past decade, sometimes causing supply disruptions, and increasing filling times for domestic cisterns. Whereas Catalina has a technical formation, runs a small street stall and has a Seguro Popular health insurance, Jorge represents a somehow higher social status, insured at IMSS due to his work as a car mechanic, and holds a higher education entrance degree. Altogether, what defines those sharing this kind of habitat biography (where water supply conditions were stable in the past and only began to deteriorate in the last 10 years or so) is not social status but the matter of being residents of the city center, or more precisely, some well-defined neighborhoods in the borough of Cuauhtémoc. This gives a strong indication that recent urbanization processes, in particular redensification, have an impact upon the quality of infrastructural conditions in these areas.

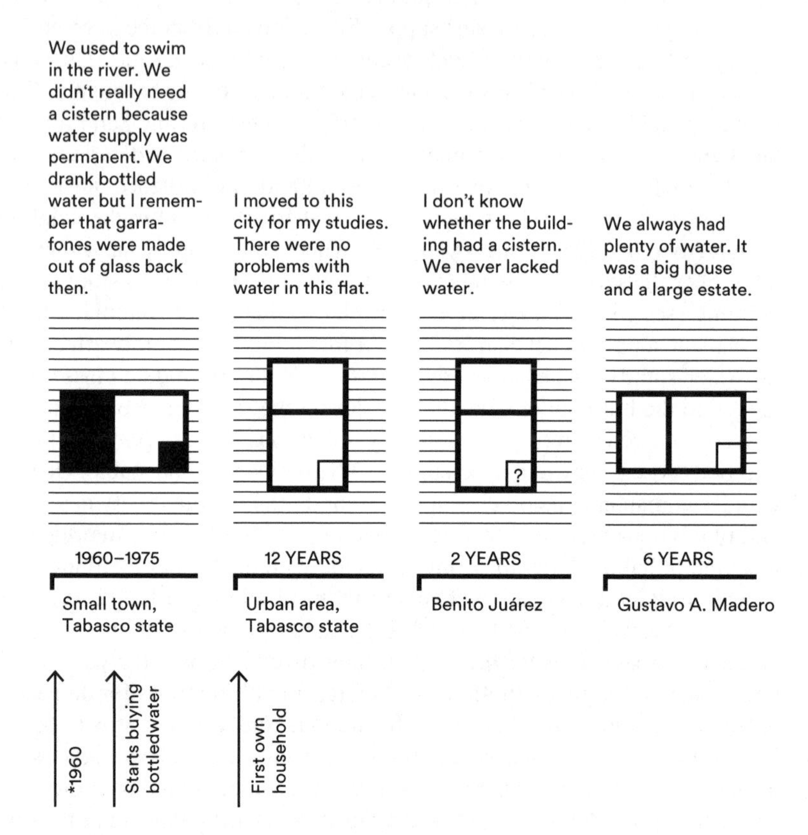

Figure 6.7: Current limitations in water supply only. Habitat biography of Carmen (E52), academic, Cuauhtémoc. Own elaboration

There was no lack of water, ever, and no cisterns. In the UK, it is common to drink tap water because it is actually potable. The infrastructure is well maintained. At first, it was strange to drink tap water – I feared that I would get sick, but nothing really ever happened.

In my building, there is no water supply in the afternoon and on weekends. The situation has worsened over the past four years, ever since they started to construct those enormous buildings along Avenida Reforma. At least twice a month we have no water at all, and there is an ongoing conflict with one of my neighbors over ordering water tankers and storage. He opposes the installation of a bigger cistern for the entire building in our yard. The old cistern we have is very small, it holds just 400 liters.

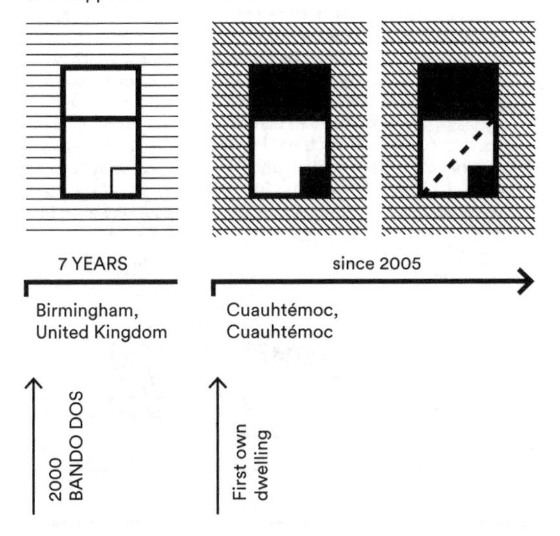

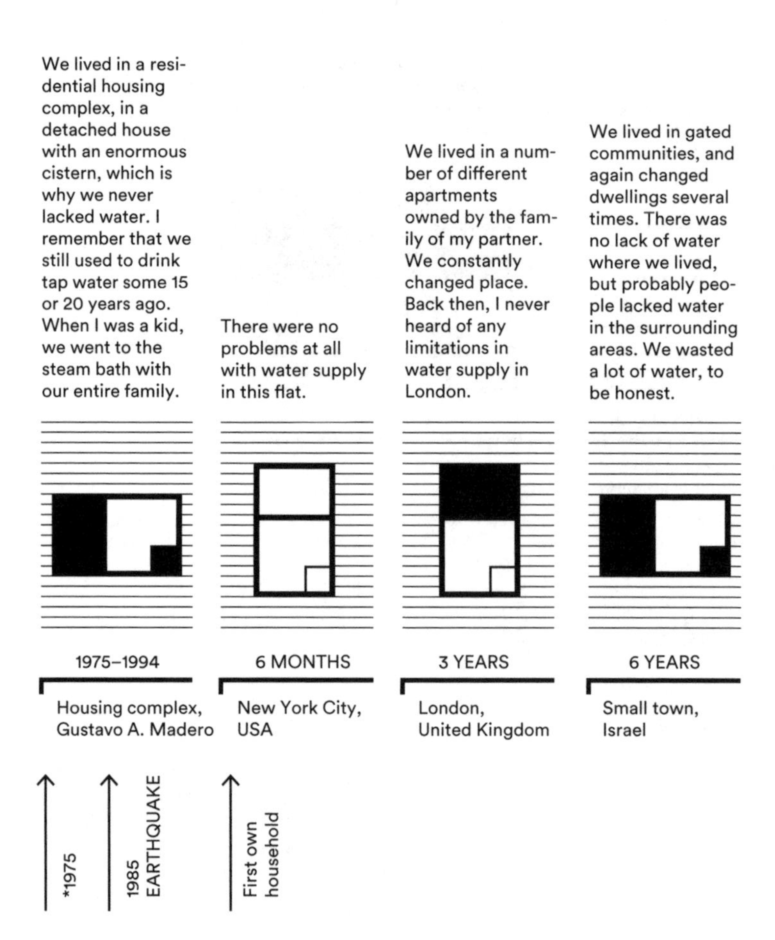

Figure 6.8: 'It was traumatic'. Habitat biography of Antonia (E20), tradeswoman, Cuauhtémoc. Own elaboration

I returned to my parents' house in Mexico when my relationship ended. Although this particular area did experience some supply disruptions, we ourselves never lacked water thanks to our big cistern. Until I moved to San Rafael, I had therefore never experienced what it means to have not enough water.

When our water supply was cut for the first time, it left a strong impression. After all, this is supposed to be a middle-class area, centrally located, where you wouldn't expect such long-lasting supply disruptions. We had no water for one week – not even for the toilet. It was very traumatic. All the dirty dishes, having to somehow clean yourself, that was very hard. Afterwards, we decided to install a bigger roof-top tank together with our neighbors.

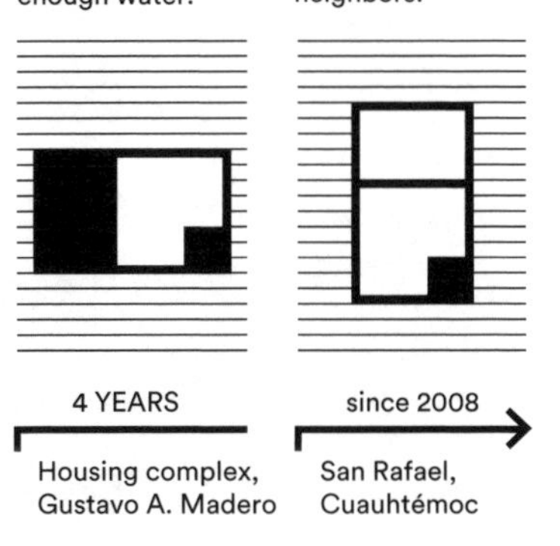

No Limitations in Water Supply

Only a small minority of five of all 44 interviewees had never experienced water supply limitations in any of their homes. At the time of the interview, they all lived in flats in the borough of Cuauhtémoc where they did not perceive any supply limitations either. They all stated that they did not experience any change in water supply conditions in their dwellings – not even during the periodic supply suspensions. This could either be an effect of stable water supply conditions in the entire neighborhood, or, as in the case of Claudia, due to additional supply by water tankers organized by the building's administration in case of a disruption of supply (see Fig. 6.9). Apart from Oscar (E14), all those who shared this kind of habitat biography could be characterized as members of the middle class: having obtained university degrees and being insured at IMSS (private sector employees' health insurance), occupied as accountants, journalists and tradeswomen, and displaying a certain material standard in their dwellings. In other words, having never experienced a lack of water at home is a privilege of few people interviewed for the present study, and unsurprisingly, one shared predominantly by the better-off.

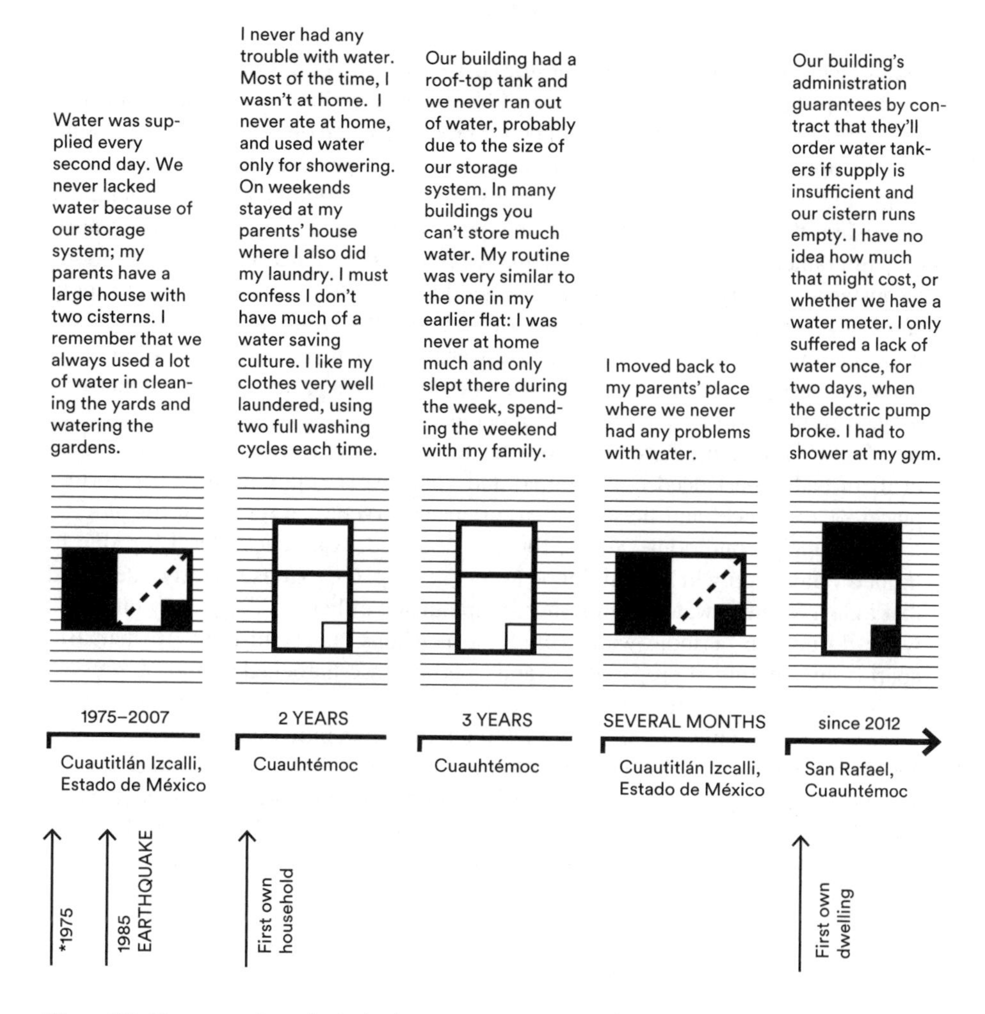

Figure 6.9: Never experienced a lack of water. Habitat biography of Claudia (E47), journalist, Cuauhtémoc. Own elaboration

Conclusions

Today, what is the influence of these experiences on people's practices of water use? First, there is the obvious: having experienced limitations in water supply in the past seemed to make people feel better prepared when once again facing a similar situation. Mexico City's periodic water supply suspensions known as *cortes de agua,* and the way people felt affected by them are a good example. As is hardly surprising, those who were entirely unexperienced with supply limitations reportedly felt rather shocked when first facing the impacts of such a suspension. Antonia, for instance, vividly recalls the lack of water during the first week-long water supply suspension she experienced at the age of 34 in her flat in the San Rafael neighborhood, calling it a 'traumatic' event. Afterwards, she was quick to adapt to the new situation: she pushed for an increase of the building's water storage capacity to prevent future shortages, and started reusing water at home in order to save water (see Fig. 6.8). In turn, there are also what could be called 'wise' water users: most of them had experienced both limited and good water supply at different former places of residence, and these past experiences served as a reference for judging today's supply conditions. Silvia, for instance, had experienced hardship due to limited water supply in her childhood home (see Fig. 6.6). Consequently, she stated that today, fetching water from a shared tap outside of her building every once in a while during the periodic water supply suspensions "ain't suffering" to her. But apart from a feeling of preparedness and a certain case-hardened attitude expressed by those who had suffered supply limitations in the past – which role does a person's dwelling history, understood as a collective rather than individual experience, play for everyday practices? Is there more to these practices than a reaction to or coping strategy based on present housing and water supply conditions as discussed in chapter 5? These questions will be answered in the following by analyzing links between these habitat biography types and a selection of water-related practices.

7. PAST EXPERIENCES AND CURRENT PRACTICES

To explore the relation between historical experience and current practices of domestic water use, two of the four types of habitat biographies identified in chapter 6 were employed in particular. Those who never suffered any water supply limitations in past or present dwellings form the first group, consisting of five cases. It is set in contrast to the second group: eleven interviewees who had suffered supply limitations in the past but are well supplied today. This second type of habitat biography is hence characterized exclusively by past limitations, whereas supply limitations are non-existent today and thus play no role for people's practices of water use. It is therefore the ideal group to be compared with those who had never suffered water supply limitations in any dwelling whatsoever. For the same reason, the two remaining habitat biography types – the majority of interviewees who experienced supply limitations in present as well as former dwellings, and those subject to supply limitations in the present dwelling only – were not studied in depth here. This approach will help to clarify whether past experiences with water supply exert any considerable influence on practices at present. The chapter at hand explores differences between the two selected types of habitat biographies for the practices of domestic stockpiling (7.1), grey water reuse (7.2) and with respect to the imagined landscapes of water supply (7.3). This covers some of the most common water-related practices analyzed in the present study, as discussed in chapter 5.

7.1 STOCKPILING WATER AND PAST

Stockpiling water in one's home is an individual strategy which aims to increase the independence from water supply frequencies and sudden glitches in water supply – whether they be caused by insufficient pressure, water supply intervals which do not match everyday routines, or asynchronous patterns of water use amongst neighbors sharing a collective storage tank. In fact, many of those who took to stockpiling in their dwelling today were subject to water supply conditions which provided a motivation to take action to prevent domestic water shortages (see 5.3.3). The question arises whether past experiences with water supply limitations in former dwellings play any role in this practice – or whether it is merely a way of coping with current supply conditions. That puts a focus on those 11 interviewees who had experienced limitations in the past and reported a stable water supply for their present dwelling. Would people continue to stockpile water after supply had become reliable? On the contrary – a vast majority of those subject to water supply limitations exclusively in the past (8 of 11 interviewees) did not keep any water stored at home beyond what was gathered in roof-top tanks and cisterns. Many had actually given up stockpiling when water supply became permanent in a new home – like

Eugenia, who moved to the city center after spending almost two decades in a neighborhood with rationed water supply in Iztapalapa. She stopped stockpiling water in a barrel as she felt this was no longer necessary in her new dwelling in the Tabacalera neighborhood where she perceived no risk of lacking of water in the future. Hence for these eight interviewees, it seems as if everyday practices were adapted as water supply (or storage capacities) became sufficient and reliable. Others who had suffered in the past would now only store water in buckets in their flat as a preventive measure when one of the general supply interruptions was announced by either the media or by the administration of their building. Having hauled water to the roof-top tank of his parent's house in Ecatepec in buckets amidst intermittent supply as a teenager, Alan, for instance, proudly recalled that today, he knew what to do when water supply in his flat in the city center was cut off temporarily (and with prior notice) to allow for a disinfection of the building's collective rooftop storage tanks.

> *I have been very lucky [here…]. The only time we didn't have water was when the roof-top tanks were cleaned. (…) I took a shower very early on the first day, gathered my four buckets. (…) I used them for the toilet, tried not to use dishes. And the next day, I used the other bucket to take a shower (…) using a bowl (…). I remember that I even had some extra water left. Two days without water, and I had kept some!* (Alan E30: 357–359, Cuauhtémoc)

Apparently, he felt well prepared for a supply interruption – this indicates those who learned to do so in the past can still recur to this practice when needed. Likewise, others took to measures such as grey water reuse when facing announced supply disruptions, as will be discussed later. However, it remains unclear whether these experiences actually 'wear off' gradually over time, or if practices such as stockpiling are given up abruptly as soon as supply conditions improve. But as a general rule, stockpiling was indeed given up when people no longer saw any need for it. What is more, being unexperienced with water supply limitations seemed to motivate some to abstain from preventive measures altogether. When he moved to a flat shared with other students and regular water supply where stockpiling was practiced, Carlos actively pushed for a change:

> *When I arrived, there was a huge container, fitting some 50–60 liters, and huge buckets precisely for when water would be scarce. But [they…] occupied a lot of space, and as we didn't use them a lot, we threw them out (…). I haven't (…) even been close to think: Híjole, what will we do if there is no water?* (Carlos E35: 16 and 42, Cuauhtémoc)

Such a feeling of security could be read partly as a reaction to what is experienced as a reliable supply in the current house, and partly as a result of his habitat biography: while Carlos had lived in a neighborhood with intermittent supply as a child, he reported that domestic storage facilities and an automatic pump always guaranteed a stable supply for his family. Thus the only time he had experienced a lack of water was on some occasions when the pump was broken and water could not be moved to the roof-top tanks to provide enough pressure for all taps.

In the end, what seemed to matter for the practice of stockpiling were not so much past experiences but whether water supply within the dwelling itself could be trusted. This is not necessarily synonymous with the perceived water supply conditions of the neighborhood – there are additional aspects revealed by interviews and habitat biographies which might not be visible if a reportedly stable water supply alone were taken into consideration. With respect to those who had experienced supply limitations exclusively in the past, there was only a small group of three interviewees who continued to store water at home despite a reportedly permanent water supply. Another three interviewees had a habitat biography coined by past and present limitations in water supply, but continued to stockpile even as current supply limitations were a matter of low water quality rather than non-permanent supply or inadequate water pressure. Take the example of Hilda, who lives in a housing complex in Iztapalapa:

> *I keep [water stockpiled in 5-liter bottles] because it can happen at any moment and without prior notice that water stops running. (…) at least, this will give me a wild card (..) so I won't remain without water for essential things like taking a shower or (…) the toilet.* (Hilda E5: 106)

Hilda's habitat biography reveals that she endured a lack of water in past dwellings, and today also faced random supply interruptions from time to time due to competition amongst neighbors over water stored in the shared cistern of the apartment block. In general, those who were stockpiling water in their dwelling despite having access to a reportedly good supply were flat dwellers who experienced a strong competition over water stored in cisterns shared with neighbors in the same building or housing complex (see also 5.3.3): of nine households stockpiling water despite a permanent supply, six shared a cistern with others, and another two had no cistern or similar domestic storage device installed at all. Such fears over water availability amidst neighbors' asynchronous patterns of water use were often (but not always) combined with limitations experienced in former dwellings. In contrast, all those who gave up stockpiling did so as they trusted the reliability of supply, in particular when it comes to steadiness and pressure, or at least felt that the installed storage capacities made any additional storage in the dwelling unnecessary.

> *There is one advantage: half of our building['s base] is the cistern, so it is mega huge, and six roof-top tanks. (…) if they would for example switch off [the supply] for (…) four, five days, we wouldn't run out of water.* (Magdalena E45: 7, Cuauhtémoc)

She lives in the Tabacalera neighborhood where water pressure is reportedly low at times, but her building has a cistern with an exceptionally high storage capacity of 10 m^3 per apartment. Yet storage capacities are not the only issue – it seems to depend on a number of aspects whether water availability for the individual household in a building with in-house piped water and permanent supply is sufficient. It is the result of a combination of supply characteristics which are difficult to measure for each and every domestic connection (steadiness, water pressure), the specific technical features of the water storage system, and the volumes and rhythms of all neighbors' water consumption in case of a shared cistern. The everyday knowledge and immediate experience of a lack of water in the dwelling during sudden disruptions or as a result of divergent patterns of consumption amongst neighbors was what motivated interviewees to keep a stockpile despite a supply that was considered sufficient in general. Therefore, stockpiling seems to be a way of securing domestic water availability rooted in conditions experienced recently rather than in past experiences. Although old strategies may prove to be of help when reactivated in the case of exceptional supply disruptions, it seems as if past experiences are only a marginal aspect of why people would keep water stockpiled in their dwelling. In other words, the habitat biography seems to exert no considerable influence on today's stockpiling, whereas current experiences with water supply and shared storage appear as stronger motivators.

7.2 REUSING WATER AND PAST

Collecting domestic grey water and reusing it for cleaning, flushing toilets or watering plants is common in Mexico City, where the down-cycling of water is organized in a cascade of water use, as discussed in 5.2.3. It was often seen as a way of saving water motivated by a combination of self-experienced water scarcity and altruism, as in the case of Guadalupe. She lives in a consolidated colonia popular in Iztapalapa, where water is provided during three hours per day.

> *I use the laundry water twice. [...because] it is logical that the more one turns on the tap, the more other people are left behind without water.* (Guadalupe E3: 237–239, Iztapalapa)

At the same time, grey water reuse is a simple means of securing water availability for the own household, and could hence easily be read as a simple reaction to present water supply conditions. Indeed, the empirical material shows that all but one of the interviewed households without a permanent water supply at the time of the interview were practicing reuse. Amongst the interviewees, personal experience of water supply limitation was thus clearly a motivation for resorting to grey-water reuse. Yet surprisingly, the same was also practiced in many other households where water supply was reported to be rather good and sufficient. Past experiences come into play here: as much as nine of 11 interviewees who had experienced water supply limitations only in the past but not in the current dwelling reportedly continued to reuse grey water at home. They had formerly lived in a dwelling were water supply was insufficient – be it as an effect of the devastating 1985 earthquake, or as an effect of lacking infrastructure and rationing in the respective neighborhoods. Isabel, for instance, started to reuse domestic grey water when she lived under hardship for several years in the Iztapalapan neighborhood of Las Minas, notorious for its lack of water.

> *I tell my kids: "(...) you haven't experienced what it means to have no water, to go and fetch a bucket [of water from a public hydrant...]". In Iztapalapa, for instance, I got up at three in the morning and we finished stockpiling water at seven, as at eight there would be no more water. [...My landlord] showed me that you use a tub [in the shower] and recycle that water for the toilet. That's what I keep doing so my children will learn how to save water.* (Isabel E21: 24–26)

Having moved to the borough of Cuauhtémoc, where she now lives in a flat with permanent water supply in the Paulino Navarro neighborhood, she continues to reuse water the way she had learned to in Iztapalapa. Therefore, Isabel is a typical example of a 'wise' water user (see 6). Despite not being reminded immediately by an everyday experience of limited supply that saving water might make sense, many interviewees displayed consolidated patterns of action when it comes to water reuse. This impression was further strengthened by the finding that four of all five interviewees who had never experienced limited water supply – neither in any past dwelling nor at present – did not practice grey water reuse at all. Some of them generally and directly objected the reuse of domestic grey water as they envisioned it as unsanitary.

> *I can't recycle [laundry] water (...) for mopping floors as it would not be hygienic to pour all that dirt coming off the clothes on the floor.* (Oscar E14: 156, Cuauhtémoc)

Interestingly, Oscar spoke vehemently in favor of saving water due to social motives, but said he took to other practices – such as a conscious water use in doing laundry, when washing dishes and showering – to achieve this goal. Thus it seems as if the practice of reuse has indeed been influenced by former living conditions, or habitats: those who lacked experience with water supply limitations would not reuse, whereas those who had experienced limitations in former dwellings or the present one would.

There are two additional observations, which complement rather than contradict these findings. First, the influence of past experiences on practices seems to wear off gradually with time. In several cases, the practice of collecting grey water for reuse became less and less common once it started to lose some of its relevance due to improving water supply conditions. In other words, the cascade of water use changed with improved supply conditions and/or improved domestic storage systems. As Maria recalls from the gradual consolidation process of her self-built dwelling in a colonia popular in Iztapalapa, this was partly a matter of convenience.

> *When we bought water [delivered by tanker trucks...], we gathered all of it (...). But since we installed the roof-top tank (...) just pull the flush on the toilet (...). Earlier, when we didn't have this kind of toilet (...) nor the roof-top tank, we were gathering the dirty laundry water or rain water for the toilet.* (Maria E2: 132, Iztapalapa)

The practice was not given up altogether, but less water is being reused in Maria's house since it became connected to the public water network. As supply was and continues to be rationed in her neighborhood, Maria's family has secured a sufficient water pressure in the dwelling by installing a cistern, roof-top tank and electric pump. Toilets are now flushed with tap water. As for reuse, the family collects water from showering to irrigate potted plants and the yard is cleaned with water from the runoff of the washing machine placed there, which is a less exhausting task than the kind of grey water reuse practiced earlier in this family. Under a similar logic, Dolores and the other interviewees from the same housing complex in Iztapalapa reportedly resorted to grey water reuse at times yet not on a regular, everyday basis. Notably, the Chinampac de Juarez neighborhood where they all live is supplied permanently, but all of them experienced supply limitations in past dwellings. Apparently, those who formerly learned how to cope with limited water supply still recur to this practice when necessary, having incorporated the coping strategy of grey water reuse though infrequently applying it under normal conditions. Dolores' family, for instance, flushes the toilet with laundry water during the periodic supply suspensions – but seldom does so on an everyday basis.

Second, there are those who experienced supply limitations in the past yet take to other kinds of coping strategies when facing limitations today. Despite receiving water in a limited way at present, the idea of reusing domestic grey water was not even considered by Eva and Pablo – both middle class residents in apartment buildings in the borough of Cuauhtémoc. Despite past experiences with supply limitations, these interviewees hardly adapted their everyday practices when similar limitations recurred. Instead, they sought to avoid the effects of insufficient water supply by what might be called class-specific strategies: delegating water-related tasks to domestic workers and concierges or turning water supply into a locational factor (see 5.3.2). But generally spoken, it is safe to say that there are links between the practice of domestic water reuse and past limitations in water supply.

7.3 IMAGINING WATER AND PAST

When asked to explain reasons for intermittence and water supply disparities in Mexico City, four main lines of thought evolve from the interview material. Whether pictured as an expression of social inequality, as a paternalist method of education, or as features of the untamed urban leviathan or of a city on the brink of collapse – the logic of Mexico City's water supply is imagined in a number of ways by its residents (see 5.4). These imaginations are a way of creating and recreating a landscape of water supply (and use) through a space of representation, as Henri Lefebvre would call it. The imagined landscapes of supply are hence not so much about each interviewee's concrete, material practices of water use but about the ways the (inter) relation between the individual and its surrounding world is experienced, imagined and lived on a daily basis. In how far are such imagined landscapes related to (or influenced by) personal experiences with domestic water supply? The analysis of the empirical material shows that imagined landscapes drawn by most interviewees are made up of a combination of some of the four above mentioned concepts. As chapter 5.4 has explored in detail, current water supply conditions in the own dwelling did not seem to provide a strong background for these imaginations of the urban. Hence the question arises whether past experiences with limited water supply are an aspect which exerts a considerable influence on these imagined landscapes. Is the logic of water supply throughout the city imagined in a different way by those who experienced supply limitations in the past? The answer is no – while it should be kept in mind that a vast majority of interviewees combined two or more of these imaginarios in their landscape of water supply, taking past experiences into account reveals that none of them is specific to any of the identified habitat biography types. Remarkably, the 11 interviewees who had suffered supply limitations only in the past and were well-supplied today unanimously referred to the logic of water supply as an expression of Mexico City's social disparities and unequal power relations, resulting in what they deemed a preferential treatment of residents with a higher social status. Here is how Elena, a free-lance teacher who had lived in more than one dwelling with water rationing and low water quality in both the Federal District and Mexico State but now receives a good and permanent supply in the Tlatelolco housing complex, imagined the logic of supply disparities in Mexico City:

> *It has to do with (…) a lot of racism and classism. (…) big business, the big companies are in the [..inner] city, and the workers are in the periphery. So a restaurant, an office, a hotel won't lack [water]. And if there is a lack [of water] for the workers, they themselves will have to see how to make do. I imagine that's why the water distribution is organized this way. (…) it's in the economically marginalized areas [where water lacks]. I don't believe that the pipes which arrive here have a magical capacity of yielding water while those arriving there have a constitution that doesn't yield. Rather, it is those who distribute the water who decide which water gets to which area.* (Elena E43: 112, Cuauhtémoc)

As chapter 5.4.3 has shown, this imagination is often tied to a set of specific places in Mexico City which serve as a kind of 'extreme cases' along a spectrum of water supply limitations. For many, it is the borough of Iztapalapa which represents those parts of the city given the lowest preference in the hegemonic logic of water supply. As a real estate agent, Magdalena based her comparison on her broad knowledge of neighborhoods and their water supply conditions throughout Mexico City, and directly tied supply disparities to questions of status and power:

> *It's more the type of (…) people who live in different parts [of the city]. I think that those who actually have economic potential are given a better [water supply]. (…) in the borough of Cuauhtémoc (…) there is the advantage of having probably (..) not the best water (..) but one that is much better than in Iztapalapa or Naucalpan or Gustavo A. Madero (…). As people living there won't complain (…) – just give them the grossest water! Not here, because here, in Lomas de Chapultepec, La Herradura (…), people can indeed raise an alarm. No, this is where you need to provide the best.* (Magdalena E45: 35, Cuauhtémoc)

Yet the imagination of differential water supply being organized along inequalities in social status and power is not the exclusive domain of those with a certain kind of habitat biography. Most strikingly, the idea of the city being an expression of social inequalities was also popular amongst those who had never suffered any supply limitations whatsoever. Indeed, four of five interviewees representing this habitat biography type drew a picture of the imagined landscape of supply which combined this imaginario with one of a lack of urban planning and untamed urban growth (see 5.4.2). A combination of socioeconomic questions and lack of infrastructural planning, for instance, was what shaped the landscape of water supply as imagined by Fabiola, a resident of a well-supplied apartment in Roma Norte:

> *Perhaps [water] is distributed in an unequal way. Maybe we don't have a good supply system. It has to do with where you live and also with your income level. Water often is something that you can access according to your position (…). I guess it has something to do with poor urban planning. (…) it's the most remote areas which suffer most, and the city really spread without being planned.* (Fabiola E37: 8 and 18, Cuauhtémoc)

However, water supply disparities were attributed to socioeconomic questions in combination with an unregulated growth of the urban leviathan by no means only by those having never experienced supply limitations – six of those 11 interviewees sharing a habitat biography with past limitations in water supply also shared this imagination. The remaining two imaginarios were even more at odds with any specific habitat biography type: the idea of water supply disparities as a form of punishment or education by a paternalist state showed no clear tendency as for interviewees' past experiences but was shared more often by those who were subject to water supply limitations today. Similarly, a logic of disparate water supply shaped by an imagined uncontrolled population growth and possible future collapse of the city (see 5.4.1) seemed widely unrelated to past experiences – it was shared by four (of eleven) of those limited only in the past and two (of five) of those who had never been limited in water supply. The imagined landscape of supply of Tlatelolco residents Martina and Esther, for instance, were in the end based on a quite complex and biased logic. While linking good water supply conditions in upper-class neighborhoods to social status, a quite different logic dominated their imagined landscape of supply when it came to the other end of the social spectrum. There, a combination of an imagined excessive population growth (see 5.4.1), and a lack of regulation of informal urbanization processes were what they saw as causes of water supply disparities. Therefore, even the minority of those who shared a habitat biography without any supply limitations did not provide an unambiguous trend when it comes to imagined landscapes of water supply. For many, the supply logic was a matter of social inequality, for others one of unregulated demographic and urban growth in Mexico City – and all this seemed rather unrelated to collective experiences with water supply. All in all, the picture is too fragmented to speak of a clear structuration of imagined landscapes based on the presence or absence of episodes of water supply limitations in the interviewees' habitat biographies. It can be stated that the imagined landscapes of water supply widely overlapped throughout the two groups of habitat biographies under scrutiny – and hence across the aspect which was initially thought to make a difference: the presence or absence of experiences with water supply limitations. What is most remarkable is the widespread awareness of socioeconomic disparities and their interlinkage with urban space – an idea that was shared by a majority of interviewees independent of their own social status and their living conditions in the past and at present.

To what extent do past experiences with water supply conditions influence current practices of water use? The present chapter has focused on two particular types of habitat biographies to answer this question, concentrating on those interviewees who were not subject to supply limitations at the time of the interview. As for the set of practices analyzed here, it can be stated that different kinds of habitat biographies hardly make a difference when it comes to water using practices. Only one practice – domestic grey water reuse – was found to be related to past experiences with water supply limitations, whereas stockpiling seemed much more linked to current supply conditions, and was given up as water provision improved. Furthermore, the way people imagine urban supply logics resulted to be entirely independent from both their past and present water supply conditions. Whether political socialization or religious values or public discourses – it seems as if other mechanisms are at play when it comes to these imagined landscapes of water supply. And even where some relation between habitat biography and practices of water use could be identified – as in the case of domestic water reuse – the finding is watered down as the influence of past experiences seems to lose relevance with the absence of such limiting conditions in the present dwelling. This strengthens the impression of a strong link of (most) practices of water use to present supply conditions rather than past experiences.

The empirical findings presented in this chapter challenge the hypothesis that the habitat plays a strong role in the co-production of habitus. Against the underlying assumption, today's supply conditions seem to exert a much stronger influence on water-using practices than past experiences. There is a dominance of the actual rather than of history in the form of personal experiences, in other words – at least when it comes to questions of everyday domestic water use. This is not to say past experiences in the sociospatial setting of the home are entirely irrelevant for current practices – yet when directly cross-tested with those practices of water use analyzed here, it seems as if they are less relevant than current conditions. Notwithstanding, shared historical experiences in a specific spatial setting can play a role on a different and somehow more indirect manner, as the discussion over the non-potability of Mexico City's tap water will show (see 8). Leaving immediate biographical experience behind, it can also be argued that the historic dimension is relevant on a different scale, as it is materialized in the very same urban infrastructures which today shape water supply conditions in Mexico City's homes. Resulting from the production and regulation of nature, urban infrastructures are necessarily a product of past spatial practices and decision-making, and a precondition for the stabilization of capitalist development (see Smith 1984).

In the way applied here, the habitat biography method as a crucial part of the analysis provides evidence that past experiences in the specific sociospatial setting of the home do afterwards not necessarily translate into practices of water use. Yet it can be argued that such a method of gathering and comparing habitat biographies, as introduced by the author specifically for the present study, contributes to a deeper understanding of a subject's practices by overcoming a solely actor-based perspective as structures of exclusion and inequality come into view. Habitat biographies draw attention to the manner in which the collective experience of dwelling history is shared by many who form part of and act within similar processes of urbanization. Based on these findings and reflections, it seems as if Bourdieu's concept of habitus constitutes a useful starting point for understanding water-using practices from a sociospatial perspective, though this approach should be broadened and re-considered (see 9).

8. REFLECTION ON THE SOCIOSPATIAL CHARACTER OF DOMESTIC WATER USE

Taking the social embeddedness of water (see Budds 2009: 420) as a starting point, the current study has explored links between the production of space and domestic practices of water use in Mexico City. This mainly refers to the sphere of the dwelling as an individually experienced yet collectively shaped habitat and its respective infrastructural conditions, assuming that both past and current water supply conditions in the home are major parameters in the pre-structuring of these domestic practices. From a theoretical point of view, the habitat is thought to be of relevance in a dual sense: regarding current living conditions, and regarding former habitats as the concrete spatial setting where habitus is incorporated. The present chapter discusses to what extent these assumptions were confirmed by the analysis of the empirical material (8.1 to 8.3) and reflects the research strategy (8.4), before showing up paths for future research in the field of domestic water use from a sociospatial perspective (8.5).

8.1 A PREDOMINANCE OF THE ACTUAL

A connection to the water network alone is not a guarantee for proper and sufficient domestic water supply in Mexico City as both the literature and the interview material reveals. When it comes to steadiness of water supply, the spatial differentiation within the metropolis is manifest not least in the sharp contrast between the two boroughs of Iztapalapa and Cuauhtémoc. Even though there are clear indications that inner-city neighborhoods are now also starting to experience intermittence or insufficient water pressure as a result of ongoing processes of redensification, these differences are still pronounced. This becomes immediately clear when one compares the water supply situation in a number of Iztapalapan neighborhoods (rationed to once a week) to the kind of limitations typical for the city center, such as low water pressure in the evening. During periodic city-wide suspensions of water supply induced by the water utility, individual stockpiling in buckets, being at stand-by and other domestic storage strategies specific to those supplied intermittently tend to become a general practice amongst Mexico City's residents, as the empirical findings have shown. In this sense, it can be argued that rationing and intermittent water supply itself is what could be called a perpetuated state of exception. Some interpret these "governmental adaptation strategies" (de Alba 2016: 2) as a reaction to the inability of providing enough water for all. In consequence, water supply limitations have turned into normality, and many residents of underserviced areas appear to believe that this situation is unlikely to change in the near future. Exemptions from paying domestic water bills due to irregularities in supply, as granted by the Federal District's government since more than a decade to some neighborhoods of

Iztapalapa, only deepen this impression. Rather than an exception, the disparate landscape of urban water supply which serves as a framing condition for these everyday practices is the product of a rationalization of social disparities materialized on different spatial scales. These involve not only a sociospatial differentiation on the urban scale but also the interconnectedness, highly unequal in terms of power, between the metropolis and its hinterland in terms of water extraction (see Swyngedouw 2006: 62). Past and current social struggles over water extraction – for instance over an extension of the Cutzamala system, providing water from another catchment area to Mexico City, and in peripheral areas within the Federal District's jurisdiction, such as Xochimilco – are a case in point (see Peña Ramírez 2004a, Castro 2007, Campos Cabral/Ávila García 2013).

Coping (Individually) with Status Quo

Regarding water supply disparities within the city itself, the empirical findings presented here leave no doubt that from domestic storage to doing laundry, most of the studied practices of domestic water use displayed a clear dominance of the actual in terms of water supply conditions. People store water at home in reaction to intermittent and unreliable supply schemes, they reuse water in order to make limited reserves last longer, and they buy bottled water for cooking and drinking as tap water is seen as unfit for these purposes. All these practices can be understood as direct coping strategies motivated by current supply conditions. Erik Swyngedouw distinguishes four different strategies[1] people employ to improve their access to water: passive acceptance, individual resistance, self-help and social protest (see Swyngedouw 2004: 150 ff.). The present findings from the Federal District indicate that most practices fall in the realm of resignation and adaptation to the status quo through individual solutions which strongly shape everyday routines (see 5). Collective protest and resistance against an unequal access to water seems to be mostly spontaneous and short-lived, even in the often systematically under-supplied parts of the metropolis. Spontaneous road blocks demanding water, hijacking of public water tankers and organized protests in front of the CONAGUA headquarters are the most common forms. Yet these social contestations tend to be heavily fragmented in temporal and spatial terms, rarely uniting with other parts of the metropolis facing the same marginalization and discriminatory water supply conditions. They thus seem widely unable to develop into an urban social struggle which would seek structural change, and get beyond the reach of clientelism (see Swyngedouw 2004: 152 ff.). Moreover, these rare contestations are often met with repression by the state. Repeated protests in Mexico City's Peñón de los Baños neighborhood amidst a situation in which private vendors usurp water from local wells even during periods of water supply suspension (see for instance Cruz Flores 21.02.2009, Jiménez/Arteaga 19.09.2013) are only one example. Other strategies of

1 Based on his own empirical findings, and Espinoza and Oliden's 1988 publication on water supply in *barrios populares* of Lima, amongst others.

gaining access to water amount to individual resistance (see Swyngedouw 2004: 151) – for instance the (outlawed) installation of private suction pumps to overcome low water pressure at the domestic tap, and the use of clientelistic networks to obtain access to water delivered by public tankers during water shortages. Tending to have a divisive effect, these strategies do not induce any change of the status quo of water supply on an urban scale.

But returning to the typical individual coping mechanisms which can be understood as a form of passive acceptance of the status quo, several practices identified in the present study show that these are also directly related to social status. The issue of bottled water consumption is a good example. With multiple bottled water use in particular being a strongly Iztapalapan practice (see 5.1.3), it seems clear that current supply conditions at the place of residence exert an influence over whether tap water is deemed fit for human consumption – and specifically for which purposes. Bottled water is hence not a means of social distinction (as is the case with the consumption of European mineral waters, see below) but as a marker of class in as far as it serves as a substitute for proper water supply conditions. Not only tended 'the poor' interviewed for the present study to consume more bottled water than better-off residents of Mexico City – higher consumption levels also directly resulted from a broader range of practices of water use in the domestic realm. Other classed coping strategies include the turning of water supply conditions into an locational factor on the middle-class housing market (see 5.3.2) and the conditioning of the popular class body for expected future living conditions through the consumption and body-related use of tap water (see 5.1.1 and 5.2.3). Though seemingly contradicting the practice of multiple bottled water use, the third strategy in fact follows the very same logic of tolerating the status quo, and even projecting it into the future. We don't expect better water supply conditions for us any time soon – following this base line, both strategies bear resemblance with what Bourdieu termed a taste for the necessary specific to those with a low social status (see Bourdieu 1982: 587 ff., and for a critique, Bennett 2011: 536).

When no water runs from the domestic tap, the widely invisible logic of supply becomes tangible, tampering with domestic rhythms of water use and a wide spectrum of purposes, from those directly related to the body to the most technical. In the face of intermittent water supply patterns, one coping strategy in particular had the capacity to alter domestic rhythms significantly: domestic storage. Storing water in the home is a means of synchronizing the rhythms of water use and supply, or as Lefebvre would have put it, mediating between the body-related practice of water use and the rational character of water supply frequencies as regulated by the distribution network.

> "Rhythm appears as regulated time, governed by rational laws, but in contact with what is least rational in human being: the lived, the carnal, the body." (Lefebvre 2010: 9)

While it can be argued that the autonomy of the modern home is itself a social construction which merely becomes evident in the moments of its failure to domesticate nature fully (for instance when water supply fails) (see Kaika 2004), domestic coping strategies can be understood as a way of regaining that autonomy. Yet even as domestic storage in general serves as a means of decreasing the heteronomy imposed by non-permanent water supply schemes and securing water availability for domestic needs despite a non-permanent supply, it often comes at the cost of additional efforts, imposing domestic schedules for filling tanks and the like (see 5.3). Where storage devices are shared with others, as is often the case in housing complexes, it is the asynchrony[2] of neighbors' water use rhythms in addition to non-permanent water supply patterns which can entail such a loss of autonomy. Strikingly, practices seeking to regain that autonomy – such as immediate water use and being at stand-by – are largely classed and gendered in that it is popular class women who resort to these strategies to cope with unsteady and unreliable water supply frequencies. Elsewhere, fully automated systems for water storage are installed, and responsibilities for water availability in the home are passed on to domestic workers and employees. The social implications of such practices and their role in the reproduction of social status in terms of class and gender were highlighted before (see Massolo 1992, González de la Rocha 1994, Bennett 1995, Crow/Sultana 2002, Dugard/Mohlakoana 2009). As becomes clear in the present work, domestic strategies to cope with insufficient and intermittent water supply continue to affect many households in Mexico City today, even as the share of dwellings without access to the water network has decreased to some 2%. But a water tap alone does not guarantee access to water, and unsteady supply can imply practices which deprive people of hours of rest, waiting for water to start running in the early morning or at any other time as schedules are often unreliable and supply disruptions can be spontaneous. It is remarkable that these coping strategies show similar effects on women (still solely responsible for domestic work in most cases), as those Clara Salazar Cruz identified in the context of the *Leche Liconsa* program. This is a state-subsidized scheme for the distribution of milk, which is offered below market rates but requires the investment of time and efforts by the beneficiaries in Mexico's urban areas (see Salazar Cruz 1999: 151 ff.). As for the storing of water, the required alertness to indicators of resuming supply and the tasks to be fulfilled once water starts running put a strong pressure not only on hours of rest but also on time for activities such as leisure, education and paid labor, and on all other activities related to what Gibson-Graham call the feminized household sector (see Gibson-Graham 2006: 261). As the wait for and storage of water occupy time and energy which people might otherwise employ to sleep, learn, relax, earn an income, attend to children or be politically active, it is water supply which dominates the rhythms of everyday life, and assists in the reproduction of social inequalities. It is in this sense that "water can become one's master rather than one's slave" (Swyngedouw 2004: 57).

As a social product always in the making, urban space not only tends to express social inequality, it can also have repercussions on the reproduction of social differences, as these examples show. This puts the role of water infrastructures as an urban commons back into the spotlight (see Gailing *et al.* 2009, Harvey 2013: 67 ff.).

2 It remains to be analyzed elsewhere why such a commoning of stored water seems to induce solidarity within the collective in some cases while leading to a more conflictuous situation in others (see 8.5).

Given their role as a major framing condition of everyday practices, which is emphasized by the findings presented here for domestic water use, it is the strong path dependence of large infrastructural networks such as water supply which has particular implications. Amidst typically long planning horizons, high initial investments, and the inflexible nature of kilometers of dug-in pipes, large technical urban infrastructures remind of Fernand Braudel's *longue durée* (see Braudel 1992). Showcasing what David Harvey termed a "spatial fix" (Harvey 1985: 33) of capital, such a long-term materialization of certain logic of water supply through technical infrastructure implies that past decisions and processes of social differentiation tend to be perpetuated through them. In other words, unequal infrastructural conditions and supply disparities could well prove to be rather stable over extended periods of time once they are materialized in an urban water network (see Graham/Marvin 2001: 190 ff.). The paradox of Mexico City's water stress in spite of its location in a (formerly) water-rich basin (see Connolly 1999: 61), rooted in colonial and modern drainage practices, and continuing until today through a highly complex hydro-technical setup undoubtedly is a case in point. Forming a rather essential part of people's living conditions, water supply infrastructures hence bear the capacity to impact on domestic practices of water use and to orchestrate domestic rhythms for extended periods of time. Altering the path is difficult, as the example of water infrastructures in shrinking (city) regions as a result of post-socialist processes of transformation has shown (see Moss 2008, Naumann 2009). This is however not purely an issue of tangible pipes and technical networks but also an institutional question as infrastructures are also materializing and thus perpetuating the path dependence of institutional logics (see Orderud/Polickova-Dobiasova 2010: 204). When it comes to the logic of water supply in Mexico City and the Federal District in particular, such an institutionalized logic would be an interesting subject of inquiry for future research, also with respect to possible changes to the hegemonic path of differential distribution.

The Emotional Geographies of Water[3]

However, it is not only in the material dimension that links between water-using practices and social status are observable; as an everyday practice, domestic water use is also situated in a complex mesh of social meaning directly involving the human body. In this context, the realm of personal hygiene is probably where "the emotional experience of being classed" (Reay 2005: 912) becomes most manifest. Strongly framed by steadiness of supply and the perceived quality of tap water, a lack of water for personal hygiene is not only a health issue but also carries strong connotations of discomfort and feelings of shame (see 5.2.1). Given the relevance of cleanliness in Mexican culture in particular, rooted in pre-colonial culture and in the pre-modernist, bourgeois *Porfiriato* period alike (see for instance Rodríguez

3 Bondi (2005), Davidson *et al.* (2005) and Smith *et al.* (2009) shall be mentioned here as an introduction into the field of emotional geographies; for an overview on literature see Sultana (2011).

Cerda *et al.* 2002, Ayala Alonso 2010), this is unsurprising. Whereas the literature is often dominated by Western conceptualizations of hygiene and purity (see Stoffer 1966, Brandt/Rozin 1997, Melosi 2000), the cultural value of water as a means of cleanliness in Mexico City is based on a dual foundation related to both modern and pre-colonial notions. In Aztec and pre-Aztec culture, water was seen as a god-like entity, not least represented through the deities *Tláloc* and *Chalchiuhtlicue* (see Hernández Andón 2007, Cruz Bárcenas 05.04.2015). Whereas bathing was attributed both health benefits and a spiritual meaning and formed part of everyday life in Tenochtitlan (see Bailey Glasco 2010: 91 ff.), an attempt to break with these traditions was imposed after the *conquista*. This was in accordance with contemporary European believes and notions of cleanliness which held that the contact of skin with water was an unhealthy practice which should be avoided at any cost; dry grooming was the norm (see Bergua Amores 2008: 784, for an overview on the development of European ideas of cleanliness). Eventually, water returned as a means of cleaning both bodies and cities in an organic conceptualization of the urban (see Gandy 2004: 181), wherein water and drainage infrastructures were conceived as arteries and veins of an imagined urban body (see Sennett 1997: 401 ff.). But it seems as if water was only again considered a fully important means of personal cleanliness in Europe with the discovery of microbiology in the late 19th century and the formation of a hygiene movement (see Bergua Amores 2008: 786). Mainly under the pressures of industrialization and capitalist urbanization and the related deterioration of urban living conditions, this eventually gave rise to the 20th century bacteriological city (see Gandy 2004: 183). With reference to Foucault, it has been argued that in a biopolitical attempt of social disciplination of the (female) body and urban society alike, cleanliness to be obtained by the use of water has gained an almost religious quality in today's Western societies:

> "Such is the importance of cleanliness today that it would not be overstating the case to say that the daily rituals associated with it have inherited the same imagery, aspiration to purity and admonishment of the real body (…) that were once the province of religion. (…) moreover, the preferred target is the female, that congenitally impure being in both messages." (Bergua Amores 2008: 790)

Upon this backdrop, the implications of a lack of water and resulting limitations in hygiene with regard to feelings of impurity and questions of self-esteem cannot be underestimated. The availability of tap water (or the lack of it) can be understood as a tangible symbol of social identity and power which is experienced by the body itself in what Farhana Sultana has termed "emotional geographies of resource access" (Sultana 2011: 163). The permanent provision of clean water suitable for all purposes is hence essentially a question of social status, or as is argued with respect to India and post-apartheid South Africa, one of citizenship (see Dawson 2010, Ranganathan 2014). Being undersupplied serves as a symbol of social exclusion – statements of interviewees supplied intermittently or with yellow, smelly water referring to themselves as 'the forgotten', or as being supplied according to their tax value support this impression (see 5.1). Two of the imagined landscapes of water supply (see 5.4) – one coined by notions of preferential treatment and the other by an imagined disciplining through supply – point in the same direction. The first imaginario pictures urban elites receiving a better water supply than others in terms of steadiness and quality as a tangible expression of social inequality; the second is apparently rooted in a self-perception in which uninformed and ignorant citizens seem to require disciplinary measures by the authorities. In this logic, water rationing appears not as a question of distribution and resource access but as a regulative measure to lower individual water consumption levels. Ultimately, it seems as if water supply in Mexico City continues to be a symbol of modernity if not 'civilization' more in general (see Peña Ramírez 2004b: 105), and as a means of social distinction of the elite, wherein good (urban) living conditions serve as a symbol of status (see also Gerdes 1994).

Passing the Buck Around: The Contested Arena of the Domestic Cistern

As a buffer between the public water network and the domestic realm, domestic cisterns are a materialization of people's reservations over actual and expected supply irregularities. That such domestic storage requires efforts and time apart from the technical installations was pointed out above. Yet of all the efforts evolving from storing water in the domestic realm, the maintenance of domestic storage devices is the most disputed due to its alleged effects on water quality. Therefore, as these private storage systems form an integral part of a significant number of Mexico City's buildings, they are now increasingly turned into an arena where responsibility for the quality of tap water is being fought over. These contestations are both discursive and material, situated between the tangible work of scrubbing residues from cisterns' walls and the voicing of opinions in the interview material as well as in the media as to who is to blame for the contamination of domestic tanks. The author wants to argue that this is a symbolic struggle over the privatization of responsibilities for the provision of water in a permanent and potable manner.

The technical literature claims that tap water quality and the cleaning of domestic storage devices are related in a two-fold way: on the one hand, pollutants and particulate matter originating from the distribution network can accumulate in domestic tanks, creating the need for periodic cleaning of all installations. On the other hand, domestic storage itself potentially further decreases water quality as domestic storage facilities were indicated as possible sources of water contamination (see for instance WHO 2011: 104). Tanks which are not covered properly are said to be particularly prone to this as residual chlorine dissipates, encouraging the growth of microorganisms (see National Research Council 1995: 47). Indeed, it seems as if domestic water tanks have a longstanding reputation as breeding grounds for disease, which was one reason why they were removed by landlords in late 19th century London (see Sheard 1994: 144, Taylor/Trentmann 2011: 226). However, it seems as if circularity is at play here, as the question remains as to what causes a (widely perceived) low tap water quality in Mexico City in the first place. For government officials and the Federal District's water utility alike, the answer seems clear: both have repeatedly stated that over 90% of water that is fed into the city's network is of potable quality, and that a lack of maintenance in domestic cisterns is the cause of limited water quality at the domestic tap (see for instance Malkin 16.07.2012, Castano 22.02.2012, Huerta Mendoza 2004: 41). However, the empirical findings contradict this claim, as domestic cisterns were generally reported to be cleaned more often in areas where tap water quality was perceived as low. There are observable, tangible contaminations in domestic cisterns – and it seems as if people act accordingly by cleaning their storage tanks more often. Whether and to what extent such contamination occurs at the source, in the distribution system or in the domestic tank – or an interplay of all these factors – hence remains an open and disputed question. At any rate, blaming those who store water in the domestic realm for limitations in water quality renders the most basic question invisible: why are there domestic cisterns in the first place? In a publication by water utility SACM itself, one expert left little doubt:

> "If you want to ensure potable water, there should be a permanent supply and people should be able to use it directly, without domestic storage." (SACM 2012: 149)

It can hence be stated that the entire range of contested issues arising over tap water's steadiness and quality crystallize in the particles that accumulate in people's domestic water tanks.

To wrap up, a dominance of the actual is manifest in most of the water-using practices explored here. Whether storage strategies linked to intermittent supply, a lack of water for domestic hygiene or the consumption of bottled water as a substitute for tap water perceived as unclean – it is current water supply conditions which most strongly frame these practices. Yet there are two significant exceptions that will be discussed in what follows: the imagined landscapes of supply (see 8.2) and the hegemonic notion of non-potability, which forms a strong foundation for the practice of bottled water consumption (see 8.3).

8.2 INSCRIBING MEANING THROUGH SPATIAL PRACTICE

Apart from the flow of water from the domestic tap, the logic of urban water supply is essentially a non-tangible concept to everyday experience. Therefore, it is particularly apt to be conceptualized and analyzed as a representational space (see Lefebvre 1991). Other than expected, in the present study these urban imaginarios (see García Canclini 1997, Lindón 2007) – or imagined landscapes of supply – were found to be widely independent from people's own living conditions, or more precisely, from their own past and current experiences with water supply. This is not to say that these imagined landscapes generally lack a material character. Instead, the identified representational spaces make reference to concrete locations within the metropolis as well as processes of urbanization, often revealing a specific concept of the urban, of society and ultimately of nature-society relations. With these imaginarios being anything but a direct abstraction from a subject's concrete material conditions it was all the more remarkable to find a vast majority of people's imagined landscapes to be characterized by a critical consciousness of sociospatial disparities throughout the city. By most interviewees, whether more or less well-off, whether themselves subject to water supply limitations or not, Mexico City's logic of water supply was thought to be shaped by a preferential treatment steered by inequalities in social status and power. Such strong awareness of the water supply situation in general and questions of distributional justice in particular can be interpreted as a hint that Mexico City's water issues are by now so pressing that they are quite high on the agenda in public discourses and people's imaginarios alike. On the other hand, there are no indications so far that this would have led to any broader kind of political mobilization on the metropolitan scale – as already mentioned, individual coping strategies and passive acceptance were predominant. While all interviewees seemed to be aware of water supply limitations in at least some parts of the city, the three remaining imaginarios they sketched out during the interviews were decidedly of a more conservative nature. An alleged link between what was imagined as an overpopulation due to excessive population growth and a depletion of natural resources would ultimately lead to a collapse of the city as a whole, some argued. As to its devastating implications for social solidarity, this Malthusian discourse of absolute (if not 'natural') limits was brought to the point by David Harvey:

> "Somebody, somewhere, is redundant and there is not enough to go round. Am I redundant? Of course not. Are you redundant? Of course not. So who is redundant? Of course! It must be them. And if there is not enough to go round, then it is only right and proper that they, who contribute so little to society, ought to bear the brunt of the burden." (Harvey 1996: 148)

Other interviewees attributed problems with water supply to what they saw as a lack of proper regulation of urban growth and poor infrastructural planning – an imaginario which falls in line with the very same developmentalist logic showcased by hegemonic water policies on the Federal and city level. Yet the most disturbing imaginario unveiled in the present study is the one wherein intermittent patterns of water supply are attributed to a disciplining state educating its ignorant citizens to save water – and to be rightly doing so. As mentioned above, this poses questions about self-perception related to social status as well as it is strangely at odds with the widely voiced mistrust against public institutions amidst allegations of corruption and the like. However, it is hardly possible to grasp Lefebvre's notion of urban space containing and implying social relations (see Lefebvre 1991: 142) more precisely than this kind of representational space does.

In conclusion, it is not necessarily people's own material-spatial experience upon which they construct imaginarios combining material and discursive elements into an imagined landscape of water supply. Whether former or current dwellings were sites of water supply limitations or not – the spatialized metaphors of each of the four identified imaginarios were shared across these types of concrete experiences with urban infrastructural conditions. Yet in linking these metaphors to specific places in the city – or as Huffschmid and Wildner (2009: 54) call it, by inscribing meaning through spatial practice – the collective subjectivity of these imaginarios is always also a manner of producing urban space.

8.3 A SYMBOL OF MISTRUST?

The most striking result – given the assumed spatial situatedness of habitus on which the present research draws – is that there were no articulate differences in the current practices of water use of those having enjoyed the privilege of regular and unlimited domestic water supply in the past, and those who were excluded from these privileges. Depending on whether interviewees had experienced water supply limitations in any of their former or current dwellings, the empirical material did provide a number of clearly distinguishable types of habitat biographies. However, apart from one significant exception which will be discussed below, these habitat biography types could generally not be linked to any particular set of practices. The original hypothesis that a person's dwelling history provides a framing condition and production context where a certain kind of habitus is incorporated and later comes to play in everyday practices (see 2.4) has therefore to be rejected – at least when it comes to (most) practices of water use. In fact, current living conditions here seem to provide a much stronger context for practices than any living conditions experienced in the past. As for the empirical failing of the theory-derived hypothesis that the production of habitus is not only socially but also spatially embedded, a reflection on the theoretical approach and on the method of habitat biographies is to be found the following chapter.

Nevertheless, there was one major exception where past experiences indeed seemed to matter for water use. Yet before going into detail, we will briefly consider why grey water reuse, the other practice which appears to be linked to a certain habitat biography, is actually no such exception to the identified rule: current water supply conditions are what matters for practices of water use. Domestic grey water reuse was identified as a way of down-cycling water based on a functional distinction between water types for body-related and for more technical purposes along a cascade-scheme of water use (see 5.2.3). Most of the interviewees who were subject to supply limitations in the past continued to reuse grey water – even when there were no problems with water in their current home. In contrast, the practice of reuse was unfamiliar to those who had always enjoyed a permanent water supply. However, the practice of domestic water reuse was not as static as it seemed to be, and in case supply improved, it tended to be given up gradually over time. For instance, water reuse might be limited to certain simple tasks such as cleaning the yard with run-off from the washing machine, while other, more complex reuse-chains were abandoned, mainly for reasons of convenience. Falling in line with what Rodríguez Cerda and colleagues identified as a 'grand principle of abundance' amongst residents of the borough of Iztapalapa – the common belief that permanent supply at the (own) domestic tap indicates a general abundance of water (see Rodríguez Cerda *et al.* 2002) – this is a further indication that in the end, it is recent experiences with water supply that hold a strong role in framing current practices, lessening the relevance of how water was provided in any of the former dwellings.

Yet this is not the case with one exceptional practice clearly indicating an influence of former living conditions on current practices: the almost universal substitution of Mexico City's tap water as a source of drinking water. The common notion of tap water being the very opposite of potable means that it is usually either pretreated, or directly substituted with bottled water. Here, it is the (perceived) quality of water in the past (rather than the steadiness of supply) which serves as a guiding principle for the way water-drinking is nowadays handled in Mexico City. The collective experience of the devastating 1985 earthquake and its aftermath, which saw a disruption of water supply and led to official health warnings over tap water consumption, seems to have turned into the persistent notion that Mexico City's tap water is generally unfit for human consumption. As the widely voiced mistrust in tap water quality, the extraordinarily high levels of bottled water consumption, and its multiple uses presented in the previous chapters indicate, this notion is highly influential in contemporary Mexico City. As to why the practice of avoiding tap water consumption has stabilized (rather than turning into an elusive principle amidst the dominance of the actual, as was the case with most other experiences potentially pre-structuring water-using practices), there are a number of possible reasons.

Disputing Potability

First, it has to be noted that what is understood here as a subjective perception contains a material base together with a dimension of social construction, linked to own experiences and social realities. Bourdieu himself understands habitus as an internalized "*modus operandi* informing all thought and action (including thought of action)" (Bourdieu 1977: 18), and hence as a guiding principle for both practices and perception. Applied to the controversial issue of tap water quality in the Federal District, there are some issues which may be identified only through proper water analysis in the laboratory[4], and there is everyday perception with all it implies. Due to its theoretical orientation, the present study treats tap water quality strictly as a matter of observation by the interviewed themselves – wherein "the subject's definition of the situation takes precedence over the objective situation" (Denzin 1989: 183) – as it is mainly interested in the impact of these (subjective) perceptions on actual practices of water use. From this point of view, the idea of undrinkable tap water is further underpinned by palpable limitations in the quality of supply, such as a visible cloudiness or unpleasant smell of water in some parts of the city. However, such differences in tap water quality throughout the Federal District are not only reported by interviewees from the boroughs of Cuauhtémoc and Iztapalapa but also by the water utility itself. In addition, the question of who defines potability is not an uncontested one. The official declaration of tap water being "100% potable" in all but some 60 neighborhoods of the Federal District (see SACM

4 Here is not the place to assess the Federal District's tap water quality from an expert point of view regarding its chemical, physical and biological characteristics and actual potability. A future study could combine a water analysis of samples taken at domestic taps in different parts of the city and at several points in the distribution network, test for compliance with drinking water norms, and combine it with people's subjective perceptions of tap water quality.

14.04.2014), for instance, is sometimes actively challenged where it contradicts people's own perceptions, as was the case in this Iztapalapan housing complex:

> Water is a bit cleaner at times, but mostly it's filthy (..). In fact, we brought a bottle of water, just as it runs from the tap, to meetings with [...representatives of SACM]. (...) and one of the water guys was told to drink it. If he claimed this water was potable, he should drink it – and he didn't want to. (...) He said: "Yes, it's potable." – "Well then, drink it, so you'll see what you provide us with." And no, he didn't bring himself to do so. (Cecilia E50: 122, Iztapalapa)

Meanwhile, even in the absence of palpable limitations of water quality particularly in the city center, only very few interviewees actually took to drinking tap water. They cited their tight economic situation and the cost of bottled water as a reason, and drank from the tap in spite of being aware that Mexico City's tap water is widely considered as not potable. In what follows, this hegemonic idea will be described as the *notion of non-potability*. It may seem paradox that several other residents of the city center whom the author spoke to casted doubt about this non-potability of tap water, describing it as a kind of 'myth' (to use their own words[5]) – in particular because the very same interviewees kept consuming bottled water as a substitute to potable tap water. In essence, even those actively challenging what they claimed to be a 'myth' continued to abstain from tap water consumption. Such robustness of the concept of non-potability as a hegemonic guiding principle for everyday practices suggests it may in fact form part of a set of incorporated rules, that is, habitus. This is also supported by the fact that both the empirical material and the existing literature on perceptions of water quality in Mexico City are largely unanimous about the non-potability of its tap water (see Tortajada 2006: 15, Jiménez Cisneros *et al.* 2011: 81). Upon this backdrop, the mere question whether tap water was considered potable was often dealt with as if it were a rather naïve one during fieldwork for the present study. Here is Marta's answer to the author inquiring why she did not drink tap water in her Iztapalapan home:

> *No. Well, well, sister, why would that be? [...We installed a water filter in the cistern so...] It doesn't taste bad, but you mistrust it.* (Marta E51: 118, Iztapalapa)

That such a self-evident nature of the answer seemed to be taken for granted by most of the interviewees further suggests that the notion of non-potability is an integral part of a set of widely unreflected, incorporated rules which seem not worth mentioning, in a manner of speaking. Yet as we will discuss below, this could also be linked to a powerful urban imaginario which may have lost some of its concrete references along the way.

Mistrusting Institutions

Second, the way trust in tap water quality is (re)negotiated could well be linked to questions of trust in public institutions in general. For the (US-)American context, it was argued that bottled water consumption and home water purification can be

5 See Ana E36: 74, Alan E30: 153 and Fabiola E37: 38 (as quoted in chapter 5.1.1).

read as symptoms of declining trust in public institutions and their (assumed) capacity to provide safe drinking water (see Hamlin 2000), not least in an environment where capital and state compete over people's trust amidst heightened risk awareness and health concerns (see Wilk 2006). In particular the existing literature on the urbanization of nature (see Graham/Marvin 2001, Swyngedouw *et al.* 2002, Gandy 2004, Kaika 2005, amongst others) pointed out how a modern infrastructural ideal of a permanent, centralized and universal urban water supply – though never fulfilled in most urban areas around the world – serves as a strong symbol of state authority while redefining the relation between nature and society. The production of modern cities was possible due to a simultaneous production and regulation of nature (see Smith 1984 (2010)) – and that refers particularly to urban infrastructures such as the provision of water of a drinkable quality[6]. The assessment of water quality was turned into a task of experts – and in succession, of public institutions – precisely during the rapid industrialization and urbanization of late 19thcentury England, when the impact of these processes on the general living conditions became more and more manifest (see Hamlin 2000).

For the Mexican context, the intrinsic link between public water infrastructures and state regulation was summarized under the term "hydrocracy" (Wester 2009: 10). The post-revolutionary constitution of 1917 declared *el agua de la nación* as a public task, and also defined that access to water could be obtained only by the way of concessions granted by the Federal state. Water policies were also a materialization of revolutionary demands, canalized into a land reform based partly on the introduction of communal land tenure and -use in the *ejidos,* and, to an even bigger extent, to the modernization of agriculture through large-scale centralized state-led irrigation projects (see Aboites 1998, Wester 2009). While urban water supply was not in the focus of the initial initiative, it nevertheless formed the base for the developmentalist concept of an ever-increasing water provision through increased extraction and import of water from other catchments through big infrastructures, so clearly materialized in Mexico City's Lerma and Cutzamala systems (see Legorreta 2006 and 4.1). Arguably, this logic still forms the backdrop of today's water policies carried out by the national water commission CONAGUA, a successor of the *Secretaría de Recursos Hidráulicos* founded in 1946. However, today's almost universal mistrust in the potability of Mexico City's tap water raises the question as to when trust in the public authorities in assessing and providing water apt for human consumption began to fade. It seems as if it was precisely the event of the 1985 quake, which not only shook Mexico's largest metropolis deeply but subsequently also led to profound political changes – not least given the obvious incapability and passivity of the state itself in its reactions to the disaster (see Davis 1994, Monsiváis 2005, Moctezuma Barragán 2012). It seems logical that this may have also affected the realm of water quality assessment, in particular given the health warnings issued by the public authorities in the aftermath of the quake. It

6 Some go so far as to argue that a lack of potable water and related health risks in the growing cities of 18th century England were a major impediment to early industrialization and the rise of capitalism itself, only to be overcome by the advent of tea from the British colonies as a first 'safe' beverage apart from alcohol (see Kemasang 2009).

seems likely that over the years, this mistrust was fostered by a general sense of the Mexican state and its agencies as being rife with clientelism and corruption (see Tejera Gaona 2003, Tuckman 2012). Carrying ambivalent meanings, the replacement of tap water by bottles is not only to be understood as a reaction to state failure in providing good living conditions for all, but also as a rejection of the water tap as a symbol of a public good. In this logic, bottled water represents a growing alienation, ultimately expressed in the consumption of water which is promoted as a highly customized product 'liberated' from the common pipe (see Strang 2004: 217). For the Mexico City context however – in light of increased levels of bottled water consumption particularly in poorer neighborhoods, and in absence of empirical evidence that bottled water serves as a means of conspicuous consumption (at least amongst the interviewed popular and middle class residents) – the mistrust in tap water and its substitution seems more a matter of lack of good universal living conditions than one of aspired social distinction[7]. Mexico City's residents are now increasingly turning to private companies when seeking seemingly trustworthy alternatives. This process seems to have been driven not least by the beverage industry capitalizing on these public doubts over tap water quality.

Commodifying Drinking Water

Third, an ever-increasing commercialization of Mexico City's drinking water – in the literal sense – has become indisputable during the last years in the context of a neoliberal framework. Bottled water is now a ubiquitous good and a daily staple for almost the entire urban population, with Mexico as a whole displaying one of the highest per capita-consumptions worldwide. What is more, many of the interviewed used bottled water not only for the purpose of drinking but also for a wide spectrum of other purposes, from cleaning teeth to preparing food (see 4.1.6 and 5.1.3). In this context, the efficacy of the social construct of tap water being generally unpotable should not be underestimated, in particular given its hegemonic position. It is not far-fetched to assume that the common belief of tap water being unfit for human consumption underpins a further commodification of drinking water in Mexico through the proliferation of its bottled form. Notably, some marketing strategies directly exploit this by taking up health concerns and fostering risk perception. A popular brand of disinfection solution, for example, widely used for food preparation and water pre-treatment is promoted as "protecting families from illnesses for more than 50 years"[8]. Another case in point are small local purification plants visited by the author in Iztapalapa, seeking to boost the sale of their product under slogans such as "Pure water – secure life" or "Don't put your family's health at risk". Mimicking marketing campaigns by the beverage industry's big players,

7 This stands in a clear contrast to the social connotation the consumption of internationally branded mineral waters, for instance, receives in Europe and elsewhere. The Mexican elite could well display similar practices of status-related consumption of certain branded bottled waters but did not form part of the empirical sample for the present study.

8 See http://microdyn.com.mx, accessed 09.01.2015.

these advertisements play upon the notion of undrinkability of both tap water and other bottled waters without naming them directly. Moreover, they represent the individualizing tendency of a neoliberal logic putting pressure upon urban public goods by shifting the responsibility for health and well-being from the public into the private realm (see Bakker 2012, Brenner *et al.* 2012). Where this discourse identifies the purchase decisions of those managing water in the household as a possible source of health risk – instead of putting those in charge of the quality of tap water under scrutiny – people's health is pictured as an individual lifestyle choice rather than a product of (socially differentiated) urban living conditions. The persistence of the belief in undrinkable tap water, in part based in the 1985 earthquake, could presumably be traced back to some extent to marketing strategies of this sort. In Mexico as elsewhere, capital has greatly benefitted from a weakening trust in public water supply. That bottled water is a product which benefits from setting itself apart from tap water is openly celebrated in industry-oriented publications:

> "the perception that bottled water meets or exceeds the standards of tap water (...) is supported by the fact that people commonly pay a higher unit cost for bottled water (...). Unlike tap water, which may be exposed to a variety of conduit materials during its (...) travel to the consumer, the residence time of bottled water in a single container made of a single material can be days, weeks, months, or years. (...) consumers of bottled water are subject to a myriad of individual labelling possibilities, whereas there are no such labels for tap water." (Bryan 2001: 122)

In other words, bottled water could well be a commodity much more 'attractive' to capital than tap water, not just in Mexico. That Vicente Fox, ex-manager of Coca Cola-FEMSA for Mexico and Latin America, took office as Mexican president in the year 2000 can be interpreted as just one of the most manifest expressions of the beverage industries' growing influence[9]. Regarding (legal) water extraction[10], it is notable that FEMSA as the biggest concession holder of this industrial sector obtained a considerable share of its water concessions during the same Fox administration (see Flores/Rosas Octavio 2009). All these are clear indications that alleged 'peripheral' regions such as the ordinary cities of the Global South, including their more marginalized parts and populations are clearly in the mind of (not just) international capital. The commodification of potability draws on such stratified markets promising a huge customer base and demand as well as on mistrust in public water supply. Indeed, such exploitation of discourses around tap water has been subject to academic debate (see Barlow/Clarke 2003, Strang 2004, Wilk 2006). These discourses are usually set in a quite specific local context and are not free of contradictions as products serve differentialized markets. From *Selters* to *Perrier,* French,

9 The significance of the beverage industries also becomes manifest in extremely elevated levels of soft drink consumption, wherein Mexico is one of the biggest consumers worldwide (as it is for bottled water). According to a recent study, the high intake of sugar-sweetened beverages, proliferating diabetes and obesity-related cardiovascular diseases, is linked to 24,000 deaths in Mexico in 2010 alone (see Singh *et al.* 2015).

10 There are indications that some beverage companies in Mexico diversified their water sources by extracting groundwater without a regular concession through work-around arrangements (see Reis 2014: 549).

Italian and German mineral waters[11] for instance were and are sold for centuries around the world as expensive lifestyle products and means of social distinction (see Wilk 2006: 312). At the same time, tap-water based bottled waters, with few added features over ordinary tap water, are commercialized in the UK, Germany and elsewhere – the case of Coca-Cola's *Bonaqa* is notorious (see Blatter *et al.* 2001: 44, Roebke 27.02.2003, Heiser 10.12.2010). All this can be understood as an attempt to introduce a lower tier to the European bottled water market, thus following the Mexican path.

Yet to the author's knowledge, the role of marketing in fostering growing bottled water markets by specifically playing upon spatial imagination and traumatic experiences in the past – such as the case of the notion of tap water being non-potable which is widely shared in Mexico City – has so far not been put under scientific scrutiny. To grasp the interrelation between the social construct of water potability and such commodification of water for human consumption on different levels, future studies could also benefit from a fine-grained distinction of purposes according to the cascade of water use (as sketched out in Fig. 5.10).

Between Concrete and Imagined

Finally, it may be argued that the idea of Mexico City's tap water being non-potable has been so persistent over the years precisely as it transcended its original scale and turned into an incorporated rule. Indeed, this is one of the few findings of the present study pointing towards a link between habitus and a tangible historical experience: the collective memory of the 1985 quake. As the entire metropolis (and in the long term, the entire country) was shaken by the repercussions of this major earthquake in a concrete, symbolic and political sense, triggering profound changes, it can be argued that the persistent notion of non-potability is strongly linked to the urban scale. The empirical evidence allows the assumption that this event gave rise to the notion of tap water being unfit for human consumption, first due to the immediate impact of the quake, destroying part of the water supply infrastructure, disrupting supply and deteriorating water quality in the short term, and soon after due to the rising mistrust in the government and public institutions given their inactivity which led to Mexico City's residents taking matters into their own hands (see 5.1). Some 30 years later, it is most remarkable how the notion of non-potability has apparently become increasingly independent from the collective experience of the seismic event itself and subsequently also transcended the urban scale. A case in point is the consumption of bottled water – widely considered an apt substitute to potable tap water – which is on the rise in Mexico since at least the early 1990s, now constituting one of the biggest markets of this product worldwide (see 4.1.6). It can hence be assumed that the notion of non-potable tap

11 Generally spoken, mineral waters stem from specific springs and receive no futher treatment prior to bottling, whereas according to the German *Mineralwasserverordnung* (federal regulation on mineral waters), table waters are not required to indicate a specific origin and may be pre-treated, including the addition of minerals.

water has become an incorporated rule which is also passed on to and reproduced by those who did not themselves experience the triggering event of the earthquake (be it younger generations or non-native population). Simultaneously, there are indications that apart from Mexico City, other urban areas throughout the country are now also displaying elevated levels of bottled water consumption (see IADB February 2010). It appears as if concerns over tap water quality evolved from being a kind of knowledge based upon a concrete spatial experience (the impact of the 1985 earthquake upon water supply) into an unspoken hegemonic rule which continues to exert an influence on people's everyday practices today, in Mexico City and elsewhere. So at some point, the notion of non-potability seems to have taken up a life of its own. By now, it appears to be as much concrete as imagined – and it might be worth the debate whether it should be called a myth, as some of the interviewees suggested.

It remains a task for future research to reveal the construction of this notion and the nature of its temporal and spatial mobility, including how (and in how far) it became independent from its initial spatial context. Interestingly, in this aspect it also closely resembles the concept of urban imaginarios as introduced in 5.4. According to the literature, imaginarios may draw upon a material base through perceptions and lived spatial practices as they assign a collectively shared meaning to urban space. Yet they may also be largely decoupled from tangible spatial features and material experience (see Lindón 2007, Hiernaux 2007). The notion of non-potability could thus also be understood as a collectively shared urban imaginario, whatever its concrete roots. On the other hand, limitations in tap water quality which do not encourage its consumption, to say the least, are undoubtedly perceptible in many of Mexico City's neighborhoods.

If we understand the contestations over tap water quality in Mexico City as at least in part symbolic, this makes them all the more of interest for their potential role in the reproduction of social inequality. As discussed above, the commodification of drinking water is by now most advanced precisely amongst poorer residents in undersupplied parts of the city such as Iztapalapa. In what appears to be a combination of the notion of non-potability and perceptible shortcomings of water running from the tap, it is in these areas that the perception of water quality is reportedly worst (see also Monroy/Montero 22.01.2014). All in all, former living conditions (as documented through habitat biographies) were found to play only a very limited role in the pre-structuring of today's practices of water use – at least when it comes to experiences made in the dwelling itself. Yet the historical becoming is by no means decoupled from its spatial context, as the notion of non-potability – rooted in an impressive past event and now widely incorporated as a collective social rule – demonstrates.

8.4 REFLECTION ON RESEARCH STRATEGY

It seems that every research project involving Bourdieu's concept of habitus inevitably faces the question "So what kind of habitus is it?" at some point. This was no different for the present study, which used habitus as part of its theoretical framework to explore space-practice relations. Rather than searching for 'the habitus of water use', 'Mexico City's habitus of water use' or anything of the like, it employed the concept of habitus merely as a theoretical vehicle and connecting rod between people's (past) sociospatial settings and everyday practices. This theoretical backing and its involvement with the production of space was highly productive and led to the development of the instrument of habitat biographies. A detailed reflection on this newly introduced method is offered in 9.2, while 9.1 is dedicated to the hypothetical relation between habitus and habitat and its implications in the view of the present study. Returning to a potential quest for habitus, any methodological approach would need to start from one common ground: due to habitus' unreflected nature as an incorporated set of social rules, it reveals itself per definition only through social practice (see Bourdieu 1977: 18). Hence any study seeking to discern habitus empirically would have to draw for instance on reflexive methods covering social practices through self-observation, and the collective (re)production of habitus through focus group discussions. The latter has become a common method in UK-based habitus research by now (see Reay 2005, Merryweather 2010, Wills *et al.* 2011: 733). A number of methods were developed in the last years to deepen the former approach. Apart from 'classical' participatory observation, they reach from visual anthropology and reflexive photography (see Collier/Collier 1986, Dirksmeier 2007, Dirksmeier 2013) to water diaries and other instruments of 'self-observation' (see DeLongis *et al.*1992, Jean-Baptiste 2013, Lahiri-Dutt 2015).

As for the sociospatially oriented, explorative and fieldwork-based research strategy of the present study, a number of things can be learned. Tackling everyday practices through a range of different methods was successful in as far as it produced multiple perspectives on the same issue: how is water used in the home, and what are the framing conditions? Each of the methods provided a different piece of the overall mosaic, which would hardly have been obtained through each of them alone.

Bring the Interview Home

First, the conduction of individual interviews at home proved to be a good method to combine the collection of empirical material on everyday practices through narration and direct questioning with a direct access to the dwelling. Being at the very place where water is used allowed for an observation of both social interactions related to these practices and of the dwelling's materiality. This observation of material features was supported by a checklist for material housing conditions which was filled in after each interview and also covered domestic devices related to the use, storage, pre-treatment and disposal of water. Moreover, being in the home during the interview allowed for an observation of interactions of household members and water use which (potentially) occurred *en passant*. Examples are the interviewees who were in the middle of doing their laundry by hand when the author showed up for the interview, or the usual offer of a glass of water, which simultaneously provided a good starting point for the exploration of domestic practices. Through the interviewer's observations, which were documented in a postscript, the interview at home helped to unveil what might be considered too usual and obvious to the interviewee to be mentioned in an interview. These non-narratable aspects, which are obscured to the quotidian view (much as the concept of habitus itself) were analyzed in the present work as the expression of practices – such as the home-made textile water filter on a domestic tap pictured in chapter 5 which nobody found worth mentioning – and, above all, the tangible sociospatial setting of practices. This applies to both the dwelling itself and the respective neighborhoods. Future studies dedicated exclusively to one specific water use (take drinking) could involve a more profound participatory observation of domestic practices with a longer duration of stay in each home (longer than 2 hours, that is). However, any such approach has its clear limitations in more intimate practices such as personal hygiene, where self-observation techniques such as water diaries may be more adequate (see Lahiri-Dutt 2015).

Cater to the Focus Group

Second, the instrument of focus group discussions proved to be challenging in terms of organization yet highly productive in not only the reproduction but also the generation of collective narratives. These would be widely inaccessible to individual interviewing; the focus group provides an opportunity for a social interaction in which the story line is widely directed by the participants themselves. Applied in four different groups, this method produced impressive results as to urban imaginarios, collective memories and shared experiences in the realm of water supply and water use. Yet the same four groups also showcased its challenges and limits. Focus groups require a degree of organization and preparation beyond comparison with the individual interview in terms of timing, location and recruitment of potential participants – all the more so if the composition of the group is sought to be widely homogenous according to some predefined characteristics. Take the hypothetical example of an all-female focus group from one neighborhood in Iztapalapa preselected for its kind of water supply. How to motivate neighbors of one and the same housing complex, for instance, to volunteer for participation, show up all in time and bring their female neighbors, but not their cousin who is visiting from Ecatepec? In other words, collective sessions are much more complex to organize and often require the support from a locally renowned facilitator who is able to motivate people to participate. It is more complex to meet the participants' needs to lower entrance barriers to focus groups than is the case with individual interviews. When conducted at home or at the workplace, the latter are more effortless for the interviewee as they require no additional travelling and can be scheduled according to the interviewee's time constraints alone. In this sense, key factors for the success of a focus group discussion were a close-by and easily accessible location (cultural center, church), a locally respected facilitator who greatly increased people's commitment to participate, the prospect of an enjoyable social atmosphere[12] and not least a topic that catches people's attention. It was also clear that discussions were more prolific if the facilitator her- or himself did not participate in the event itself. Introducing additional material, in particular photo elicitation, was found to be a useful supplement to the topic guide as it was able to trigger debate amongst participants. By lessening the need for the researcher to pose direct questions, it led to a higher degree of self-directedness of the debate.

Interacting and Exchanging

Third, the interview situation and the focus group discussion are not mere tools to obtain information or inductively build theoretical concepts – they are always also an arena of interaction. A narration is produced – individually during the interview, or collectively in the focus group. (Not only) through the habitat biography method, a process of active remembering is induced by the interview process (see 9.2). As others emphasized before, there is always also a dynamic (though mostly unequal) interaction between interviewer and interviewee, and an almost (psycho)therapeutical function of interviews targeting personal experiences which allow people to voice their concerns, sorrows and opinions potentially kept in private otherwise (see Bourdieu 2010: 400 ff., Bondi 2005). This seems to be all the more the case in interviews with socially marginalized voices. Particularly when the interview is able to unfold a situation of mutual respect and trust – for instance through what Bourdieu calls a "companionship amongst women" (Bourdieu 2010: 397), or similar contexts where interviewee and interviewer are (or seem to be) linked by some kind of unspoken affinity or corresponding identity potentially allowing to overcome social differences – it can nurture this effect. Take Antonia, who after commenting that her partner did not share enough of the burden of domestic work, laughed and remarked:

> *It's also somehow therapeutic, this interview. I get rid of...* (Antonia E20: 242)

Moreover, the attention provided through the fact that a (foreign) researcher is studying one's very own everyday life is apparently understood by some as a kind of recognition that their own subjective experiences are actually of relevance. Several interviewees also used the interview situation to actively inquire about water-related issues in Europe, for instance regarding water tariffs, irrigation methods and rain water use in Hamburg. Others requested a detailed explanation of the

12 It was a former interviewee who pointed out that offering the prospect of sandwiches and refreshments after the debate would motivate people to participate – which it actually did, after some 25 sandwiches were home-made prior to two of the focus group discussions. Such a move is of course not entirely uncontroversial in ethical terms in a social context where such catering is a typical feature of political campaigns on the neighborhood level, and where votes are reportedly bought during election periods for the prospect of a free plasma TV screen or a supermarket voucher over 160€ in Mexican pesos.

study's purpose, and how it might be of use with respect to their own situation. Due to their collective character, it was particularly during focus group discussions that a perspective for change could be opened up through sharing of problems and experiences of infrastructural discrimination, which people might aspire to overcome collectively once they recognize their non-individual nature. Such an empowering character clearly emerged during one of the discussions when the all-female participants from a housing complex in Iztapalapa eagerly debated possible future options for actions involving all neighbors given the problems with water quality they were (all) facing. However, this does not entirely neutralize what continues to be an unequal position between researcher and interview partners in the production process of knowledge. The same focus group inquired about the purpose of the present study in detail before uttering the following:

> Alba: *What you said is very important, but you have a better idea of the problem whereas we for instance don't – we [know] that we receive dirty water, that's it. (…) if you already [spoke] to various groups, maybe here in Iztapalapa and the center, well, you have a different perspective.*
>
> Interviewer: *Yes, the idea is to share (…) the results I get. Sure, it's not just so I can put my book into the shelf and that's it, but so that people…*
>
> Sonia: *Send a book to Ramón Aguirre!* (FG III: 354–356, Iztapalapa)

Sonia's final remark regarding the director of the Federal District's water utility SACM ended in collective laughter. Calling attention to the problematic water supply conditions they experience in their everyday life but also get them related to the 'bigger picture' was hence an important aim for participants of this focus group, and the request to inform official representatives of the water utility can be interpreted as a kind of collateral the researcher is expected to bring up in exchange for retrieving the collective narration of that very afternoon.

In this context, there was an interesting situation during the recruitment of interview partners in one Iztapalapan neighborhood. At a certain point of the fieldwork process, I was explicitly seeking to interview locals devoid of a domestic cistern – a matter which was understood to impact strongly on domestic water use, as this neighborhood receives water only once a week due to rationing. As the author had worked in this part of the city before, field access through local contacts was relatively good to begin with. However, finding households without domestic storage systems was a challenge, as some kind of cistern was apparently installed in almost all dwellings in this particular, more or less consolidated colonia popular by now[13]. Having luckily identified several potential 'cases' to add to the list of interview partners, and being introduced to each of them by a neighbor or a former interview partner, the author was met with a remarkably assertive, negative reaction by one woman in particular. She lived in a relatively unconsolidated dwelling that was supposed, according to her neighbors, to lack a domestic cistern. After knocking at her door, explaining the aim and scope of the study and asking whether she would volunteer to be interviewed, the woman told the author that she had no inten-

13 Though that does not mean that the storage system is necessarily automated, with everything this implies for domestic storage practices (see 5.3).

tion whatsoever to participate in a project that, from her point of view, was designed only to showcase her family's poverty and precarious living conditions. In a way, she could not have marked her status as a subject rather than an object of inquiry more clearly and thus make the unequal relations between (potential) interview partners and researcher more explicit. This can be understood as an act of resistance against what she expected to be the symbolic violence of a highly unequal interview situation, and also against her potential objectification in the course of the research (see Bourdieu 2010: 395 ff.). The (often weak) reciprocity between the social scientist and the subjects he (or she) inquires is of course a longstanding issue, especially in qualitative empirical research[14] (see Shapiro 1980). These differences between the researcher and the people she interacts with during an empirical research project can hardly be 'wished away' but need to be made explicit while trying to minimize them, and by following a research agenda aimed at propagating social change not least through activating empirical methods – for instance those aimed at unearthing collective memories and (re)writing histories (see 9.2).

The Thorny Issue of Trust

Fourth, there is the question of trust and confidentiality: considering that interviews were deliberately located in people's homes and also exploring some quite intimate topics (take the hygiene-related questions), mutual trust was of the essence. It could be argued that it was precisely the (seemingly) neutral position of an outsider (rather than, say, a neighbor or family member) which allowed people to voice relatively private or tabooed topics such as whether a lack of water affects personal hygiene, and whether domestic grey water is reused for flushing toilets. Especially questions calling for an act of remembrance were found to be able to unearth emotional reactions not necessarily of sad nostalgia only, but of joyful childhood experiences with water, or endured and overcome hardships. Regarding the question of trust, the author's gender was also a crucial factor during fieldwork in Mexico City. As already reported in an earlier study (see Schwarz 2009: 12), it is likely that often hour-long visits to people's dwellings by a male rather than female researcher would have been met with suspicion, potentially even getting into conflict with moral standards upheld by the local community. Enjoying a certain level of trust in the studied neighborhood through some key person (reached through snowballing), or any previous field contact is also of the essence – all the more in a social context such as the Mexican one, being generally coined by strong feelings of insecurity amidst corruption, impunity and devastating levels of extreme violence (see Huffschmid *et al.* 2015). This is highlighted by cases of harassment of individuals not forming part of a local community who engage in collecting data or taking

14 It can be argued that this is all the more so in a postcolonial context as the very act of colonialization is assisted by geographical mapping excercises, the ethnological gaze in empirical field work, and the establishment of power through de- and reterritorializations and specific ways of knowledge production (see for instance Harley 1989, Zimmerer 2004, Crampton/Krygier 2006, Porto Gonçalves 2009).

pictures, only to be accused of preparing kidnappings, robberies or other crimes[15] (see for instance Duhau/Giglia 2008: 353 f.). A sensible establishment of reliable field access through local residents who introduce the researcher to the local community and the trust created through this can hence prove crucial for both the security of the research team as well as the success of the research itself.

Enter through Comparison, Exit by Complexity

Finally, inspired by comparative approaches, the present study took to the pre-selection of two boroughs of the Federal District as a starting point to explore potential differences in everyday practices under differing urban living conditions, especially regarding the water supply situation. Further down the line, the empirical results show that the picture is more complex than official water provision data and public discourse might indicate. In particular, the water supply situation in the borough of Cuauhtémoc was found to be not as homogeneous and without limitations as both these sources claim. There are areas with low water pressure and others where interviewees reported experiencing intermittent supply at the domestic tap. The borough of Iztapalapa, though generally more disadvantaged than the city center in terms of infrastructures, is not as homogeneous in its water supply conditions either. In particular with respect to water quality, there are strong internal differences between Iztapalapan neighborhoods. So when it comes to water supply conditions, the picture of each borough is both internally complex and also intertwined in similar processes and showcasing similar phenomena. The same accounts for practices of water use, which were found to be organized along and influenced by aspects such as housing types, the availability of certain domestic devices such as cisterns, supply frequencies and perceived tap water quality (see 8.1) rather than referring to people's residency in certain neighborhoods or boroughs alone. The challenge of any approach which takes a comparative stance is to adequately grasp the internal contradictions and intertwined processes in and between both entities rather than oversimplifying them to achieve a clear dichotomy of two contrasting cases as expected from the very beginning. From the author's perspective, taking two assumedly similar yet different boroughs as starting points for the present study was instructional, as it illustrates (again) how the complexity of urban processes requires a sensitive approach open to the processual and relational character of urban space.

15 Particularily from the periphery of Mexico City but also from other Mexican states, there are reports of violence against and even killings of investigators (though not necessarily academics). One gruesome case was the lynching of two men conducting market research on tortilla production in Ajalpan, Puebla, who after being accused of being kidnappers were tortured and killed by an angry mob of local residents in October 2015 (see Gil Olmos 21.10.2015, Marcial Pérez 21.10.2015). Equally unsettling is the growing number of journalists disappeared and murdered each year, making Mexico one of the most dangerous countries for journalistic work. In sharp contrast to the lynchings, most reports blame the massive violence against the media on both drug cartels and governmental agents (see https://rsf.org/en/mexico, accessed 23.11.2015). All in all, this makes Mexico a rather difficult field for empirical investigation in particular.

8.5 OPEN TASKS FOR FUTURE RESEARCH

"Señálale sus traumas a quien no compre"
(Monsiváis 2012: 212)

With respect to the sociospatial character of Mexico City's tap water, the notion of non-potability and its commercialization through bottled water which is so thoroughly grasped by Carlos Monsiváis' phrase is the major line of thought to be followed in the future, along with two other issues arising from the present work. Both the question of commoning water resources in the domestic realm and beyond, and the emotional geographies of water sharing and saving deserve further research. While the following passages focus on issues which are mainly at play in the everyday use of water on the micro scale, it should nevertheless be mentioned that there is also clear need to deepen the analysis of the hegemonic logic and sociospatial patterns of water distribution, provision and consumption in Mexico City and its implications for the reproduction of social inequality from a more quantitatively oriented perspective or through a mixed-method approach involving engineers, sociologists, political scientists, economists and geographers alike. One necessary precondition for such venture would undoubtedly be a better access to water supply data to begin with.

Commercializing Emotions?

The character of the widespread discourse that Mexico City's tap water is undrinkable was laid out in detail in chapter 8.3. It is clear that this notion, in its impact and hegemonic character but also its origins, deserves further attention in future studies. What precisely remains of the post-earthquake experiences with water when it comes to drinking? When and how exactly did bottled water become hegemonic over the practice of boiling? But most crucially: in how far has the idea of Mexico City's tap water being undrinkable become as much imagined as concrete? Was it somehow decoupled from actual supply conditions or any specific memory of supply limitations in the aftermath of the 1985 quake? An empirical inquiry in the direction of these questions could be methodologically based on a thorough discourse analysis of various types of texts in combination with focus group discussions, in which participants are also asked about their habitat biographies in the realm of the 1985 earthquake, putting specific scrutiny on drinking-water issues. This will allow deepening the systematical uncovering of past experiences with tap water quality in Mexico City and their relation to the practice of drinking from a historical perspective. When it comes to reading spatial practices in such a (discursive) sense (see Huffschmid/Wildner 2009) to fully grasp the dialectic between material spatial practices and spaces of representation (see Harvey 1989: 220), the writings of Doreen Massey and Michel Foucault on space, power and discourse should be taken into account. The aim would be to deconstruct the notion of non-potability as a discourse infused with power relations, exploring its arbitrariness. Where this notion relates to the trauma of the 1985 earthquake, bottled water promises a secure and healthy solution. The beverage industries' advertisement strategies also increasingly play upon hydration as a healthy lifestyle choice – a narration which particularly targets women's body image in a region where obesity is widespread. Consequently, it seems that in Mexico, bottled water is not only merchandized with reference to a widespread fear of waterborne diseases and health concerns but it also sells a desirable image of the (female) body – and even mind. Bonafont's 2013 Deshidratas anónimas campaign[16] showed stressed-out women on the brink of nervous breakdown finding relief in drinking two liters of bottled water a day. This highly emotionalized context showcases how the construction of identity under capitalist conditions quite substantially draws upon practices of consumption. In the words of Eva Illouz,

> "'Emotions' are an essential, albeit unsufficiently acknowledged mechanism explaining how consumer needs and desires connect to the system of *production of wants*." (Illouz 2009: 379; emphasis added)[17]

In this sense, it appears as if Mexico's thriving sale of drinking water in plastic bottles draws upon the ongoing commercialization of emotions, also those linked to tap water supply and imaginations regarding the public water network. In this context it seems promising to enter in conversation with experiences from Italian cities, where demand for bottled water has been high for several decades despite the provision of potable tap water, or cities such as Beijing, where bottled water sales are on the rise upon the backdrop of a growing middle class striving for distinction through consumption beyond 'ordinary' tap water. At the same time, boiling and filtering are the common practices of domestic water pre-treatment until now, and debates around tap water's potability similar to the Mexican ones are rife (see Zihan 22.07.2012). Such issues are to be juxtaposed with questions as to why residents of Leipzig or Vienna would turn to bottled water consumption despite a longstanding 'tradition' of the consumption of water provided through the domestic tap – and thus a seemingly hegemonic trust in (mostly public) water suppliers. A comparative stance would allow these experiences to engage with each other (see Robinson 2011), and to do so from an explicitly sociospatial perspective.

Competition over Commons?

Storing water within a building, usually in a cistern, is a straight-forward means of creating a buffer to regain a certain domestic autonomy of water use. Yet another layer of meaning and social interaction is added to this practice when storage

16 To add to the controversity of this campaign, the cartoon-like characters on the labels of Bonafont's water bottles were drawn by the cartoonist José Trinidad Camacho alias *Trino*, renowned in Mexico for his satirical stance.

17 Illouz hereby refers to emotions in a socio-cultural rather than individualistic sense: "Emotion is less a psychological entity than it is a cultural and social one: through emotion we enact cultural definitions of personhood as they are expressed in concrete and immediate relationships with others. Emotion is thus about where one stands in a web of social relationships" (Illouz 2009: 384).

devices are installed in multi-family buildings and housing complexes, where stored water is then shared amongst neighbors: the flow of the public, state-regulated good of water, as obtained from the municipal network, is turned into highly localized commons[18]. Such a "social practice of commoning" (Harvey 2013: 73) water in the domestic cistern turns into a ground where questions of commons are negotiated and struggled over on two different scales. First, collectively stored water could turn into an exclusionary commons due to the fact that other users along the line, not connected to the same cistern, are deprived of water as only a certain amount of the liquid is provided through the public network, and water pressure might sink as a consequence of increased domestic storing. Redensifing neighborhoods in Mexico City's center like San Rafael and Cuauhtémoc, where residents complain about a decrease of water pressure in the context of a building boom involving office towers and apartment complexes with enormous cisterns provide a good example – as do repeated claims by the water utility and the Federal District's government that the refilling of the entire network and restoring of full service after the total supply suspensions applied every third month or so take up to two days due to the ubiquity of domestic storage tanks (see Romero Sánchez 07.04.2012, Animal Político 02.11.2012). Moreover, it can be argued that on an urban scale "you need pressure to make water flow" (Anand 2011: 543), both in physical and social terms. In the negotiation over public water policies and access to water, local communities (such as certain neighborhoods, gated communities or organizations of business owners) are likely to possess a different 'voice' as well as social and financial capital to negotiate their access to water, for instance regarding questions such as which areas are subjected to rationing, who is cut off during supply suspensions and who is not, or who gets supplied by public tankers first. Of course these are often questions of the social status of the members of the respective local community to begin with (see Ranganathan 2014). What Stephen Graham coined as "premium network spaces" (Graham 2000) – bypassing less-affluent or less-powerful users of infrastructure through customized urban infrastructures – is another expression of this.

Second, back to the question of shared storage at the building level, the empirical material also reveals mechanisms of exclusion and competition within the (often involuntary) collective that shares the stored water, in particular due to different domestic everyday rhythms. As storage capacities are always limited at some point, this often means the creation of co-dependent rhythms of water use which can directly affect domestic practices. Take the examples of Hilda and Ana (see 5.3), employees returning home at night only to find that their neighbors had used up all water stored in the domestic cistern. As they lived alone, there was no one else to take care of domestic water management during the day, and secure access to water in this competitive neighborly situation. In other cases, however, neighbors expressed solidarity with those connected to the same storage tank, and reduced their own consumption to a certain level so everybody would obtain a sufficient

18 For the present purpose, by *commons* we can roughly understand an entity under "collective management by communities" (Bakker 2012: 30).

share of the stored water. It remains to be studied more in detail what exactly influences these two opposed manners of dealing with shared water. Frank Parkin's Weberian concept of social closure, based on mechanisms of either social exclusion or solidarization, might be of help here (see Parkin 2009: 158 ff.). Indeed, these contradictions form an integral part of the debate over the concept of commons (see Bakker 2012: 33).

Thus it should have become clear that the commons of stored water can bear an exclusionary character both in the internal and external sense. On the other hand, in potentially posing questions on property relations, on access to public goods and good living conditions (and for whom), it is precisely in such commons that an option for progressive urban politics shines through. Far from romanticizing the notion, it remains to be debated in detail to which extent and how a collective self-organization of the provision of public (urban) goods such as water could provide a viable alternative in a commodifying world (see Harvey 2013: 87). Urban water supply with its typically large technical infrastructure would seem a particular challenge, but there are promising proposals such as Thessaloniki's K136 initiative[19]. From the urban down to the local scale of the block and the building, these questions deserve attention in future research which could engage from a critical stance with a range of theoretical approaches to the commons, from Elinor Ostrom (Ostrom 1990) over Michael Hardt and Antonio Negri (Hardt/Negri 2011) to Gibson-Grahams' community economies (Gibson-Graham 2006: xi) and Murray Bookchin's libertarian municipalism (Bookchin 1992). Amidst its growing popularity, it is also worth noting the vagueness and multiple connotations of the concept of the commons including the more neoliberal ones, which Gehrig (2015) elaborates upon in some detail.

In this vein, the widespread awareness of supply disparities amongst Mexico City's residents, as displayed in the often cited imaginario of 'preferential treatment' (see 5.4.3), could serve as a starting point for a collective re-thinking of urban living conditions, and the collective organization of and access to public goods for all. In which way is the distribution of water in the city currently organized according to these imaginarios – and what might be other ways? It is therefore crucial to deepen the understanding of what shapes urban imaginarios as (potentially utopian) horizons of (political) action (see also Hiernaux 2007: 26). While current water supply conditions seemed to make hardly a difference when it comes to people's imagined landscapes of water supply, public discourses could well play a role, as could political socialization, religious and moral values, or the participation in

19 After years of struggle against a (full) privatization of Thessaloniki's water utility, the K136 initiative (see http://www.136.gr/article/what-initiative-136) launched a succesfull referendum against water supply privatization in early 2014, proposing that each resident buy a share of 136€ in the water utility to turn it into a company owned directly by the population of Thessaloniki and managed by cooperatives on the municipal level (see Steinfort 03.06.2014). Harsh austerity measures subsequently imposed on Greece by the EU under the leadership of Germany, most promimently from 2014 on, and their push for further privatization of key infrastructures render a rather bleak outlook. Yet at the same time, they provide the very backdrop upon which progressive approaches for urban water supply in the cracks of the state-or-capital dichotomy need to be developed.

certain urbanization processes – for instance the collective organization of neighbors in housing projects under the roof of the *movimiento urbano popular,* or the shared struggle and collective work typically applied during the consolidation process of irregular settlements. Returning to collective water storage as a commoning coping strategy which is obviously on the rise in Mexico City, such a need to better understand the involved imaginarios also applies, and future studies could explore the imaginary dimension of this (often quite involuntary) practice of sharing in its sociospatial character. What kind of collective is imagined, for instance, around the domestic cistern, and how do I see my own role in it? Who is included and excluded from these collectives? What are its unspoken rules? Do I appropriate all available water for my own private use, or are common interests to be considered – both from an internal perspective of the collective (me and my neighbors), and on the urban scale and beyond? Exploring imaginarios hence also requires an adequate definition of their situatedness in a specific scale, and the way it is intertwined with other scales. In this context, the call to pay more attention to the spatiality of the commons (see Moss 2014) is of particular interest.

Similar questions sprout on the contested field of water saving, in particular upon the question which common interest, (imagined) collective or egocentric motives someone is saving for, and on which spatial scale(s) these imaginarios are located. This matters for a range of academic disciplines, from ethnography to environmental sociology and economics, concerned with the use of public goods and socially constructed 'natural' resources. Yet apart from the practice of domestic grey water reuse, which provided the empirical base for the induction of the already mentioned concept of a cascade of water use, the entire issue of motivations for water saving was deliberately excluded from the present study. This was decided mainly due to the strong positivistic and economistic streak which dominates much of the research in that particular field of domestic water use as it lingers on the limits between consumption and use. Any future research on water saving from a sociospatial perspective would hence need to actively enter in debate with and aim to be dissociated from the hegemonic behaviorist and rational-choice approaches of that field. Again, Eva Illouz' work on the commercialization of emotions is of interest in this context (see Illouz 2009), as is the growing field of emotional geography (see for instance Thrift 2004, Davidson *et al.* 2005, Bondi 2005, Smith *et al.* 2009, Sultana 2011).

There is no doubt that the urban imaginarios related to questions of sharing and saving water, as well as the widespread notion of non-potability deserve more attention in future research. They could be systematically explored through a combination of personal narratives, focus group discussions and discourse analysis, including the reconstruction of the discursive character of metaphors in biographical narratives as proposed by Ute Karl with reference to Michel Foucault (see Karl 2007). Moreover, such studies can also benefit from the concept of a cascade of water use (as developed in 5.2.3) by providing a deeper understanding of the perceived and always also imagined characteristics of water designated to specific purposes, from body-related to technical. Following Farhana Sultana's call to "explore how emotions and embodied subjectivities play a role in the ways that natural resources come to influence everyday life" (Sultana 2011: 171), I would argue that only such a situated studying of everyday practices which are sensitive to subjective needs, perceptions and (socially embedded) emotions can fully grasp the meaning of domestic practices of water use.

9. REFLECTION ON THE RELATION BETWEEN HABITAT AND HABITUS

According to what may well be one of Bourdieu's most popular quotes, the habitus makes the habitat (see Bourdieu 1991: 32) – but, at least, regarding the empirical findings of the present work, there are no indications that this concept is of a reciprocal nature (see 6 and 7). Generally speaking, the idea of some kind of direct relation between past habitats and habitus can be rejected with respect to domestic water use in Mexico City. This is clearly indicated by the results of an in-depth analysis of the sociospatial setting assumedly providing a framing condition for the formation of habitus. Upon this backdrop, there is a need to reflect the conceptual approach employed in the present study from both a theoretical (9.1) and a methodological perspective (9.2), also with respect to its potential value for future research.

9.1 REFLECTION ON THE CONCEPTUAL APPROACH

Habitats do not make habitus at all. As good as this may sound if we strive to reject a fetishization or reification of space[1], some questions remain regarding the interplay between space and social practices. To begin with, I wish to shortly recall the motivation for drafting the hypothesis that habitats (or concrete living conditions in a sociospatial sense) might exert a potential influence upon the generation of habitus. As developed in detail in chapter 2, the motivation for the present work to empirically analyze possible repercussions of differentiated urban living conditions on the reproduction of social inequalities via everyday practices is based on two theoretical assumptions: First, these living conditions form an integral part of an urban space understood as a social product constantly in the making and being shaped by social relations. Second, it was assumed that precisely in its *always socially produced* character (and it is crucial to stress this), space might serve as some kind of framing condition for social praxis (in particular everyday practices). That is not to say that habitat is drafted as a necessary or exclusive pre-condition for social reproduction. Rather, it seems to be one amongst a complex host of categories such as class, gender, ethnicity or color, along which social differences are constructed. Introducing Bourdieu's notion of habitus as an incorporated set of social dispositions, and applying these assumptions to form a conceptual approach under the heading of *spatializing habitus,* lead to the following reasoning: If habitus were to 'make' the habitat, could it then be assumed that the socially produced space of the habitat, in turn, may also, in some way, serve as a framing condition in the emergence and development of (future) habitus? This question brought forth

1 Reification is an objectification of social relations; in Georg Lukács' terms, it "degrades time to the dimension of space" (Lukács 1971: 89).

the hypothesis that the habitat might play a discernable role in the co-production of habitus, as advanced by the habitus-habitat-practice model in chapter 2.3. Hence the need to clarify the relation between past habitats (or sociospatial settings) and habitus, in particular with respect to the making of the latter. This falls in line with some of the existing geographical literature drawing upon Bourdieu's concept of habitus, which specifically pointed out a need to further explore that relation (see Haferburg 2007: 342), also with respect to a possible reciprocity (see Schroer 2006: 88 f.).

While Bourdieu himself explicitly called for an analysis of the interrelations between social structures and physical space (see Bourdieu 2010a: 117), it seems as if in his writings he never directly considered reversing the habitus-makes-habitat dictum (see Schroer 2006: 89). Given that Bourdieu's conceptualization of 'social space' follows a Durkheimian tradition in understanding it as a material expression of social structures (or in Bourdieu's words, of abstract social space), this seems logical and consistent. However, there are a number of indications throughout his texts that Bourdieu might not have entirely dismissed the idea of some kind of feedback from spatial structures to social status. At several points, he refers to 'spatial profits' and 'localization profits' which simultaneously were to represent and reinforce a hierarchical spatial distribution according to people's social status (see Bourdieu 1991: 30). But the idea that urban space (and hence habitats) might also be framing the reproduction of social structure – rather than simply objectifying and reflecting them – is perhaps most explicitly expressed in this quote from *The Weight of the World,* in which Bourdieu hints towards an inviting character of space:

> "In general, the part of the mediator, through which social structures are gradually transformed into structures of thought and predispositions, is played by the *covert commands and silent calls to order of the structures of appropriated space*. More precisely, the imperceptible incorporation of the structures of social order is without doubt taking place (…) through (…) repeated experience of spatial distances, wherein social experiences sustain their position (…) but also (…) through the movements (…) of the body [whereby they] convert to spatial structures and such *naturalized social structures organize and qualify social* ones." (Bourdieu 2010a: 119; emphasis added)[2]

Referring to what could be called a kind of spatial stimuli he proceeds to outline what this could imply when applied to urban space, or more precisely, to the interplay between space and body:

> "Architectonic spaces, whose *silent bids are directly addressed to the body, demand (…) reverence*. Thanks to their substantial invisibility, they are without doubt the most important components of a symbolism of power and of the entirely real effects of symbolic power" (Bourdieu 2010a: 120; emphasis added)

Physical space, Bourdieu argues, symbolizes power as social relations are inscribed in it – and this symbolic domination can indeed have repercussions upon sociospatial practices, and the reproduction of social inequality. He also mentions the examples of better-off and of marginalized neighborhoods, which hold the capacity to symbolically (de)valuate their inhabitants by means of a 'good' or 'ill-reputed' address –

2 All translations from the 2010 edition of *Das Elend der Welt* were performed by the author.

with quite tangible effects, for instance regarding their access to the labor market. Yet these spatial differences are also of a material nature – herein rests the Bourdieuian concept of 'localization profits', for instance when it comes to the proximity and quality of public facilities such as schools and clinics, or urban infrastructures.

Notably, this is where a relational understanding of space comes back into the picture. Whereas Bourdieu linked society and space in his habitus-makes-habitat dictum, and in the idea that urban space represents a reflection or expression of social structure, he also gave the above mentioned hints on a potential feedback between urban space and social status. I would argue that it is precisely the concept of a relational space (as a product of ongoing sociospatial processes) which is able to fill the gap by helping to avoid what John Agnew famously termed the "territorial trap" (Agnew 1994): A reification of space through its conceptualization as a mere container of social relations[3]. Yet in the light of, for instance, the recent debate in German-spoken urban sociology on an alleged 'intrinsic logic of cities', which heavily draws upon Bourdieu's concept of habitus in particular, it seems inevitable for the present work to give up the initial wording that it aims to somehow 'spatialize habitus'. Instead, it seems more appropriate to return to the idea that "time and space are produced and reconstituted through the enactment of practices" (Shove *et al.* 2012: 133), or what Nigel Thrift (again with reference to Bourdieu) calls a situatedness of practices in both time and space:

> "Human action must be seen for what it is: a continuous flow of conduct in time and space constantly interpolating social structure. Such a view of human agency is necessarily contextual. (…) human action takes place in time as a continual time- (and space-)budgeting process and as an irreversible sequence of actions. (…) it is the *ad hoc* improvisatory strategy imposed on people's practices by the fact that they have a limited time in which to carry out particular activities (…) that is a crucial part of practice (…). Practice, therefore, is always situated in time and space. (…) the places at which activity is situated are the result of institutions which themselves reflect structure." (Thrift 1996: 71)

Yet even as such a contextual approach to human agency takes social structures and power relations as a starting point for any sociospatial analysis, some questions remain as to the spatiality of social becoming. Being characterized by their heterogeneity, how can the complexity of both the processes of social becoming and of production of space be fully grasped in their interrelatedness?

From a relational perspective, it could be argued that space, understood as a social product always in the making (see Massey 1999), also produces relatively permanent or stable structures in a number of possible ways. It is in that respect that space might serve as a framing condition for ongoing spatial practices, on the one hand, and for the reproduction of social relations, in part also through spatial practices, on the other. Infrastructural networks and, to a lesser degree, other features of the built environment are usually relatively stable as they display a strong path-dependency and serve as a 'spatial fix' to capital (see Harvey 1985: 83, Graham/Marvin 2001). The previous chapters have shown how social practices are spatially situated, as those practices of domestic water use in Mexico City studied in the

3 Though John Agnew does so with reference to the concept of the territorial nation state.

present work were clearly framed by water supply conditions. A similar framing is also performed by the production of spatial knowledge and discourses (for instance a map displaying areas subject to water rationing), and by the social meaning attributed to space via everyday experience. This is the case when, for example, unclean tap water (see 5) or severely limited patterns of everyday mobility (see Schwarz 2013) come to symbolize social marginality for those affected. However, it was found that collective memories forming part of these spatial imaginations were not necessarily linked to immediate subjective experiences. The most widespread *imaginarios* regarding the logic of water supply in Mexico City, for instance (see 5.4), seemed to be widely independent of people's own water supply conditions in both past and present. Nevertheless, these imagined spaces rested heavily on Mexico City's actual supply disparities – of which people seem to be widely aware. The notion of non-potability, as analyzed in 8, provides another case in point, though it also allows for some interesting considerations as to whether and how this hegemonic notion has by now transcended its spatial and temporal situatedness by leaving behind Mexico's largest metropolis and spreading to other regions. In this sense, the following section traces some of the remaining questions emerging from a reflection upon the concepts of habitus and habitat.

Habitus on Stand-by?

In a similar vein as relational space, the concept of (socially produced) habitus is based on one consideration in particular: its processual character. In other words: How flexible is habitus, once incorporated? And to which extent can these sets of incorporated rules be discarded, ignored or adapted under changing circumstances? For a start, this is precisely the strength of the habitus concept as the author understands it: Its relative stability, in the sense of an incorporated set of guiding principles or social rules, which operates in the realm of the unconscious[4]. If habitus is a principle not applied consciously, it can hardly be subject to a conscious decision for change. In line with Bourdieu's own writings and Bourdieuian scholars, this is precisely what makes habitus such a strong tool of social distinction. When set in the 'wrong' or unexpected social context, it is habitus that is most likely to immediately render the person new to the field an outsider, as Bourdieu put it, through "that experience, which you expose yourself to every time you enter a room without fulfilling all pre-requirements it demands implicitly from everybody who occupies it" (Bourdieu 1991: 32). The other way around, it is habitus which allows people to easily conform to and integrate into a field to which they possess the corresponding habitus. Therefore, habitus is conceived as a powerful tool to exacerbate social advancement by reinforcing social distinction, and to reproduce and stabilize social relations by concealing their arbitrariness and socially constructed character (see

4 The literature emphasizes that habitus in the sense of Bourdieu is by no means 'unconscious' in Freudian terms. Instead, its 'unconsciousness' refers to the idea that its contents are widely inaccessible to the subject him- or herself as the circumstances of its incorporation are forgotten (see Fuchs-Heinritz/König 2011: 115).

Bourdieu 1977: 164). This leads to the question as to what would happen if someone with a certain habitus were confronted with a social situation where this precise set of rules is not of avail or results inapplicable. How soon and under which circumstances are incorporated social rules subject to change? How fast may a habitus be discarded or adapted once put under scrutiny in a new environment? To return to the empirical findings of the present study, let us assume the notion of non-potability (see chapter 8) were a disposition forming part of a specific habitus unconsciously incorporated by (many) residents of Mexico City. They would, in other words, be unable to actually provide any particular reason as to why tap water should not be drunk. What would happen once people moved to somewhere else where water supply conditions are different and other social rules as to the potability of tap water might apply? Some of the interviewees from Mexico City who had lived abroad for a period of time recalled their initial surprise over the practice of drinking water directly from the tap. A case in point is the interview with Elena, who allegedly "almost fainted" when she first witnessed her colleagues in Bogotá drinking water straight from the tap. Like others, she reported to have eventually adapted to this custom, considering it highly convenient in comparison to bottled water consumption. However, once returning to Mexico, these interviewees reportedly returned to avoiding tap water as a drinking water source. Are certain dispositions hence set 'on stand-by' as long as the person is situated in an 'unfitting' field? And, following the same vein, how (fast) are such incorporated rules unlearned, given their astonishing perseverance in Mexico City itself? Clearly, the very concept of habitus rests on a certain robustness and tenacity in the face of change. Yet Bourdieu later stressed that such a mismatch between incorporated rules and actual conditions is indeed a moment wherein the long-lasting yet not permanent nature of habitus comes to the fore:

> "in all the cases where dispositions encounter conditions (including fields) different from those in which they were constructed and assembled, there is a *dialectical confrontation* between habitus, as structured structure, and objective structures." (Bourdieu 2005: 46)

In line with this, it has been argued that it is precisely due to its social producedness that habitus is by no means determined permanently. Gail Weiss, for instance, made the point that habitus expands with experience and thus changes with time (see Weiss 2008: 229 ff.). In what could well be applied to the notion of non-potability which seems to stem from both the collective shock resulting from Mexico City's 1985 earthquake and the subsequent social transformations, Weiss refers to the events of 9/11 to make her point that extraordinary situations in particular seem to possess the ability to alter and transform existing habitus in a radical way[5] (see ibid.: 231–232). (Furthermore, these extreme situations might be directly instrumentalized to implement new regulations or alter existing power relations). To return to the Mexico City case, it can be argued that in the aftermath of the earth-

5 This should by no means imply that terrorist attacks and earthquakes were somehow comparable events *per se*. Though what may need further clarification is the potential impact of such extreme events as triggers of processes of social transformation (see for instance Wisner *et al*. 2004).

quake, a new social rule was created and eventually incorporated: Do not drink tap water, at least not without pre-treatment! This social disposition (as part of a certain habitus) appeared to be quite steady, most likely due to a range of additional factors outlined in chapter 8 which supported its stabilization over time. The deep rupture which the 1985 earthquake presented for everyday life in Mexico City may hence have provided the base for such a radical change of the guiding principle when it comes to drinking water. Notably, this falls in line with Néstor García Canclini's indication that it is "social or natural shocks" in particular through which urban *imaginarios* as a sedimentation of collective experiences may be altered (see García Canclini 2013: 40). However, Elizabeth B. Silva also has a point in arguing that transformations of social norms may also be "recurrently generated" in the realm of daily routines (rather than only during sudden ruptures, or a lack of 'fit' between habitus and field) as long as these norms are subject to a critical awareness[6] (see Silva 2005: 95 ff.).

In the end, such considerations over the durability of habitus (or some of its dispositions) bring us back to a core question on Bourdieu's Theory of Practice: How exactly is habitus created – that is, how are social rules incorporated? This is certainly not the place to retrace the long-standing debate on this matter, which mainly seems to be nurtured by the vagueness displayed by Bourdieu's own writings in this respect, and by diverging interpretations in the Bourdieu-based literature (see Fröhlich 1994, Lau 2004, Steinrücke 2006, Bennett 2011). Apparently, further research is needed to tackle the issue of habitus' (in)flexibility and, more importantly, to push for a stronger theoretical conceptualization regarding the way habitus is generated (see Dirksmeier 2007). The present work sought to contribute to that by (conceptually) situating the production of habitus in a sociospatial setting via the instrument of habitat biographies.

Habitus Makes Habitat?

In reflection, if we are to take a relational understanding of space seriously, Bourdieu's statement that the habitus makes the habitat needs to be put under scrutiny. To actually 'spatialize' Bourdieu, it may not be enough to look at people's past spatial experiences in an attempt to grasp the sociospatial conditions under which a certain habitus is incorporated. More importantly, it can be argued that there is a need for future studies resting on a similar conceptual approach to reflect upon the making of habitat to specifically tackle this process in its social and spatial situatedness. A potential interplay between habitus and habitat formed a crucial part of the theoretical conceptualization behind the present work, yet it has to be stressed that habitats are of course never simply a product of habitus to begin with – and even less of just one, singular habitus. This immediately gives rise to the question as to who makes habitat, and whose habitus might be at play in this production process.

6 In one of his last texts, Pierre Bourdieu also highlighted the potential transformability of habitus through conscious training, which he likens to overcoming a certain pronunciation of a language (see Bourdieu 2005: 45).

Assuming that the production of urban space (including habitats) is an ongoing social process in which a whole range of producers is involved on different scales, whose habitus in particular are we talking about? Take the hypothetical example of someone living in an intermittently supplied flat in one of Iztapalapa's modernist housing complexes – a short recapitulation of the number of actors potentially involved reminds us that the generation of habitat can by no means be conceived as an individualistic procedure in which one subject would simply play out his or her incorporated habitus. First and foremost is the inhabitant of the flat, who may have installed a private roof-top tank due to irregularities in water supply. Hereby, she seeks to secure her ability to do laundry when returning home from work in the evening. She may have also got rid of the bathtub in her flat. Thus she appropriated and shaped a certain aspect of her habitat. Apart from her own actions, there are her neighbors, also making use of water and of the housing complex in a certain way; there are urban planners who set up the urban development plan (whether applied or not), architects who projected and bricklayers, plumbers and carpenters who actually built the housing complex; there are those producing water tanks and building domestic cisterns; there are bottled water vendors from a local purification plant, as there are, crucially those who manage and those who maintain the municipal water infrastructure, applying a certain logic of distribution, and so forth. As this example demonstrates, not only would a considerable number of people be involved in the production of this (hypothetical) habitat, but it is also likely that they come from various social groups and (professional) fields. Yet habitus, in the sense of Bourdieu, is precisely the feature of a certain social group or field, so that several different and interfering habitus would be at play in the production of any particular habitat, or in other words, of concrete living conditions. Moreover, these are not simply the product of multiple, combined individual actions but of a certain mode of spatial production. The modernist housing complex – with its rational concept and often repetitive materiality striving to both represent and proliferate a certain (modern) way of life and a certain household type (the family) which is imagined as the fundamental unit of society – is only an especially telling example.

Finally, it is precisely in its quality as a social product that urban space becomes a site and stake of social struggle:

> "*Space is at once result and cause, product and producer*; it is also a *stake*, the locus of projects and actions deployed as part of specific strategies, and hence also the object of wagers on the future" (Lefebvre 1991: 142; emphasis added).

This regulatory character of space derives from its homology to dominant social relations; the production of space is, in other words, a process which generates one of "those 'permanences' that can give order to social being and directions to social becoming" (Harvey 1996: 347). In the end, this renders the production of space a moment stabilizing the hegemonic order while simultaneously, and at least potentially, opening up a source and site of social contestation and radical transformation of both space and society (see Brenner/Elden 2009: 367, Goonewardena 2012: 86 ff.). It is in this sense that "the city is both a product of the past and (...) assists in creating the now and tomorrow" (Ward 1998: 232) – though one might want to add: potentially yet not necessarily in a progressive sense.

In conclusion, if space is understood as always a social construct and social process to begin with, a coherent analysis of the making of habitat needs to take different modes of (spatial) production into account, and also make reference to different scales. While the sphere of everyday life, involving particularly (but not exclusively) the domestic realm and public space, is arguably where a (seemingly) individual habitat is arranged and appropriated, its material bases are largely produced elsewhere (in other fields and locations) and by others – and ever more so in a capitalist society based on a pronounced and increasingly globalized division of labor. This involves different dimensions of production of space, from institutional logics and regimes of regulation over land and housing markets to infrastructures as 'enabling' parameters of urban development and capitalist accumulation through urbanization (see Smith 1984, Harvey 1996). In this sense, the micro scale of the dwelling and the urban scale are intrinsically linked and entangled to a point of inseparability – but still require a careful distinction during the analysis of their production process. The habitat is an arena where (dominant) social relations are at play (see Haferburg 2007), but at the same time, it is also a social product always in the making, and this process is (at least potentially) open to contestations of the *status quo* in terms of living conditions. Reading people's everyday practices from a sociospatial perspective may reveal these contradictions.

9.2 HABITAT BIOGRAPHIES: METHODOLOGICAL REFLECTION

Habitat biographies, as developed and applied in the present work, are an instrument which was inspired by the habitus concept and by calls for a stronger focus on the role of urban space in its formation. The method was explicitly developed with the aim of analyzing the conditions of the historical becoming of habitus from a sociospatial perspective. Habitat biography types were then constructed with respect to interviewees' experiences with water supply in former dwellings and analyzed for their possible relation to current practices of domestic water use. As such, the instrument of habitat biographies, though inspired by the Bourdieuian Theory of Practice, is not to be confused with a method to grasp habitus on its own right. In the way developed here, it seems of avail to complement a habitus analysis in future studies by offering a way to obtain crucial data on one of the most relevant biographical experiences: The manner in which a person dwelled. Moreover, it is able to show how the experienced living conditions are of a collective rather than an individual nature. For this reason, the method seems promising – in combination with focus group discussions, reflexive photography, and other methods of non-directive and autobiographical interviewing – for future research involving the concept of habitus in social geography and urban studies. Precisely as it aims to unveil both the historicity of spatial practices and the spatiality of social becoming, the method could also be of use to other academic fields – such an approach is rooted in Emile Durkheim's proposal to overcome the limits between the disciplines of sociology and history, which was significantly advanced by Fernand

Braudel (see Steinmetz 2011, and Parnreiter 2007). It also hopes to contribute to what Walter Benjamin outlined as a task of any materialist historiography:

> "Materialistic historiography (...) is based on a constructive principle. (...) A historical materialist approaches a historical subject only where he encounters it as a monad. In this structure he recognizes the sign of a Messianic cessation of happening, or, put differently, a revolutionary chance in the fight for the oppressed past. He takes cognizance of it in order to blast a specific era out of the homogeneous course of history – blasting a specific life out of the era or a specific work out of the lifework. As a result of this method *the lifework is preserved in this work and at the same time cancelled; in the lifework, the era; and in the era, the entire course of history.*" (Benjamin 1968: 262–263; emphasis added)

In this sense, the habitat biography method can also be understood as a contribution to a stream of research that seeks to meet two of the major concerns of the 'structurationist school'[7]: To employ a theory of practice, and herein, to analyze social relations and temporal as well as spatial structures together rather than in a separate manner (see Thrift 1996: 68–72). Following these reflections, this section is dedicated to the method of habitat biographies as developed here, its benefits and possible future applications.

Grasping the Collectiveness of Everyday Experience

When embedded in a focused interview, as in the present study, the instrument of habitat biographies was found to encourage a historical narrative, allowing the interviewee to organize this narrative along past dwelling experiences. In the present case, these narratives were typified and then employed to analyze current practices of water use for a potentially shared history. This is one of the strongest qualities of the method in the way it was applied here: To allow for a generalization, or in a sense, to allow zooming out from the individual narrative without losing the empirical groundedness. The construction of empirically saturated types of habitat biographies along empirical regularities and meaningful relationships (see Kluge 2000) was a major step towards a generalization of the seemingly individual experiences with water supply limitations: It allowed linking them back to processes of social and spatial differentiation. The analysis of habitat biographies during the present study, for instance, produced first indications that in Mexico City, a limited water supply was not only a matter of location and/or social status but also linked to certain biographical periods, more specifically: the process of entering into homeownership. To what extent people face a (temporal or more persistent) deterioration of infrastructural conditions as an effect of buying a dwelling or plot and thus ending their status as tenants is another question that could be studied more in detail in future investigations involving the instrument of habitat biographies.

Arguably, the exercise can be taken further in future applications of the method through the introduction of a more detailed periodization during the comparative analysis of these narratives, thus improving the identification of what Fritz Schütze

7 Represented by Roy Bhaskar, (early) Anthony Giddens and Pierre Bourdieu, amongst others.

called the processual character of biographies (see Schütze 1983). Strengthening the historical analysis through a stronger periodization of habitat biographies would allow deepening the abstraction from the individual case during the analysis and facilitate a comparison across different times and places (see for instance Bentley 1996). The exact character of such periods could either be based upon theoretical considerations[8] and/or derived inductively from the empirical material itself. Organized along a timeline, such a periodization allows for a documentation and classification of both collectively shared biographical events (e.g. forming the first independent household after leaving the parents' house) and those of relevance on the urban scale (e.g. the privatization of the municipal water utility or a new urban plan) and beyond (e.g. national bankruptcy or a free trade agreement). Such paralleling of subjective and at the same time social experiences documented in the dwelling biographies with collective moments of crisis (such as done here with the 1985 earthquake) and ongoing societal processes of transformation serves to overcome a solely actor-based perspective by shedding light on the collectively shared nature and social situatedness of everyday experiences and biographies. In this sense, the different habitat biography types based on people's experiences with water supply limitations presented in chapter 6 are just a first exemplary application of this empirically grounded tool. All in all, the instrument of habitat biography seems promising for future empirical research interested in the field of collective experiences in a sociospatial context, that is, the historicity of spatial experiences and practices.

Active Remembrance

But habitat biographies are more than an instrument which allows for a documentation and profound analysis of the sociospatial situatedness of collective experiences. Apart from its ability to set a seemingly subjective experience into a collective context through the interpretation process, the invitation to share a narrative also has a potential impact directly on the studied field, more precisely for the interviewed person her- or himself. In *Understanding,* Bourdieu argued that (narrative) interviews, as an extraordinary form of communication and social interaction, provide a particular occasion for interviewees to unearth experiences and to voice problems, needs and wishes. Therefore, an interview situation characterized by mutual trust and respect may provide

> "the preconditions for the formation of an extraordinary discourse (...) that existed all along but would never have been delivered otherwise, and was only waiting for favorable circumstances to exert itself." (Bourdieu 2010b: 400)

Enabling the development of narrative, the interview situation itself can hence be understood as a process of social construction – or in the words of John Law and

8 While Immanuel Wallerstein's world-systems analysis is a prominent example (see Wallerstein 2004), more recent approaches developed in the realm of regulation theory also seem promising (see e.g. Jessop/Sum 2006).

John Urry, as an enactment of the social world: By studying the social world, social sciences both make and remake it (see Law/Urry 2004). The research process itself is a social interaction rather than a documentation of social realities. Upon this backdrop, applying the habitat biography method as part of a narrative interview may well be able to induce what can be understood as a process of active remembrance in the sense of Marcel Proust and Walter Benjamin: By provoking a recollection and reappropriation of personal experiences.

> "According to Proust, it is a matter of chance whether an individual forms an image of himself, whether he can take hold of his experience. But there is nothing inevitable about the dependence on chance in this matter. A person's inner concerns are not by nature of an inescapably private character. They attain this character only after the likelihood decreases that one's external concerns will be assimilated to one's experience. [...Proust] coined the phrase *mémoire involontaire*. This concept bears the traces of the situation that engendered it; it is part of the inventory of the individual who is isolated in various ways. Where there is experience *[Erfahrung]* in the strict sense of the word, certain contents of the individual past combine in the memory *[Gedächtnis]* with material from the collective past. Rituals, with their ceremonies and their festivals (...) kept producing the amalgamation of these two elements of memory over and over again. They triggered recollection at certain times and remained available to memory throughout people's lives." (Benjamin 2003: 315–316)

Similar to such rituals, the interview situation provides an occasion during which such a recollection of memories is possible. In effect, the earnings of such an act of active remembrance triggered by the interview situation surfaced from time to time during the field work for the present study. By recalling memorable experiences[9] and by producing their habitat biography when asked for a narrative on experiences with water supply in past dwellings, interviewees did more than just provide information of interest for the researcher: They undertook an act of active remembrance. At least one interviewee took advantage of the interview situation to comment about this directly. After having finished the 'formal' part of the interview, voice recorder switched off, one woman, Gabriela, lead the interviewer to the door telling her how she had enjoyed recalling her childhood memories when she played with her siblings at a river in rural Oaxaca, and water was abundant and potable. If the interview situation is understood as an instance where the social is enacted, the point of this exercise cannot be to verify whether such memories are 'consistent' or 'valid' but to uncover their social meaning. What is more, where such memories are set in relation by the interviewee – potentially also at a later moment in time – to his or her currently experienced water supply conditions, they have the potential to raise consciousness for a currently deprived situation by unveiling it under the layer of apparent normality it might be covered with as part of daily routines. Remembering might thus open up an opportunity to reflect upon what could be different and how, by undermining the naturalization of the established order and "exposing the arbitrariness of the taken for granted" (Bourdieu 1977: 169). Such a perspective for change could also be opened up by a sort of empowerment over the sharing of problems and experiences of infrastructural discrimination, which people might

9 The interview guideline employed in the present study included one question in particular which aimed at a personal, memorable experience with respect to water.

aspire to overcome collectively once they recognize their non-individualized nature – something that appears to have happened during one of the focus group discussions, and was also strongly present in the imaginario of a preferential treatment inscribed in the water supply logic. In this sense, research based on habitat biographies (and other methods aimed at activating memories) can serve as an instrument dedicated to an unveiling of the collectiveness of historical experience by taking subjective perspectives seriously while also introducing a sociospatial perspective. In this manner, collective experiences and imaginarios can turn into relevant horizons for change, giving sense and direction to future actions (see García Canclini 2013: 39). Though, of course, not necessarily of a progressive nature – as the imaginario linking potential urban collapse to population growth has shown – such an approach could nevertheless be able to trigger a consciousness for shared experiences and reactivate collective memories, thus opening up possible new paths of thought on urban futures.

Conclusions

To summarize, the value of the instrument of habitat biographies lays in what it offers to other fields interested in the dialectics between society and space: To establish a process of active remembrance, and to grasp (everyday) experiences in their collective yet simultaneously subjective nature. This is of avail for future research in the realm of everyday practices from a sociospatial perspective. To return once again to the example of a notion of non-potability of tap water apparently hegemonic amongst residents of Mexico City, which the present study was able to identify through an explorative approach: A mistrust in tap water quality was indeed reported previously in the literature – rather than discovering this mistrust anew, what the present study was able to do was link it back to past sociospatial experiences, notably the 1985 earthquake. This finding was directly induced from the interview material during the process of coding and the subsequent analysis. Future research in this direction could benefit from the instrument of habitat biographies by concentrating exclusively on questions of potability and water drinking over the entire life course. This will provide material which allows identifying very specific habitat biography types as regards the conditions and manners of past water drinking in particular, which could then be cross-tested with people's current practices. Such research would provide deeper insights as to when, where and under which specific circumstances bottled water became hegemonic over the practice of tap water consumption and boiling. Hereby, habitat biographies could help to clarify when and how the notion of non-potability started to take up a life of its own.

By providing a historical contextualization from a sociospatial perspective, including questions tracing an interviewee's habitat biography into interview guidelines also seems promising for future research in other fields such as geography, sociology, ethnology and urban studies. Precisely for their potential role in pre-structuring motivations and perceptions of the involved actors, there have been, for instance, calls for a more thorough inclusion of biographical factors into research on processes of social transformation and their spatial implications (see Naumann 2009: 241). Focusing these biographies on the habitat might be indicated in cases such as the present study – nevertheless, a similar relational-spatial approach to biographies could be adapted to cover not only dwellings but also workplaces and places of leisure, for instance for use in research on mobility, multilocality and migration (see for instance Rolshoven 2008). A comparison of habitat biographies within or between certain professional fields is another possible application. The academic field (in Western European societies) with its nomadic mid-level faculty staff and conference-hopping upper ranks seems highly promising for such a venture, as do seasonal workers, leased laborers and the unemployed. With respect to Mexico City, tackling the spatiality of everyday life through people's dwelling biographies was able to once more call attention to the interconnectedness between alleged rural hinterlands and the metropolis as well as residential and everyday mobility patterns within the city and the often long-running processes of up- and downgrading of urban living conditions (see for instance Azuela de la Cueva 1989, Ward 1998, Duhau/Giglia 2008, Salazar Cruz 2010). This also points out that by adopting a historical perspective, the method might prove valuable when studying processes of urban in- and exclusion and gentrification – be it in Mexico City, Leipzig, or Beijing – through a close observation of residents' biographies beyond the area subject to the most visible upgrading processes. Rather than painting displacement of low-income residents and tenants as a matter of individual or 'life-style' decisions, an analysis of habitat biographies involving a clear periodization would allow to link them back to development processes on the urban scale, thus highlighting the collective character of displacement (see Bernt/Holm 2009). Finally, I want to argue that whenever people's spatial practices are concerned, adopting a habitat biography perspective could be used to understand different spheres or fields in their interrelatedness rather than slicing them apart. An adapted version of the method could be employed to not only grasp people's experiences of the place of residence or the work place, the school and so on, but also to analyze these conjointly and from a perspective that takes the situatedness of these experiences in time and space into account (see Thrift 1996). The applicability of the method in a range of different research contexts could well contribute to a more transdisciplinary approach to the urban as well as support the project of a more global urban studies that takes to "thinking cities from elsewhere" (Robinson 2015) by starting from multiple yet subjective experiences and weaving them together as a collective history of (urban) space.

10. CONCLUDING REMARKS

This book is about water, and the manner in which people's water use is linked to their habitat, or home. In more abstract terms, it tackles the spatial relevance of social practices on the micro scale. Its conceptual framework infuses the Bourdieuian praxeological approach with a relational understanding of space, and I hope that it has opened up a different perspective on everyday practices by grasping them in their past and present spatiality. I believe that the present work is of relevance well beyond the Mexico City context in providing a theory-derived sociospatial approach to everyday practices, and developing new empirical methods. When it comes to domestic practices of water use, putting an emphasis on questions of habitat and widening the research horizon to include material, conceived and imagined spatial practices provided a set of new insights. I will now shortly revisit the reflection on conceptual approach and empirical findings, as discussed in chapters 8 and 9.

The present work highlighted the relevance of past and present spatialities of everyday practices such as water use. Conceptually, it sought to explore the relation between everyday practices and urban space through a habitus-habitat-practice model based on the Bourdieuian praxeological approach combined with a relational understanding of space. Other than expected, the influence of past experiences with supply limitations on current practices of water use was found to be minor in comparison with the predisposition provided by current conditions of water supply. Former practices seemed to be given up under changing circumstances. The habitat hence appears to frame people's everyday practices primarily in its actual condition, at least when it comes to domestic water use in Mexico City. This runs contrary to the initial, theory-derived assumption that habitats might in some way predispose the generation of habitus. The results show that habitats have a much more immediate potential of regulating everyday practices rather than exerting an influence through past dwelling experiences (though these are by no means irrelevant, as are other potential influences). In spite of this rejection of an assumed reciprocity between habitat and habitus, the theoretical approach was of great avail particularly as it inspired the development of a method to grasp past spatialities from a subject-centered perspective. These habitat biographies should be of use for future studies in several fields, as they provide an instrument to understand the urban condition as an ongoing process in its historicity while unveiling the collectiveness of historical experience by taking subjective perspectives seriously (see 9.2). It is also clear that the reproduction and potential transformation of habitus need to be studied further, specifically when it comes to the role of those spatial processes which parallel the incorporation of social dispositions.

As for the employed research design, it can be argued that, along with habitat biographies, two of the here developed empirical methods in particular contribute to a further 'down-scaling' of research on domestic water use as proposed by Dena Fam (2015) and colleagues. First, the cascade of water use (see 5.3) is a tool able to grasp the fine-grained distinctions between body-related and technical use of water from different sources and of different (perceived) qualities in the home. Second, deliberately conducting the interviews directly in the interviewee's home allowed for a participatory observation of both social practices and the dwelling's materiality, thus enabling the researcher to grasp the spatiality of the studied everyday practices more comprehensively. All in all, by having developed a sociospatial, subject-centered approach to everyday practices, the present work hopes to contribute empirical instruments suitable for an application in various contexts, thus allowing to set cities in conversation over specific urban experiences and phenomena (see Robinson 2011), not just in the realm of water.

With respect to the empirical results, questions of urban space and everyday practices of water use in Mexico City were found to be linked in a number of ways. The present work traced the sociospatial nature of practices of domestic water use, and was able to reveal how supply limitations influence daily routines. These findings are of particular relevance amidst a logic of supply rife with sociospatial disparities. The relatively poor borough of Iztapalapa continues to be one of the worst supplied parts of the Federal District in terms of steadiness, water pressure and water quality as perceived by its residents. Yet just as in the centrally located and socially more heterogeneous borough of Cuauhtémoc, there are also marked differences within the borough and between its neighborhoods; it would be misleading to assume a spatial homogeneity on the borough level. The overall picture seems clear: not only are water supply conditions in Mexico City generally imagined as an expression of social inequality, as the prominent urban imaginario of preferential treatment indicate – those affected by supply limitations also take to domestic practices which appear more than likely to further deepen this inequality. Poor households in Iztapalapa consuming elevated levels of bottled water for multiple purposes due to low tap water quality, and women being on stand-by at night in areas subject to water rationing, waiting for water supply to resume, are cases in point. In this context, the everyday practices of water use of the Federal District's residents from the middle and popular class – who stand for an ordinary everyday experience in this metropolis[1] – are apparently framed much stronger by current living conditions (including water supply) than by people's past experiences with supply, as chapters 5 to 8 have shown.

In addition, many of the here discussed everyday practices involving water in the home shed a light on the silent encroachment of neoliberal principles in the Mexican water sector, not just when it comes to questions of ownership of urban infrastructure. The propagation of a logic of (partial) self-supply observable in Mexico City's homes parallels the ongoing infrastructural unbundling through the privatization of tasks of the Federal District's public water supplier[2]. While it remains to be clarified elsewhere to what extent these processes are directly related,

1 Rather than, say, the pool-owning elites in their lavish villas or condominiums, or the most marginalized urban poor living in newer informal settlements in the far periphery of Mexico City without access to piped water whatsoever.

2 While the Federal District's water utility remains in public ownership, tasks such as billing, metering, and the maintenance of the secondary supply network were privatized through

the present work called attention to the proliferation of individual responses to water supply limitations in Mexico City's households. Keeping water in domestic storage tanks, waiting for water provision to resume, reusing domestic grey water, and the ubiquitous consumption of bottled water have all become essential everyday practices in Mexico City amidst unequal and often unreliable patterns of water supply. And women in poorer parts of the metropolis are the ones bearing the brunt of such exclusionary configurations of urban space. Along with a commodification of water especially in its bottled form, this reflects a trend towards an individualization wherein everyone becomes a consumer responsible for his or her own wellbeing. It goes without saying that such a notion lies at the heart of neoliberal ideology, "rooted in an idealized conception of competitive individualism and a deep antipathy to sources of social solidarity" (Brenner *et al.* 2012: 30). Under these conditions, securing water availability for the own household turns into a kind of self-regulation, as several of the domestic practices of water use identified here illustrate. Amidst unsteady and at times unreliable water supply conditions, domestic storage tanks symbolize an endeavor to maintain the autonomy of domestic rhythms of water use, and there may be competition over sharing stored water amongst neighbors. At the same time, a logic of self-supply as a reaction to limitations in water provision has led to a sophisticated system of water use cascaded along different qualities for specific purposes, from those directly related to the human body to the most technical, in Mexico City's households. The commodification of water is perhaps most visible in the realm of drinking water, where bottled water has now become the most common if not exclusive option for Mexico City's residents. Selling plastic bottles containing a product declared as 100% natural as the only healthy option for human consumption can be read as a remarkably contradictory strategy. While dams, pre-treatment facilities and urban water networks tamed and urbanized the flow of water (in terms of amount, timing, pressure and quality) in a modernizing stance (see Kaika 2006), bottling plants seem to be the new instrument through which water's (alleged) contents, composition and its usability for predefined purposes are commercialized. When studying water use from an everyday perspective, as in the present work, this indirect process of privatization – by shifting some of the responsibilities of the water utility itself to the micro scale, into the domestic sphere, rather than selling off the pipes and ducts of the public supply network – lies in plain sight. Households take over tasks apparently abandoned by the water utility (which sure enough provided a differential service to begin with). Water supply limitations in terms of steadiness of supply and water pressure are an everyday reality in many neighborhoods of the Federal District, countered on the micro scale through the installation of domestic cisterns, electric pumps and roof-top tanks. At the same time, (perceived) limitations in tap water quality, along with other factors, now increasingly motivate the purchase of bottled water not solely for purposes of drinking. Using water in the home turns into a demanding task as water supply disrupts and shapes daily routines, resulting from a supply logic wherein social

temporary concessions in 1994, which were renewed and expanded in 2004 and 2014 respectively (see CEDS 2010: 125 f., Romero Lankao 2011).

relations are embedded (see Budds 2009: 420), and which tends to enact unequal power relations in a spatial manner. It is through this very supply logic that water infrastructure holds the power to limit people's command over even the personal space of the home. Such heteronomy in the everyday practices of Mexico City's residents is a bold expression of the demanding character of water, and implicitly, water infrastructures. Pointing out the regulatory, power-infused character of infrastructure, this is precisely how "the medium of infrastructure space makes certain things possible and other things impossible" (see Easterling 2014: 14).

Nevertheless, the act of demanding water, particularly better water supply conditions, is often highly individualized in Mexico City today. The city's residents certainly aspire to a certain quality and quantity of water, but social contestation over water appears limited in scale and scope and often heavily fragmented in terms of time and space. In Mexico City, there are a host of NGOs and academics working on the problem of water, and access to water as a human right, for instance. Yet it would seem to me, given the scale of the city's water problems, as if only a relatively small share of the population directly affected by exclusionary water policies and limitations in supply actively organizes collectively around these issues. Even amidst a discriminatory and sociospatially differentiated logic of supply, people now seem to adapt rather than protest. This could be read as a sign of general resignation over Mexico's hegemonic political and economic situation, indicating that people may have little hope for political change. Currently, political contestation over water seems largely limited to individual acts of resistance[3] which are often of a divisive nature and thus inherently conservative as regards the *status quo* (see also Swyngedouw 2004: 150 f.). What little collective protest over an unequal access to water there is seems to be mostly spontaneous and short-lived, even in the often systematically under-supplied parts of the metropolis. It typically takes the form of road blocks near neighborhoods affected by extraordinary supply disruptions, which end once local water supply is restored. Apparently, these one-dimensional protests can easily be coopted and resolved in a fragmented and divisive way by providing water to protesters.

Fortunately, and perhaps more importantly from a geographical perspective, a different picture emerges when we abandon our myopic focus on supply disruptions at the domestic tap. Today's emancipatory struggles over water in and around Mexico City mainly involve questions of control over water and water services. The most prominent example by far is the resistance by the indigenous Mazahua women's movement against a continuous enlargement of the Cutzamala system, aiming to exploit additional water sources for Mexico City (see Legorreta 2006: 91 ff., Gómez Fuentes 2009, Campos Cabral/Ávila García 2013). Similar struggles over water extraction for the Cutzamala system are ongoing in the Temascaltepec area (see Fernández 09.08.2012) and elsewhere. The Federal District itself has also seen such protests, whether in peripheral parts of the borough of Cuajimalpa or in

3 Such as the installation of suction pumps to increase pressure at the domestic tap, conflicts between those queuing for water, or the hijacking of public tanker trucks during water supply suspensions.

the Peñon de los Baños neighborhood near the airport[4]. And these movements stand in the tradition of earlier water struggles in the Mexico City area[5] (see Castro 2004, 2006). What distinguishes them from, say, short-lived actions demanding water for one neighborhood, is, I would argue, their socio-territorial character. They oppose the expropriation or destruction of collectively produced spaces on different scales, which symbolize and materialize power relations[6]. Protests over water extraction and control over water resources could, in other words, be read as a form of social struggle over processes of de- and reterritorialization (see Zibechi 2008, Porto Gonçalves 2009, Raffestin 2010), similar to contestations over the privatization of ejidal land (see Salazar Cruz 2014a, Schwarz/Streule 2016). Yet one should take care not to confuse them with conflicts between 'the rural' and 'the urban' (both of which have become essentially undistinguishable).This socio-territorial character of emancipatory struggles over the ownership and organization of water suppliers and over water as a resource needs to be considered more in detail in future research. That includes the question whether and how other contestations regarding the control over urban infrastructures and water services – such as civil disobedience in the Federal District against the partial privatization in 1994 through the non-payment of bills and sabotage of meters (see Castro 2004), and public protest over a proposed federal water law in 2014/2015, criticized for pursuing a further neoliberalization of the Mexican water sector[7] – fit in this perspective.

Other open questions for further research were indicated in chapter 8.5, in particular regarding the generation and stabilization of the disputed notion of non-potability of Mexico City's tap water, which displays an undeniable degree of "genesis amnesia" (Bourdieu 1977: 79), and the sharing of and competing over collectively stored water.

4 Recent examples range from protests in San Lorenzo Acopilco – a village in the borough of Cuajimalpa, where the local community wants to maintain its right to use water from own wells on communal land instead of being forcefully connected to the Cutzamala system (see Salgado 08.06.2013) – to repeated conflicts over water extraction from wells in the neighborhoods of Peñon de los Baños (see Cruz Flores 21.02.2009, Jiménez/Arteaga 19.09.2013) and San Bartolo Ameyalco (see Desinformémonos 02.03.2014, González Alvarado 24.05.2014).

5 In the late 1980s, for instance, water protests erupted in peripheral municipalities in the Estado de México such as Ecatepec and Chimalhuacán when local, community-run water suppliers were threatened to be put under federal state-control or privatized (see Castro 2004: 337 ff.).

6 Though it should be stressed that such socio-territorial movements are not *per se* progressive, and may also be of an exclusionary nature (see Harvey 1996: 209).

7 The original proposal by the federal government for the *Ley General de Agua* allowed a further opening of the water sector to private capital, a penalization of unauthorized water-related research in the academic and non-academic realm, and was criticized for violating the constitutional human right to water, amongst other points. Since 2012, the law-making process was accompanied by an organized debate involving various indigenous and political organizations, critical academics and NGOs across Mexico, resulting in an independent proposal for a new water law. This *Propuesta Ciudadana para una nueva Ley General de Agua* (see http://aguaparatodos.org.mx/la-iniciativa-ciudadana-de-ley-general-de-aguas for a full text), was largely ignored during the official legislative initiative in 2015, triggering protest marches and an online-petition receiving support from several thousand people in a few days, after which the government announced some changes to the law (see Campuzano 05.03.2015, Olivares Alonso 06.03.2015, Becerril 08.03.2015, González Alvarado 08.03.2015).

With respect to water supply in Mexico City, the present work called attention to the effects of supply limitations on the household level, while it remains a task for experts in the field to develop progressive urban water policies able to induce a change on the supply side, particularly regarding the technical implementation. Yet if water infrastructure is considered an influential parameter of common welfare (see Moss/Hüesker 2010: 11), questions of ownership and resource allocation arise, and it can be argued that these are primarily political rather than technological issues (even though water suppliers may paint a different picture). In order to enable a broad public debate, there is a need to improve public access to information on water supply conditions and tap water quality in the Federal District[8]. In terms of ownership, there are interesting proposals for a collectively organized provisioning of public goods such as water, for instance by Thessaloniki's K136 initiative proposing a workers' collective to run a municipal water utility owned by all residents of the city. These can serve as an alternative to the direction currently pursued by the Federal District's government for its own water utility, absurdly following the beaten path of a (further) unbundling of infrastructures and services, and marketization of the resource – even as the negative effects of such ventures could be observed over the last two decades around the globe, and effectively lead to a re-communalization of municipal water utilities from Buenos Aires to Berlin. From a technical point of view, tackling the issues of maintenance of the existing water supply network, particularly with respect to leakage and quality of tap water seems paramount, though again, both risk to be used as an argument for a further opening towards private capital. With respect to the domestic realm, low-tech solutions for rainwater harvesting may be an easy way of decreasing the dependency on water supply patterns and provide water for some of the more technical purposes in Mexico City's households[9], at least during the rainy season. Such a decentralized rain water harvesting would provide an additional source not a substitute for tap water. Though rain water harvesting may constitute an additional domestic chore, it would differ from other domestic practices aimed at securing supply as it is based on a non-exclusive *usufruct*. Hence it is a non-competitive manner of increasing water availability for one household without affecting others. In the end, however, urban water supply remains, in its entire complexity in a technical as well as social and political sense, a collective issue calling for a collective rather than individual

8 In 2012, it was announced that the water utility SACM would install a daily monitoring system on water supply (in terms of quantity, quality and supply frequencies on the borough level) and publish the results (see Asamblea Legislativa del Distrito Federal 26.04.2012). However, not much of this was achieved by 2015. The current situation is a far cry from these stated intentions, as the author's own Kafkaesque experiences in undertaking the task of obtaining specific data sets on public water supply through the Federal District's so-called 'transparency portal' in 2013 illustrate (see 3.2). Furthermore, planned limitations on public access to water data in the 2015 proposal for a new Mexican water law on the federal level give little reason for hope.

9 Rainwater-fed cisterns for the purpose of flushing toilets were installed in several schools in Iztapalapa as part of a pilot project (see El Universal 12.06.2013). This is all the more relevant as schools in the local area were forced on several prior occasions to close temporarily amidst a lack of water for precisely this purpose (see Molina Ramírez 26.06.2005, González Alvarado 14.03.2012).

solution. We need to pose questions of distribution and access, and have questions of space – from water's sources to its daily use in each and every dwelling – clearly in mind while doing so. Sketching "alternatives to the undesired present" (Marcuse 2012: 37) in terms of water supply is a matter which may well start by exposing the effects of differential and unequal patterns of supply upon everyday life, as this book does, and, as Mexico City's Right to the City charta (see Zárate 2010) demands, a means of building the city we dream about.

REFERENCES

Aboites, Luis (1998): *El agua de la nación. Una história política de México (1888 – 1946)*. México, D.F.: CIESAS.

Abrahams, Jessica/Ingram, Nicola (2013): The Chameleon Habitus: Exploring Local Students' Negotiations of Multiple Fields. *Sociological Research Online* 18, 4.

Adler Lomnitz, Larissa (1993): *Cómo sobreviven los marginados*. México, D.F.: Siglo XXI.

Adler, Ilan (2011): Domestic water demand management: implications for Mexico City. *International Journal of Urban Sustainable Development* 3, 1: 93–105.

Adler, Patricia A./Adler, Peter (2000): Observational techniques. In: Denzin, N. K./Lincoln, Y. S. (Eds.): *Handbook of qualitative research*. Thousand Oaks: Sage, 377–392.

Agnew, John (1994): The territorial trap: The geographical assumptions of international relations theory. *Review of International Political Economy* 1, 1: 53–80.

de Alba, Felipe/Noiseux, Yanick/Nava, Luzma F. (2006): Neoliberalismo y privatización del agua en México. *Mundo Urbano*, 30.

de Alba, Felipe (2016): *Challenging state modernity: Governmental adaptation and informal water politics in Mexico City*. In: Current Sociology, online first (23.09.2016), 1–13.

Alegre, Helena/Baptista, Jaime M./Cabrera Jr., Enrique/Cubillo, Francisco/Duarte, Patrícia/Hirner, Wolfram/Merkel, Wolf/Parena, Renato (Eds.) (2007): *Performance indicators for water supply services*. London: IWA Publishing.

Allen, Adriana (07.11.2014): *'Everyday infrastructural planning' in the urban global south: Urbanisation without or beyond large infrastructure?* Conference paper. UGRG Annual Conference 2014: Critical Geographies of Urban Infrastructure.

American Life Lines Alliance (April 2001): *Seismic Fragility Formulations For Water Systems. Part 2 - Appendices*. Baltimore: American Society of Civil Engineers.

Anand, Nikhil (2011): Pressure: The PoliTechnics of Water Supply in Mumbai. *Cultural Anthropology* 26, 4: 542–564.

Anand, P. B. (2007): *Scarcity, entitlements and the economics of water in developing countries. Sharing water peacefully*. Cheltenham: Edward Elgar.

Animal Político (02.11.2012): En cuanto al suministro de agua… GDF y CONAGUA se hacen bolas. *Animal Político*.

Animal Político (30.03.2014): Federación garantizará abasto de agua, GDF decide suministro: CONAGUA. *Animal Político*.

Aridjis, Homero (1994): *La leyenda de los soles*. México, D.F.: Fondo de Cultura Económica.

Arredondo Brun, Juan Carlos (2007): *Adapting to Impacts of Climate Change on Water Supply in Mexico City*. United Nations Development Programme, Human Development Report Office, Occasional Paper, 2007/42. México, D.F.

Asamblea Legislativa del Distrito Federal (26.04.2012): *Reformas a Ley de Aguas del Distrito Federal modifica suministro a deudores*. México, D.F.

Askew, Louise E./McGuirk, Pauline M. (2004): Watering the suburbs: distinction, conformity and the suburban garden. *Australian Geographer* 35, 1: 17–37.

Atkinson, Will (2011): From sociological fictions to social fictions: Some Bourdieusian reflections on the concepts of 'institutional habitus' and 'family habitus'. *British Journal of Sociology of Education* 32, 3: 331–347.

de Ávila, Mónica (11.03.2013): Problemas de abasto de agua en el DF, para 2014. *Noticieros Televisa*.

Ayala Alonso, Enrique (2010): Privacidad, higiene y comodidad en la casa colonial. In: Ayala Alonso, E. (Ed.): *Habitar la casa. Historia, actualidad y prospectiva*. México, D.F.: Universidad Autónoma Metropolitana, 36–61.

Ayala, Gustavo/O'Rourke, Michael (1989): *Effects of the 1985 Michoacan Earthquake on Water Systems and other Buried Lifelines in Mexico,* 89-0009. Buffalo: National Center for Earthquake Engineering.

Azuela de la Cueva, Antonio (1989): *La ciudad, la propiedad privada y el derecho*. México, D.F.: El Colegio de México.

Bahrdt, Hans P. (1975): Erzählte Lebensgeschichten von Arbeitern. In: Osterland, M. (Ed.): *Arbeitssituation, Lebenslage und Konfliktpotential*. Frankfurt am Main; Köln: Europäische Verlagsanstalt, 9–37.

Bailey Glasco, Sharon (2010): *Constructing Mexico City. Colonial conflicts over culture, space, and authority*. New York: Palgrave Macmillan.

Bakker, Karen (2003): Archipelagos and networks: urbanization and water privatization in the South. *The Geographical Journal* 169, 4: 328–341.

Bakker, Karen (2012): Commons versus commodities. Debating the human right to water. In: Sultana, F./Loftus, A. (Eds.): *The right to water. Politics, governance and social struggles*. Milton Park, Abingdon, Oxon, New York: Earthscan/Routledge, 19–44.

Bapat, Meera/Agarwal, Indu (2003): Our needs, our priorities; women and men from the slums in Mumbai and Pune talk about their needs for water and sanitation. *Environment and Urbanization* 15, 2: 71–86.

Barkin, David (Ed.) (2006): *La gestión del agua urbana en México. Retos, debates y bienestar.* Guadalajara: Universidad de Guadalajara.

Barlow, Maude/Clarke, Tony (2003): *Blue gold. The battle against corporate theft of the world's water.* London: Earthscan.

Barnes, Trevor J./Minca, Claudio (2012): Nazi Spatial Theory: The Dark Geographies of Carl Schmitt and Walter Christaller. *Annals of the Association of American Geographers* 103, 3: 669–687.

Becerril, Andrea (08.03.2015): El dictamen de Ley General de Aguas tiene visión privatizadora y represiva: expertos. *La Jornada*.

Becker, Anne/Burkert, Olga/Doose, Anne/Jachnow, Alexander/Poppitz, Marianna (Eds.) (2008): *Verhandlungssache Mexiko Stadt. Umkämpfte Räume, Stadtaneignungen, imaginarios urbanos*. Berlin: b_books.

Beilin, Ruth/Sysak, Tamara/Hill, Serenity (2012): Farmers and perverse outcomes: The quest for food and energy security, emissions reduction and climate adaptation. *Global Environmental Change* 22, 2: 463–471.

Belina, Bernd/Miggelbrink, Judith (Eds.) (2010): *Hier so, dort anders. Raumbezogene Vergleiche in der Wissenschaft und anderswo*. Münster: Westfälisches Dampfboot.

Benjamin, Walter (1968): Theses on the Philosophy of History. In: Benjamin, W.: *Illuminations. Essays and reflections*. New York: Schocken Books, 253–264.

Benjamin, Walter (2003): On Some Motifs in Baudelaire. In: Eiland, H./Jennings, M. W. (Eds.): *Selected writings, 1938–1940*. Cambridge: Harvard University Press, 313–355.

Bennett, Tony (2011): Culture, choice, necessity: A political critique of Bourdieu's aesthetic. *Poetics* 39, 6: 530–546.

Bennett, Vivienne (1995a): Gender, Class, and Water: Women and the Politics of Water Service in Monterrey, Mexico. *Latin American Perspectives* 22, 2: 76–99.

Bennett, Vivienne (1995b): *Politics of water. Urban protest, gender, and power in Monterrey, Mexico*. Pittsburgh: University of Pittsburgh Press.

Bentley, Jerry H. (1996): Cross-Cultural Interaction and Periodization in World History. *The American Historical Review* 101, 3: 749.

Bergua Amores, José A. (2008): Ideology, Magic and Spectres: Towards a Cultural Analysis of Water Use. *Current Sociology* 56, 5: 779–797.

Berker, Thomas (2013): "In the morning I just need a long, hot shower". *Sustainability: Science, Practice & Policy* 9, 1: 57–63.

Berking, Helmuth (2008): "Städte lassen sich an ihrem Gang erkennen wie Menschen" – Skizzen zur Erforschung der Stadt und der Städte. In: Berking, H./Löw, M. (Eds.): *Die Eigenlogik der Städte. Neue Wege für die Stadtforschung*. Frankfurt am Main: Campus, 15–31.

Berking, Helmuth/Löw, Martina (Eds.) (2008): *Die Eigenlogik der Städte. Neue Wege für die Stadtforschung*. Frankfurt am Main: Campus.

Bernt, Matthias/Holm, Andrej (2009): Is it, or is not? The conceptualisation of gentrification and displacement and its political implications in the case of Berlin-Prenzlauer Berg. *City* 13, 2: 312–324.

Bescherer, Peter/Liebig, Steffen/Schmalz, Stefan (2014): Editorial: Klassentheorien. *PROKLA* 175: 152–162.

Bhaskar, Roy (1978): On the Possibility of Social Scientific Knowledge and the Limits of Naturalism. *Journal for the Theory of Social Behaviour* 8, 1: 1–28.

Bhaskar, Roy (1998): *The possibility of naturalism. A philosophical critique of the contemporary human sciences*. London, New York: Routledge.

Blatter, Joachim/Ingram, Helen M./Lorton Levesque, Suzanne (2001): Expanding Perspectives on Transboundary Water. In: Blatter, J./Ingram, H. M. (Eds.): *Reflections on water. New approaches to transboundary conflicts and cooperation*. Cambridge: MIT Press, 31–53.

Böhme, Hartmut (Ed.) (1988): *Kulturgeschichte des Wassers*. Frankfurt am Main: Suhrkamp.

Bohnsack, Ralf (2003): *Rekonstruktive Sozialforschung. Einführung in qualitative Methoden*. Opladen: Leske + Budrich.

Bohnsack, Ralf (2007): Gruppendiskussion. In: Flick, U./Kardorff, E. von/Steinke, I. (Eds.): *Qualitative Forschung. Ein Handbuch*. Reinbek bei Hamburg: Rowohlt-Taschenbuch-Verlag, 369–383.

Bolaños, Ángel (23.01.2013): Abasto de agua en el DF bajará 10%. *La Jornada*.

Bolaños, Claudia (28.03.2013): Listas delegaciones para atender desabasto de agua. *El Universal*.

Boltvinik, Julio (1997): Aspectos conceptuales y metodológicos para el estudio de la pobreza. In: Schteingart, M./Boltvinik, J./Duhau López, E./Castillejas, M. (Eds.): *Pobreza, condiciones de vida y salud en la ciudad de México*. México, D.F.: El Colegio de México.

Boltvinik, Julio (2010): Principios de medición multidimensional de la pobreza. In: Pichardo Hernández, H./Hurtado Martín, S. (Eds.): *(In)justicia social, identidad e (in)equidad. Retos de la modernidad*. México, D.F.: Universidad Autónoma Metropolitana, 55–93.

Boltvinik, Julio (2012): Evaluación de la Pobreza en México y el Distrito Federal, 1992–2010. Valoración Crítica de las Metodologías de Medición, las Fuentes y las Interpretaciones. In: Ordóñez Barba, G. (Ed.): *La pobreza urbana en México. Nuevos enfoques y retos emergentes para la acción pública*. Tijuana; México, D.F.: El Colegio de la Frontera, 23–90.

Bondi, Liz (2005): Making connections and thinking through emotions: between geography and psychotherapy. *Transactions of the Institute of British Geographers* 30, 4: 433–448.

Bookchin, Murray (1992): *Urbanization without cities. The rise and decline of citizenship*. Montreal, New York: Black Rose Books.

Botton, Sarah/Gouvello, Bernard de (2008): Water and sanitation in the Buenos Aires metropolitan region: Fragmented markets, splintering effects? *Geoforum* 39, 6: 1859–1870.

Bourdieu, Pierre (1977): *Outline of a theory of practice*. Cambridge: Cambridge University Press.

Bourdieu, Pierre (1982): *Die feinen Unterschiede. Kritik der gesellschaftlichen Urteilskraft*. Frankfurt am Main: Suhrkamp.

Bourdieu, Pierre (1985): *Sozialer Raum und "Klassen". Leçon sur la leçon. Zwei Vorlesungen*. Frankfurt am Main: Suhrkamp.

Bourdieu, Pierre (1991): Physischer, sozialer und angeeigneter physischer Raum. In: Wentz, M. (Ed.): *Stadt-Räume*. Frankfurt am Main: Campus, 25–34.

Bourdieu, Pierre (1997): *Der Tote packt den Lebenden*. Hamburg: VSA Verlag.

Bourdieu, Pierre (2005): Habitus. In: Hillier, J./Rooksby, E. (Eds.): *Habitus: A Sense of Place*. Burlington: Ashgate, 43–49.

Bourdieu, Pierre (2006): Sozialer Raum, symbolischer Raum. In: Dünne, J./Doetsch, H. (Eds.): *Raumtheorie. Grundlagentexte aus Philosophie und Kulturwissenschaften*. Frankfurt am Main: Suhrkamp. 354–368.

Bourdieu, Pierre (2010a): Ortseffekte. In: Bourdieu, P./Accardo, A./Balazs, G. B. S./Bourdieu, E./Broccolichi, S./Champagne, P./Christin, R./Faguer, J.-P./Garcia, S./Pialoux, M./Pinto, L./Podalydès, D./Sayad, A./Soulié, C./Wacquant, L. J. D. (Eds.): *Das Elend der Welt*. Konstanz: UVK, 117–123.
Bourdieu, Pierre (2010b): Verstehen. In: Bourdieu, P. *et al.* (Eds.): *Das Elend der Welt*. Konstanz: UVK, 393–410.
Braig, Marianne (2004): Fragmentierte Gesellschaft und Grenzen sozialer Politiken. In: Bernecker/Walther/Braig, M./Hölz, K./Zimmermann, K. (Eds.): *Mexiko heute. Politik, Wirtschaft, Kultur.* Frankfurt am Main: Vervuert, 271–308.
Brandt, Allan M./Rozin, Paul (Eds.) (1997): *Morality and health*. New York: Routledge.
Braudel, Fernand (1992): *Gesellschaften und Zeitstrukturen*. Stuttgart: Klett-Cotta.
Breña Puyol, Agustín F. (2007): La problemática del agua en zonas urbanas. In: Morales Novelo, J. A./Rodríguez Tapia, L. (Eds.): *Economía del agua. Escasez del agua y su demanda doméstica e industrial en áreas urbanas*. México, D.F.: Cámara de Diputados, LX Legislatura; Universidad Autónoma Metropolitana; Miguel Ángel Porrúa, 69–89.
Brenner, Neil/Elden, Stuart (2009): Henri Lefebvre on State, Space, Territory. *International Political Sociology* 3, 4: 353–377.
Brenner, Neil/Peck, Jamie/Theodore, Nikolas (2012): *Afterlives of neoliberalism*. London: Bedford Press.
Brown, D. F. (2006): Social Class and Status. In: Brown, K. (Ed.): *Encyclopedia of Language and Linguistics*. Amsterdam: Elsevier, 440–446.
Bryan, Karen L. (2001): In: Search of a Uniform Standard for Bottled Water. In: LaMoreaux, P. E./Tanner, J. T. (Eds.): *Springs and Bottled Waters of the World. Ancient History, Source, Occurrence, Quality and Use*. Berlin, Heidelberg: Springer.
Budds, Jessica (2009): Contested H2O: Science, policy and politics in water resources management in Chile. *Geoforum* 40, 3: 418–430.
Burdett, Richard/Sudjic, Deyan (2007): *The endless city*. London: Phaidon.
Butler, John F. (2008): The Family Diagram and Genogram: Comparisons and Contrasts. *The American Journal of Family Therapy* 36, 3: 169–180.
Butler, Judith (2005): Photography, War, Outrage. *PMLA* 120, 3: 822–827.
Calhoun, Craig (1993): Habitus, Field and Capital: The Question of Historical Specificity. In: Calhoun, C./LiPuma, E./Postone, M. (Eds.): *Bourdieu: Critical Perspectives*. Chicago: The University of Chicago Press.
Campos Cabral, Valentina/Ávila García, Patricia (2013): Entre ciudades y presas. *Revista de Estudios Sociales,* Agosto 2013: 120–133.
Campuzano, Margarita (05.03.2015): *Aprueban diputados en comisiones iniciativa de Ley General de Aguas que atenta contra el derecho humano al agua*. México, D.F.: Centro Mexicano de Derecho Ambiental.
Castano, Ivan (22.02.2012): Mexico's Water War. *Forbes Magazine*.
Castro, José E. (2004): Urban water and the politics of citizenship: the case of the Mexico City Metropolitan Area during the 1980s and 1990s. *Environment and Planning A* 36, 2: 327–346.
Castro, José E. (2006): *Water, power and citizenship. Social struggle in the Basin of Mexico*. Basingstoke, New York, Oxford: Palgrave Macmillan.
Castro, José E. (2007): Water Governance in the twentieth-first century. *Ambiente & Sociedade* 10, 2: 97–118.
Cerda García, Alejandro (2011): *Imaginando zapatismo. Multiculturalidad y autonomía indígena en Chiapas desde un municipio autónomo*. México, D.F.: Universidad Autónoma Metropolitana.
de Certeau, Michel (1988): *The practice of everyday life*. Berkeley: University of California Press.
Chappells, Heather/Medd, Will/Shove, Elizabeth (2011): Disruption and change: drought and the inconspicuous dynamics of garden lives. *Social & Cultural Geography* 12, 7: 701–715.
Cifuentes, Enrique/Suárez, Leticia/Solano, Maritsa/Santos, René (2002): Diarrheal Diseases in Children from a Water Reclamation Site in Mexico City. *Environmental Health Perspectives* 110, 10: A619–A624.

Cleaver, Frances (1998): Choice, complexity, and change: Gendered livelihoods and the management of water. *Agriculture and Human Values* 15, 4: 293–299.

Cole, Ardra L./Knowles, Gary J. (Eds.) (2001): *Lives in Context. The Art of Life History Research*. Lanham: AltaMira Press.

Collier, John/Collier, Malcolm (1986): *Visual anthropology. Photography as a research method*. Albuquerque: University of New Mexico Press.

Comisión Económica para América Latina y el Caribe (CEPAL) (2011): *Anuario estadístico de América Latina y el Caribe, 2011*. Santiago de Chile: Naciones Unidas.

Comisión Nacional de los Salarios Mínimos (CONASAMI) (April 2015): *Salarios Mínimos Generales por Áreas Geográficas 1992-2015*. México, D.F.

Comisión Nacional del Agua (CONAGUA) (2009a): *Estadísticas del Agua de la Región Hidrológico-Administrativa XIII, Aguas del Valle de México. Edición 2009*. México, D.F.: Comisión Nacional del Agua.

Comisión Nacional del Agua (CONAGUA) (2009b): *Situación del Subsector Agua Potable, Alcantarillado y Saneamiento. Edición 2009*. México, D.F.: Comisión Nacional del Agua.

Comisión Nacional del Agua (CONAGUA) (02.09.2009): *Suspenden extracción de agua de la presa Villa Victoria del Sistema Cutzamala. Comunicado de Prensa No.140-09*. México, D.F.: Comisión Nacional del Agua.

Comisión Nacional del Agua (CONAGUA) (2010): *Compendio del Agua, Región Hidrológico-Administrativa XIII. Lo que se debe saber del Organismo de Cuenca Aguas del Valle de México. Edición 2010*. México, D.F.: SEMARNAT; Comisión Nacional del Agua.

Comisión Nacional del Agua (CONAGUA) (2011): *Estadísticas del Agua en México. Edición 2011*. México, D.F.: SEMARNAT; Comisión Nacional del Agua.

Comisión Nacional del Agua (CONAGUA) (2012): *Programa Hídrico Regional, Visión 2030. Región Hidrológico-Administrativa XIII: Aguas del Valle de México*. México, D.F.: Comisión Nacional del Agua.

Comisión Nacional del Agua (CONAGUA) (16.07.2012): *El agua que se entrega mediante el Sistema Cutzamala es pura, transparente y apta para el consumo humano: José Luis Luege. Comunicado de Prensa No. 154-12*. México, D.F.: Comisión Nacional del Agua.

Comisión Nacional del Agua (CONAGUA) (22.08.2012): *El mantenimiento al Sistema Cutzamala no genera desabasto de agua al Distrito Federal. Comunicado de Prensa No. 237-12*. México, D.F.

Comisión Nacional del Agua (CONAGUA) (2013): *Estadísticas del Agua en México. Edición 2013*. México, D.F.: Comisión Nacional del Agua

Comisión Nacional del Agua (CONAGUA) (30.01.2013): *Mantenimiento de rutina en el Cutzamala del 1 al 3 de Febrero. Comunicado de Prensa No. 035-13*. México, D.F.

Comisión Nacional del Agua (CONAGUA) (26.03.2013): *Mantenimiento mayor al Sistema Cutzamala, del 28 al 30 de Marzo. Comunicado de Prensa No 143-13*. México, D.F.

Comisión Nacional del Agua (CONAGUA) (31.03.2013): *Culmina mantenimiento mayor al Sistema Cutzamala. Comunicado de Prensa No. 150-13*. México, D.F.

Comisión Nacional del Agua (CONAGUA) (22.05.2013): *Almacenamiento en Presas del Sistema Cutzamala, Mayo 2013. Comunicado de Prensa*. Toluca.

Comisión Nacional del Agua (CONAGUA) (21.08.2013): *Almacenamiento en Presas del Sistema Cutzamala, Agosto 2013. Comunicado de Prensa*. Toluca.

Comisión Nacional del Agua (CONAGUA) (2015): Strategic Projects: Drinking Water, Sewerage, Sanitation. http://www.conagua.gob.mx/english07/publications/StrategicProjects.pdf (accessed 28 January 2015).

Conan, Rebecca (13.03.2013): Conagua to tender Cutzamala US$403mn water line 3 in the coming weeks. *BNAmericas*.

Connolly, Priscilla (1999): Mexico City: our common future? *Environment and Urbanization* 11, 1: 53–78.

Connolly, Priscilla (2004): The Mexican National Popular Housing Fund. In: Satterthwaite, D./ Mitlin, D. (Eds.): *Empowering Squatter Citizen. Local Government, Civil Society and Urban Poverty Reduction*. Sterling: Taylor and Francis.

Consejo de Evaluación del Desarrollo Social del Distrito Federal (CEDS) (2010): *Evaluación externa del diseño e implementación de la política de acceso al agua potable del Gobierno del Distrito Federal. Informe final*. México, D.F.: Gobierno del Distrito Federal; Universidad Nacional Autónoma de México.
Consejo de Evaluación del Desarrollo Social del Distrito Federal (CEDS) (2011): *Índice de Desarrollo Social de las Unidades Territoriales del Distrito Federal. Delegación, Colonia y Manzana*. México, D.F.: Gobierno del Distrito Federal.
Consejo Nacional de Población (CONAPO) (2013): *Índice de marginación urbana, 2010*. México, D.F.: Consejo Nacional de Población.
Constantino Toto, Roberto M./Salazar Vargas, Pilar/Barrios Fernández, Laura/Morales Santos, Eduardo (2010): La sostenibilidad de la ciudad de México y la conquista de la cuenca hídrica del altiplano. Los límites de viejas soluciones y la emergencia de nuevos problemas. In: Flores Salgado, J. (Ed.): *Crecimiento y desarrollo económico de México*. México, D.F.: Universidad Autónoma Metropolitana, 231–268.
Contreras, Cintya (02.04.2013): Riesgo para la población, purificadoras de agua tienen problemas sanitarios. *Excelsior*.
Coulomb, René/Schteingart, Martha (2006): *Entre el Estado y el mercado. La vivienda en el México de hoy*. México, D.F.: Miguel Ángel Porrúa.
Coutard, Olivier (2008): Placing splintering urbanism: Introduction. *Geoforum* 39, 6: 1815–1820.
Crampton, Jeremy W./Elden, Stuart/Foucault, Michel (Eds.) (2007): *Space, knowledge and power. Foucault and geography*. Aldershot: Ashgate.
Crampton, Jeremy W./Krygier, John (2006): An Introduction to Critical Cartography. *ACME: An International E-Journal for Critical Geographies* 4, 1: 11–33.
Crow, Ben/Sultana, Farhana (2002): Gender, Class, and Access to Water: Three Cases in a Poor and Crowded Delta. *Society & Natural Resources* 15, 8: 709–724.
Cruz Bárcenas, Arturo (05.04.2015): Tláloc sigue enterrado aquí; se llevaron a Chalchiuhtlicue: Guadalupe Villarreal. *La Jornada*.
Cruz Flores, Alejandro (21.02.2009): Cinco policías lesionados al intentar dispersar protesta en Circuito Interior. *La Jornada*.
Cruz, Alejandro (21.12.2012): Acuerdan el GDF y CONAGUA una agenda de trabajo conjunta. *La Jornada*.
Cymet, David (1992): *From ejido to metropolis, another path. An evaluation on ejido property rights and informal land development in Mexico City*. New York: Peter Lang.
Davidson, Joyce/Bondi, Liz/Smith, Mick (2005): *Emotional geographies*. Aldershot: Ashgate.
Davis, Diane E. (1994): *Urban leviathan. Mexico City in the twentieth century*. Philadelphia: Temple University Press.
Davis, Mike (2006): *Planet of slums*. London: Verso.
Dawson, Marcelle (2010): The cost of belonging: exploring class and citizenship in Soweto's water war. *Citizenship Studies* 14, 4: 381–394.
de la Luz González, María (20.01.2009): Luege: Se avecina gran crisis del agua. *El Universal*.
Delgado Ramos, Gian C. (2015): Water and the political ecology of urban metabolism: the case of Mexico City. *Journal of Political Ecology* 22: 98–114.
DeLongis, Anita/Hemphill, Kenneth J./Lehman, Darrin R. (1992): A Structured Diary Methodology for the Study of Daily Events. In: Bryant, F./Edwards, J./Tindale, R. S./Posavac, E./Heath, L./Henderson, E./Suarez-Balcazar, Y. (Eds.): *Methodological issues in applied social psychology*. New York: Springer, 83–109.
Denzin, Norman K. (1989): *The research act. A theoretical introduction to sociological methods*. Englewood Cliffs: Prentice Hall.
Departamento del Distrito Federal, Comisión de Aguas del Distrito Federal (May 1994): *Agua. Una nueva estrategia para el Distrito Federal*. México, D.F.: Departamento del Distrito Federal.
Departamento del Distrito Federal, Comisión de Aguas del Distrito Federal (1996): *Agua para la Ciudad más grande del Mundo*. México, D.F.: Departamento del Distrito Federal.
Desinformémonos (02.03.2014): Ameyalco detiene el robo de su agua. *Desinformémonos*.

Devine, Fiona/Savage, Mike (2005): The Cultural Turn, Sociology and Class Analysis. In: Devine, F./Savage, M./Scott, J./Crompton, R. (Eds.): *Rethinking class. Culture, identities and lifestyles*. Basingstoke, New York: Palgrave Macmillan, 1–23.

Di Nucci, Josefina (2011): Divisiones territoriales del trabajo en Mar del Plata (Argentina). *Mundo Urbano*, 37.

Dirksmeier, Peter (2007): Mit Bourdieu gegen Bourdieu empirisch denken: Habitusanalyse mittels reflexiver Fotografie. *ACME: An International E-Journal for Critical Geographies* 6: 73–97.

Dirksmeier, Peter (2009): *Urbanität als Habitus. Zur Sozialgeographie städtischen Lebens auf dem Land*. Bielefeld: Transcript.

Dirksmeier, Peter (2013): Zur Methodologie und Performativität qualitativer visueller Methoden. Die Beispiele der Autofotografie und reflexiven Fotografie. In: Rothfuß, E./Dörfler, T. (Eds.): *Raumbezogene qualitative Sozialforschung*. Wiesbaden: Springer, 83–102.

Dovey, Kim (2005): The Silent Complicity of Architecture. In: Hillier, J./Rooksby, E. (Eds.): *Habitus: A Sense of Place*. Burlington: Ashgate, 283–295.

Dugard, Jackie/Mohlakoana, Nthabiseng (2009): More work for women: A rights-based analysis of women's access to basic services in South Africa. *South African Journal on Human Rights* 25: 546–572.

Duhau, Emilio (1997): Políticas de Suelo y Vivienda Popular. Aplicaciones en las colonias estudiadas. In: Schteingart, M./Boltvinik, J./Duhau López, E./Castillejas, M. (Eds.): *Pobreza, condiciones de vida y salud en la ciudad de México*. México, D.F.: El Colegio de México, 93–128.

Duhau, Emilio/Giglia, Angela (2008): *Las reglas del desorden. Habitar la metrópoli*. México, D.F.: Universidad Autónoma Metropolitana, Siglo XXI.

Easterling, Keller (2014): *Extrastatecraft. The power of infrastructure space*. London: Verso.

Eibenschutz Hartman, Roberto (Ed.) (1997): *Bases para la planeacíon del desarrollo urbano en la Ciudad de México*. México, D.F.: Miguel Ángel Porrúa; Universidad Autónoma Metropolitana.

Eichholz, Michael (2012): Regimes and Niches of the Water Supply Governance in La Paz, Bolivia. In: Le Sandner Gall, V./Wehrhahn, R. (Eds.): *Geographies of Inequality in Latin America*. Kiel: Selbstverlag des Geographischen Institus der Universität Kiel, 211–236.

Eichholz, Michael/van Assche, Kristof/Oberkircher, Lisa/Hornidge, Anna-Katharina (2012): Trading capitals? Bourdieu, land and water in rural Uzbekistan. *Journal of Environmental Planning and Management:* 1–25.

El Universal (30.03.2013): Continúa abastecimiento de agua mediante pipas. *El Universal*.

El Universal (02.05.2013): Pipas antiguas abastecen agua de la capital. *El Universal*.

El Universal (12.06.2013): Prevén resolver problema del agua en Iztapalapa. *El Universal*.

Escolero Fuentes, O./Martinez, S./Kralisch, S./Perevochtchikova, M. (2009): *Vulnerabilidad de las fuentes de abastecimiento de agua potable de la Ciudad de México en el contexto de cambio climático*. México, D.F.: Centro Virtual Cambio Climático.

Espino Bucio, Manuel/Cruz López, Héctor (02.04.2013): Espera al DF una crítica temporada de estiaje, advierte el Sistema de Aguas. *La Crónica de Hoy*.

Falkenmark, Malin/Lindh, Gunnar (1976): *Water for a starving world*. Boulder: Westview Press.

Fam, Dena/Lahiri-Dutt, Kuntala/Sofoulis, Zoë (2015): Scaling Down: Researching Household Water Practices. *ACME: An International E-Journal for Critical Geographies* 14, 3: 639–651.

Farfán Mendoza, Guillermo (2010): Comentarios sobre la descentralización de las políticas sociales en el México actual. In: Pichardo Hernández, H./Hurtado Martín, S. (Eds.): *(In)justicia social, identidad e (in)equidad. Retos de la modernidad*. México, D.F.: Universidad Autónoma Metropolitana, 147–161.

Farías, Ignacio (2010): Adieu à Bourdieu? Asimetrías, límites y paradojas en la noción de habitus. *Convergencia. Revista de Ciencias Sociales* 17, 54: 11–34.

Fernández, Emilio (09.08.2012): Temascaltepec se opone a extracción de agua del Cutzamala. *El Universal*.

Flick, Uwe (2007): Triangulation in der qualitativen Forschung. In: Flick, U./Kardorff, E. von/Steinke, I. (Eds.): *Qualitative Forschung. Ein Handbuch*. Reinbek bei Hamburg: Rowohlt-Taschenbuch-Verlag, 309–318.

Flick, Uwe (2012): *Qualitative Sozialforschung. Eine Einführung*. Reinbek bei Hamburg: Rowohlt-Taschenbuch-Verlag.
Flores, Gonzalo/Rosas Octavio (2009): Coca-Cola FEMSA contra México y América Latina. In: Clarke, T. (Ed.): *Embotellados. El turbio negocio del agua embotellada y la lucha por la defensa del agua*. México, D.F.: Centro de Análisis Social, Información y Formación Popular; Itaca.
Frank, Sybille/Schwenk, Jochen/Steets, Silke/Weidenhaus, Gunter (2013): Der aktuelle Perspektivenstreit in der Stadtsoziologie. *Leviathan* 41, 2: 197–223.
Frias, Sonia M. (2008): Measuring Structural Gender Equality in Mexico: A State Level Analysis. *Social Indicators Research* 88, 2: 215–246.
Friedman, Herbert/Rohrbaugh, Michael/Krakauer, Sarah (1988): The Time-Line Genogram: Highlighting Temporal Aspects of Family Relationships. *Family Process* 27, 3: 293–303.
Fröhlich, Gerhard (1994): Kapital, Habitus, Feld, Symbol. Grundbegriffe der Kulturtheorie bei Pierre Bourdieu. In: Mörth, I./Fröhlich, G. (Eds.): *Das symbolische Kapital der Lebensstile. Zur Kultursoziologie der Moderne nach Pierre Bourdieu*. Frankfurt am Main, New York: Campus, 31–54.
Fuchs-Heinritz, Werner/König, Alexandra (2011): *Pierre Bourdieu. Eine Einführung*. Konstanz: UVK.
Fuerte Celis, Maria D. P. (2013): El bando dos y la administración y dotación del servicio de agua potable en las cuatro delegaciones centrales 2000–2005. Tesis para obtener el grado de Doctorado en Geografía. México, D.F.: Universidad Nacional Autónoma de México.
Furlong, Kathryn (2006): Hidden theories, troubled waters: International relations, the 'territorial trap', and the Southern African Development Community's transboundary waters. *Political Geography* 25, 4: 438–458.
Furlong, Kathryn (2010): Small technologies, big change: Rethinking infrastructure through STS and geography. *Progress in Human Geography* 35, 4: 460–482.
Furlong, Kathryn (2014): STS beyond the "modern infrastructure ideal": Extending theory by engaging with infrastructure challenges in the South. *Technology in Society* 38: 139–147.
Gailing, Ludger/Moss, Timothy/Röhring, Andreas (2009): Infrastruktursysteme und Kulturlandschaften – Gemeinschaftsgut- und Gemeinwohlfunktionen. In: Bernhardt, C./Kilper, H./Moss, T. (Eds.): *Im Interesse des Gemeinwohls. Regionale Gemeinschaftsgüter in Geschichte, Politik und Planung*. Frankfurt am Main: Campus, 51–73.
Gale, Fay (2005): The Endurance of Aboriginal Women in Australia. In: Hillier, J./Rooksby, E. (Eds.): *Habitus: A Sense of Place*. Burlington: Ashgate, 356–369.
Gandy, Matthew (2004): Water, Modernity and Emancipatory Urbanism. In: Lees, L. (Ed.): *The emancipatory city? Paradoxes and possibilities*. London, Thousand Oaks: Sage, 178–191.
Gandy, Matthew (2008): Landscapes of disaster: water, modernity, and urban fragmentation in Mumbai. *Environment and Planning A*, 40 (1), 108–130.
García Canclini, Néstor (1997): *Imaginarios urbanos*. Buenos Aires: Editorial Eudeba.
García Canclini, Néstor (2013): Zur Metamorphose der lateinamerikanischen Stadtanthropologie. Ein Gespräch. In: Huffschmid, A./Wildner, K. (Eds.): *Stadtforschung aus Lateinamerika. Neue urbane Szenarien: Öffentlichkeit – Territorialität – Imaginarios*. Bielefeld: Transcript, 33–44.
Geertz, Clifford (1973): Thick Description. Toward an Interpretive Theory of Culture: *The interpretation of cultures. Selected essays*. New York: Basic Books, 3–30.
Gehrig, Thomas (2015): Commons – zwischen Marktliberalismus und Utopie. *Widersprüche* 35, 137: 9–24.
General Electric: Pentair Residential Filtration, L. L. (2014): Pro Elite: Water problems. http://www.proelitesystems.com/waterproblems/ (accessed 27 November 2014).
Gerdes, Claudia (1994): Fremde im eigenen Land. Kulturelle Distinktion lateinamerikanischer Eliten. In: Mörth, I./Fröhlich, G. (Eds.): *Das symbolische Kapital der Lebensstile. Zur Kultursoziologie der Moderne nach Pierre Bourdieu*. Frankfurt am Main, New York: Campus, 263–270.
Gestring, Norbert (2011): Habitus, Handeln, Stadt – Eine soziologische Kritik der „Eigenlogik der Städte". In: Kemper, J./Vogelpohl, A. (Eds.): *Lokalistische Stadtforschung, kulturalisierte Städte. Zur Kritik einer „Eigenlogik der Städte"*. Münster: Westfälisches Dampfboot, 40–53.

Gibson-Graham, J. K. (2006): *The end of capitalism (as we knew it). A feminist critique of political economy.* Minneapolis: University of Minnesota Press.

Giddens, Anthony (1984): *The constitution of society. Outline of the theory of structuration.* Cambridge: Polity Press.

Giglia, Angela (2013): Habitat und Habitus. Eine anthropologische Lektüre des Wohnens in Mexiko-Stadt. In: Huffschmid, A./Wildner, K. (Eds.): *Stadtforschung aus Lateinamerika. Neue urbane Szenarien: Öffentlichkeit – Territorialität – Imaginarios.* Bielefeld: Transcript, 171–184.

Gil Olmos, José (21.10.2015): El linchamiento en Ajalpan. *Proceso.*

Gilbert, Alan/Varley, Ann (1991): *Landlord and tenant. Housing the poor in urban Mexico.* London, New York: Routledge.

Gläser, Jochen/Laudel, Grit (2009): *Experteninterviews und qualitative Inhaltsanalyse. Als Instrumente rekonstruierender Untersuchungen.* Wiesbaden: VS Verlag für Sozialwissenschaften.

Gleick, Peter H. (Ed.) (1993): *Water in crisis. A guide to the world's fresh water resources.* New York: Oxford University Press.

Gobierno del Distrito Federal (01.11.2007): *Manual Administrativo, Delegación Iztapalapa.* México, D.F.: Gobierno del Distrito Federal.

Gobierno del Distrito Federal (2013): *Código Fiscal del Distrito Federal 2013.* México, D.F.: Gobierno del Distrito Federal.

Gobierno del Distrito Federal, Secretaría de Finanzas (n.d.a): Índice de Desarrollo Social, Delegación Cuauhtémoc. http://www.finanzas.df.gob.mx/IDS/docs/CUAUHTEMOC.pdf (accessed 20 February 2013).

Gobierno del Distrito Federal, Secretaría de Finanzas (n.d.b): Índice de Desarrollo Social, Delegación Iztapalapa. http://www.finanzas.df.gob.mx/IDS/docs/IZTAPALAPA.pdf (accessed 20 February 2013).

Gobierno del Distrito Federal, Secretaría de Finanzas (2010): Índice de Desarrollo: Cómo se construyó. http://www.finanzas.df.gob.mx/IDS/comoseConstruyo.html (accessed 7 October 2015).

Gobierno del Distrito Federal (15.03.2013): Resolución de carácter general mediante la cuál se determinan y se dan a conocer las zonas en las que los contribuyentes de los derechos por el suministro de agua en sistema medido, de uso doméstico o mixto, reciben el servicio por tandeo. *Gaceta Oficial del Distrito Federal* 1564, 8–14.

Gobierno del Distrito Federal/Sistema de Aguas de la Ciudad de México (SACM) (December 2004): *Elaboración e Integración del Diagnóstico y Estragegías para el Plan Hidráulico 2005–2015.* México, D.F.: Sistema de Aguas de la Ciudad de México.

Gobierno del Distrito Federal/Sistema de Aguas de la Ciudad de México (SACM) (October 2012): *Programa de gestión integral de los recursos hídricos. Visión 20 años.* México, D.F.: Sistema de Aguas de la Ciudad de México.

Gómez Fuentes, Anahí C. (2009): Un ejército de mujeres. Un ejército por el agua. *Agricultura, Sociedad y Desarrollo* 6, 3.

González Alvarado, Rocío (14.03.2012): Escasez de agua obliga a suspender clases en la sierra de Santa Catarina. *La Jornada.*

González Alvarado, Rocío (24.05. 2014): Reparto en pipas, negocio de $6.5 millones cada mes. *La Jornada.*

González Alvarado, Rocío (08.03.2015): La nueva norma coartaría la libertad de investigación en materia hídrica. *La Jornada.*

González de la Rocha, Mercedes (1994): *The resources of poverty. Women and survival in a Mexican city.* Oxford, Cambridge: Blackwell.

Goonewardena, Kanishka (2012): Space and revolution in theory and practice. Eight theses. In: Brenner, N./Marcuse, P./Mayer, M. (Eds.): *Cities for people, not for profit. Critical urban theory and the right to the city.* London: Routledge, 86–101.

Gottdiener, Mark (1994): *The social production of urban space.* Austin: University of Texas Press.

Graham, Stephen (2000): Constructing premium network spaces: reflections on infrastructure networks and contemporary urban development. *International Journal of Urban and Regional Research* 24, 1: 183–200.

Graham, Stephen/Marvin, Simon (2001): *Splintering urbanism. Networked infrastructures, technological mobilites and the urban condition*. London: Routledge.

Haferburg, Christoph (2003): Cape Town between apartheid and post-apartheid: the example of the Wetton-Lansdowne-Corridor. In: Haferburg, C./Oßenbrügge, J. (Eds.): *Ambigous Restructurings of Post-Apartheid Cape Town*. Münster: Lit, 65–85.

Haferburg, Christoph (2007): Umbruch oder Persistenz? Hamburg: Universität Hamburg.

Halbwachs, Maurice (1991): *Das kollektive Gedächtnis*. Frankfurt am Main: Fischer.

Halkier, Bente/Katz-Gerro, Tally/Martens, Lydia (2011): Applying practice theory to the study of consumption: Theoretical and methodological considerations. *Journal of Consumer Culture* 11, 1: 3–13.

Hamlin, Christopher (2000): Water. In: Kiple, K. F./Ornelas, K. C. (Eds.): *The Cambridge world history of food*. Cambridge, New York: Cambridge University Press.

Hand, Martin/Shove, Elizabeth/Southerton, Dale (2005): Explaining Showering: a Discussion of the Material, Conventional, and Temporal Dimensions of Practice. *Sociological Research Online* 10, 2.

Hardt, Michael/Negri, Antonio (2011): *Commonwealth*. Cambridge: Harvard University Press.

Harley, John B. (1989): Deconstructing the map. *Cartographica* 26, 2: 1–20.

Harper, Douglas (2000): Reimagining visual methdos. From Galileo to neoromancer. In: Denzin, N. K./ Lincoln, Y. S. (Eds.): *Handbook of qualitative research*. Thousand Oaks: Sage, 717–732.

Harper, Douglas (2002): Talking about pictures: A case for photo elicitation. *Visual Studies* 17, 1: 13–26.

Harvey, David (1985): *The urban experience*. Baltimore: Johns Hopkins University Press.

Harvey, David (1989): *The condition of postmodernity. An enquiry into the origins of cultural change*. Malden: Blackwell.

Harvey, David (1996): *Justice, nature and the geography of difference*. Cambridge: Blackwell.

Harvey, David (2013): *Rebel cities. From the right to the city to the urban revolution*. London: Verso.

Hastings García, Isadora (2007): Habitabilidad. Un análisis cualtitativo de la vivienda popular en la Ciudad de México. Master thesis. México, D.F.: Universidad Nacional Autónoma de México.

Heiser, Sebastian (10.12.2010): Ein Wasser, zwei Preise. *taz*.

Helbrecht, Ilse (2003): Der Wille zur „totalen Gestaltung“: Zur Kulturgeographie der Dinge. In: Gebhardt, H./Reuber, P./Wolkersdorfer, G./Bathelt, H. (Eds.): *Kulturgeographie. Aktuelle Ansätze und Entwicklungen*. Heidelberg: Spektrum Akademischer Verlag, 149–170.

Hennink, Monique M. (2007): *International focus group research. A handbook for the health and social sciences*. Cambridge: Cambridge University Press.

Hernández Andón, Elia R. (2007): La representación pictográfica del agua. *Ciudades* 18, 73: 3–14.

Hiernaux, Daniel (2007): Los imaginarios urbanos: de la teoría y los aterrizajes en los estudios urbanos. *EURE* 33, 99.

Hillier, Jean/Rooksby, Emma (Eds.) (2005): *Habitus: A Sense of Place*. Burlington: Ashgate.

Hollander, Kurt (05.02.2014): Mexico City: water torture on a grand and ludicrous scale. *The Guardian*.

Horbarth Corredor, Jorge E. (2003): Problemas urbanos del Distrito Federal para el nuevo siglo: la vivienda en los grupos populares de la ciudad. *Scripta Nova – Revista Electrónica de Geografía y Ciencias Sociales* VII, 146 (041).

Huerta Mendoza, Leonardo (2004): Métodos para purificar agua. *Revista del Consumidor* 03/2004, 40–43.

Huffschmid, Anne/Vogel, Wolf-Dieter/Heidhues, Nana/Krämer, Michael (Eds.) (2015): *TerrorZones. Gewalt und Gegenwehr in Lateinamerika*. Berlin; Hamburg: Assoziation A.

Huffschmid, Anne/Wildner, Kathrin (2009): Räume sprechen, Diskurse verorten? Überlegungen zu einer transdisziplinären Ethnografie. *FQS (Forum Qualitative Sozialforschung)* 9, 3.

Illouz, Eva (2009): Emotions, Imagination and Consumption: A new research agenda. *Journal of Consumer Culture* 9, 3: 377–413.
Instituto Nacional de Estadística y Geografía (INEGI) (1990): *Censo de Población y Vivienda 1990*. Aguascalientes: Instituto Nacional de Estadística y Geografía.
Instituto Nacional de Estadística y Geografía (INEGI) (2000): *XII Censo de Población y Vivienda 2000*. Aguascalientes: Instituto Nacional de Estadística y Geografía.
Instituto Nacional de Estadística y Geografía (INEGI) (2005): *Estadísticas del Medio Ambiente del Distrito Federal y Zona Metropolitana 2002*. Aguascalientes; México, D.F: Instituto Nacional de Estadística y Geografía; Secretaría del Medio Ambiente; Gobierno del Distrito Federal.
Instituto Nacional de Estadística y Geografía (INEGI) (2010): *Censo de Población y Vivienda 2010*. Aguascalientes: Instituto Nacional de Estadística y Geografía.
Instituto Nacional de Estadística y Geografía (INEGI) (09.04.2013): *Estadística básica sobre Medio Ambiente. Agua potable y saneamiento. Residuos sólidos. Hogares y Medio Ambiente*. Aguascalientes: Instituto Nacional de Estadística y Geografía.
Instituto Nacional de Estadística y Geografía (INEGI) (10.04.2013): *Estadística básica sobre Medio Ambiente. Datos del Distrito Federal*. México, D.F. Aguascalientes: Instituto Nacional de Estadística y Geografía.
Inter-American Development Bank (IADB) (February 2010): *Determinantes de consumo de agua embotellada en México. Informe Ejecutivo*. Washington, D.C.: Inter-American Development Bank.
International Bottled Water Association (IBWA) (2003): *Bottled Water Market Report Findings*. Alexandria: International Bottled Water Association.
International Bottled Water Association (IBWA) (2006): *Bottled Water Market Report Findings*. Alexandria: International Bottled Water Association.
International Bottled Water Association (IBWA) (2007): *Bottled Water Market Report Findings*. Alexandria: International Bottled Water Association.
Jaglin, Sylvy (2008): Differentiating networked services in Cape Town: Echoes of splintering urbanism? *Geoforum* 39, 6: 1897–1906.
Jean-Baptiste, Nathalie (2013): *People centered approach towards food waste management in the urban environment of Mexico*. PhD thesis. Weimar: Bauhaus-Universität.
Jefatura de Gobierno del Distrito Federal (05.04.2013): Resolución de carácter general mediante la cual se condona totalmente el pago de los Derechos por el Suministro de Agua, correspondientes a los Ejercicios Fiscales 2008, 2009, 2010, 2011, 2012 y 2013, Así como los Recargos y Sanciones a los Contribuyentes cuyos Inmuebles se encuentren en las Colonias que se indican. *Gaceta Oficial del Distrito Federal* 1577.
Jenkins, Richard (1992): *Pierre Bourdieu*. London: Routledge.
Jensen, Jesper (2008): Measuring consumption in households: Interpretations and strategies. *Ecological Economics* 68, 1-2: 353–361.
Jessop, Bob/Sum, Ngai-Ling (2006): *Beyond the regulation approach. Putting capitalist economies in their place*. Cheltenham: Edward Elgar.
Jiménez Cisneros, Blanca/Gutiérrez Rivas, Rodrigo/Marañón Pimentel, Boris/González Reynoso, Arsenio (Eds.) (2011): *Evaluación de la política de acceso al agua potable en el Distrito Federal*. México, D.F.: Universidad Nacional Autónoma de México.
Jiménez, Gerardo/Arteaga, Carlos (19.09.2013): Hieren a 13 policías en protesta por agua. *Excelsior*.
Kaika, Maria (2004): Interrogating the geographies of the familiar: domesticating nature and constructing the autonomy of the modern home. *International Journal of Urban and Regional Research* 28, 2: 265–286.
Kaika, Maria (2005): *City of flows. Modernity, nature, and the city*. New York: Routledge.
Kaika, Maria (2006): Dams as Symbols of Modernization: The Urbanization of Nature Between Geographical Imagination and Materiality. *Annals of the Association of American Geographers* 96, 2: 276–301.
Karl, Ute (2007): Metaphern als Spuren von Diskursen in biographischen Texten. *FQS (Forum Qualitative Sozialforschung)* 8, 1.

Kelle, Udo (1995): *Empirisch begründete Theoriebildung. zur Logik und Methodologie interpretativer Sozialforschung*. Weinheim: Deutscher Studien-Verlag.

Kelle, Udo/Kluge, Susann (2010): *Vom Einzelfall zum Typus. Fallvergleich und Fallkontrastierung in der qualitativen Sozialforschung*. Wiesbaden: VS Verlag für Sozialwissenschaften.

Kemasang, A.R.Taunus (2009): Tea – midwife and nurse to capitalism. *Race & Class* 51, 1: 69–83.

Kemper, Jan/Vogelpohl, Anne (2011): Zur Einleitung. In: Kemper, J./Vogelpohl, A. (Eds.): *Lokalistische Stadtforschung, kulturalisierte Städte. Zur Kritik einer „Eigenlogik der Städte"*. Münster: Westfälisches Dampfboot, 7–13.

Kenney, Douglas S./Goemans, Christopher/Klein, Roberta/Lowrey, Jessica/Reidy, Kevin (2008): Residential Water Demand Management: Lessons from Aurora, Colorado. *Journal of the American Water Resources Association* 44, 1: 192–207.

Kimmelman, Michael (17.02.2017): Mexico City, Parched and Sinking, Faces a Water Crisis. *The New York Times*.

Kitzinger, Jenny (1994): The methodology of Focus Groups: the importance of interaction between research participants. *Sociology of Health and Illness* 16, 1: 103–121.

Kitzinger, Jenny/Barbour, Rosaline S. (1999): *Developing focus group research. Politics, theory, and practice*. London, Thousand Oaks: Sage.

Kluge, Susann (2000): Empirically Grounded Construction of Types and Typologies in Qualitative Social Research. *FQS (Forum Qualitative Sozialforschung)* 1, 1.

Kooy, Michelle/Bakker, Karen (2008): Splintered networks: The colonial and contemporary waters of Jakarta. *Geoforum* 39, 6: 1843–1858.

Krais, Beate/Gebauer, Gunter (2002): *Habitus*. Bielefeld: Transcript.

Kreckel, Reinhard (2009): Dimensionen sozialer Ungleichheit heute. In: Solga, H./Powell, J./Berger, P. A. (Eds.): *Soziale Ungleichheit. Klassische Texte zur Sozialstrukturanalyse*. Frankfurt am Main: Campus, 143–154.

Kuckartz, Udo (2010): *Einführung in die computergestützte Analyse qualitativer Daten*. Wiesbaden: VS Verlag für Sozialwissenschaften.

Lahiri-Dutt, Kuntala (2015): Counting (Gendered) Water Use At Home: Feminist Approaches In: Practice. *ACME: An International E-Journal for Critical Geographies* 14, 3: 652–672.

Larochelle, Jeremy G. (2013): A City on the Brink of Apocalypse: Mexico City's Urban Ecology in Works by Vicente Leñero and Homero Aridjis. *Hispania* 96, 4: 640–656.

Lau, Raymond W. (2004): Habitus and the Practical Logic of Practice: An Interpretation. *Sociology* 38, 2: 369–387.

Law, John/Urry, John (2004): Enacting the social. *Economy and Society* 33, 3: 390–410.

Lefebvre, Henri (1974): *Kritik des Alltagslebens. Grundrisse einer Soziologie der Alltäglichkeit*. Frankfurt am Main: Fischer Taschenbuch-Verlag.

Lefebvre, Henri (1990): *Die Revolution der Städte*. Frankfurt am Main: Hain.

Lefebvre, Henri (1991): *The Production of Space*. Malden: Blackwell.

Lefebvre, Henri (2010): *Rhythmanalysis. Space, time and everyday life*. London: Continuum.

Legorreta, Jorge (2006): *El Agua y la Ciudad de México. De Tenochtitlán a la megalópolis del siglo XXI*. México, D.F.: Universidad Autónoma Metropolitana.

Leñero, Vicente (1983): *La gota de agua*. México, D.F.: Plaza y Janes.

Libreros Muñoz, Héctor V. (2004): El Abastecimiento y Consumo de Agua en el Distrito Federal: Una reflexión del vínculo entre la población y el medio ambiente. In: Lozano Ascencio, F. (Ed.): *El amanecer del siglo y la población mexicana*. México, D.F., Cuernavaca: Sociedad Mexicana de Demografía; Universidad Nacional Autónoma de México, 673–698.

Lindner, Rolf (2003): Der Habitus der Stadt – ein kulturgeographischer Versuch. *Petermanns Geographische Mitteilungen* 147, 2: 46–53.

Lindón, Alicia (2007): La ciudad y la vida urbana a través de los imaginarios urbanos. *EURE* 33, 99: 7–16.

López, Flor M./Hernández Lozano, Josefina (2013): La gestión y autogestión del agua como limitantes para la sustentabilidad socioambiental local. El caso de la delegación Magdalena Contreras del Distrito Federal. In: Aguilar, A. G./Escamilla Herrera, I. (Eds.): *La sustentabilidad en la Ciudad de México. El suelo de conservacvión en el Distrito Federal*. México, D.F.: Universidad Nacional Autónoma de México, 321–355.

Lukács, Georg (1971): *History and class consciousness. Studies in Marxist dialects*. Cambridge: MIT Press.

Macías, Verónica (03.02.2013): DF registrará la peor temporada de estiaje desde 2009. *El Economista*.

MacKillop, Fionn/Boudreau, Julie-Anne (2008): Water and power networks and urban fragmentation in Los Angeles: Rethinking assumed mechanisms. *Geoforum* 39, 6: 1833–1842.

Malkin, Elisabeth (16.07.2012): Bottled-Water Habit Keeps Tight Grip On Mexicans. *The New York Times*.

Marcial Pérez, David (21.10.2015): Dos hombres son linchados y quemados en el este de México. *El País*.

Marcuse, Peter (2012): Whose right(s) to what city? In: Brenner, N./Marcuse, P./Mayer, M. (Eds.): *Cities for people, not for profit. Critical urban theory and the right to the city*. London: Routledge, 24–41.

Márquez, Francisca (2007): Imaginarios urbanos en el Gran Santiago: huellas de una metamorfosis. *EURE* 33, 99: 79–88.

Martínez, Edith (13.02.2009): Prometen agua limpia para Iztapalapa. *El Universal*.

Marx, Karl (1978): Thesen über Feuerbach. In: Marx, K./Engels/Friedrich (Eds.): *Werke. Band 3*. Berlin: Dietz Verlag, 5–7.

Massey, Doreen B. (1994): *Space, place, and gender*. Minneapolis: University of Minnesota Press.

Massey, Doreen B. (1999): *Power-geometries and the politics of space-time. Hettner-Lecture 1998*. Heidelberg: University of Heidelberg, Department of Geography.

Massey, Doreen B. (2005): *For space*. Los Angeles: Sage.

Massolo, Alejandra (1992): *Mujeres y ciudades. Participación social, vivienda y vida cotidiana*. México, D.F.: Colegio de México.

Matthes, Joachim (Ed.) (1983): *Biographie in handlungswissenschaftlicher Perspektive*. Nürnberg: Verlag der Nürnberger Forschungsvereinigung.

Mayring, Philipp (2010): *Qualitative Inhaltsanalyse. Grundlagen und Techniken*. Weinheim: Beltz.

Mazari Hiriart, Marisa/Cifuentes, Enrique/Velázquez, Elia/Calva, Juan J. (2000): Microbiological groundwater quality and health indicators in México City. *Urban Ecosystems* 4: 91–103.

McGoldrick, Monica/Gerson, Randy (1985): *Genograms in Family Assessment*. New York: Norton.

Mehta, Lyla (Ed.) (2010): *The Limits to Scarcity. Contesting the Politics of Allocation*. London, Washington, D.C.: Earthscan Publications.

Mejía Montes de Oca, Pablo (2010): El sentido de equidad en las políticas de educación superior. In: Pichardo Hernández, H./Hurtado Martín, S. (Eds.): *(In)justicia social, identidad e (in)equidad. Retos de la modernidad*. México, D.F.: Universidad Autónoma Metropolitana, 225–240.

Melosi, Martin V. (2000): *The sanitary city. Urban infrastructure in America from colonial times to the present*. Baltimore: Johns Hopkins University Press.

Merrett, Stephen (1997): *Introduction to the economics of water resources. An international perspective*. London: UCL Press.

Merrifield, Andy (2000): Henri Lefebvre. A socialist in space. In: Crang, M./Thrift, N. (Eds.): *Thinking space*. London: Routledge, 167–182.

Merrifield, Andy/Swyngedouw, Erik (Eds.) (1997): *The urbanization of injustice*. New York: New York University Press.

Merryweather, Dave (2010): Using Focus Group Research in Exploring the Relationships Between Youth, Risk and Social Position. *Sociological Research Online* 15, 1.

Merton, Robert K./Kendall, Patricia L. (1946): The Focused Interview. *American Journal of Sociology* 51, 6: 541–557.

Minca, Claudio/Rowan, Roy (2015): The question of space in Carl Schmitt. *Progress in Human Geography* 39, 3: 268–289.

Moctezuma Barragán, Pedro (2012): *La Chispa. Orígenes del Movimiento Urbano Popular en el Valle de México*. México, D.F.: Fundación Rosa Luxemburg; Para Leer en Libertad A.C.

Molina Ramírez, Tania (26.06.2005): Iztapalapa paga el pato. *La Jornada*.

Monroy Hermosillo, Oscar/Montero Contreras, Delia (22.01.2014): *México, consumidor número uno de agua embotellada a nivel mundial. Calidad y disminución de la demanda de agua en el Distrito Federal.Comunicado de Prensa*. México, D.F.: Universidad Autónoma Metropolitana.

Monsiváis, Carlos (2005): *No sin nosotros. Los días del terremoto 1985–2005*. México, D.F.: Ediciones Era.

Monsiváis, Carlos (2012): *Los rituales del caos*. México, D.F.: Ediciones Era.

Montalvo, Tania L. (27.03.2014): Beber o no agua de la llave, ¿sabes qué tomas? *Animal Político*.

Montero Contreras, Delia (2009): El sistema de concesiones del agua en México y la participación de los grandes consorcios internacionales. In: Montero Contreras, D./Gómez Reyes, E./Carrillo González, G./Rodríguez Tapia, L. (Eds.): *Innovación tecnológica, cultura y gestión del agua. Nuevos retos del agua en el Valle de México*. México, D.F.: Universidad Autónoma Metropolitana; Miguel Ángel Porrúa, 93–118.

Mora, Karla (21.05.2013): Iztapalapa: Falta de agua, el verdadero 'vía crucis'. *El Universal*.

Morales Novelo, Jorge A./Rodríguez Tapia, Lilia (Eds.) (2007): *Economía del agua. Escasez del agua y su demanda doméstica e industrial en áreas urbanas*. México, D.F.: Cámara de Diputados, LX Legislatura; Universidad Autónoma Metropolitana; Miguel Ángel Porrúa.

Moss, Timothy (2000): Unearthing Water Flows, Uncovering Social Relations: Introducing New Waste Water Technologies in Berlin. *Journal of Urban Technology* 7, 1: 63–84.

Moss, Timothy (2008a): 'Cold spots' of Urban Infrastructure: 'Shrinking' Processes in Eastern Germany and the Modern Infrastructural Ideal. *International Journal of Urban and Regional Research* 32, 2: 436–451.

Moss, Timothy (Ed.) (2008b): *Infrastrukturnetze und Raumentwicklung. Zwischen Universalisierung und Differenzierung*. München: Oekom Verlag.

Moss, Timothy/Hüesker, Frank (2010): *Wasserinfrastrukturen als Gemeinwohlträger zwischen globalem Wandel und regionaler Entwicklung – institutionelle Erwiderungen in Berlin-Brandenburg*. Berlin: Berlin-Brandenburgische Akademie der Wissenschaften.

Moss, Timothy (2014): Spatiality of the Commons. *International Journal of the Commons* 8, 2: 457–471.

Mouzelis, Nicos (2007): Habitus and Reflexivity: Restructuring Bourdieu's Theory of Practice. *Sociological Research Online* 12, 6.

Myers, Greg (1998): Displaying opinions: Topics and disagreement in focus groups. *Language in Society* 27, 01: 85–111.

Mylopoulos, Yannis A./Mentes, Alexandros K./Theodossiou, Ioannis (2004): Modeling Residential Water Demand Using Household Data: A Cubic Approach. *Water International* 29, 1: 105–113.

Narsiah, Sagie (2011): Urban Pulse –The Struggle for Water, Life, and Dignity In South African Cities: The Case of Johannesburg. *Urban Geography* 32, 2: 149–155.

Nascimento, José Rente (1997): *Programa de abastecimiento y manejo de agua en la zona metropolitana del Valle de México*. Washington, D.C.: Banco Interamericano de Desarrollo.

National Research Council (1995): *Mexico City's Water Supply. Improving the Outlook for Sustainability*. Washington, D.C: The National Academies Press.

Naumann, Matthias (2009): *Neue Disparitäten durch Infrastruktur? Der Wandel der Wasserwirtschaft in ländlich-peripheren Räumen*. München: Oekom Verlag.

Nestlé Waters: Home: About us: Our story: Key dates in history. http://www.nestle-waters.com/aboutus/Pages/key-dates-in-history.aspx (accessed 17 January 2013).

Nieswiadomy, Michael L. (1992): Estimating urban residential water demand: Effects of price structure, conservation, and education. *Water Resources Research* 28, 3: 609.

Olivares Alonso, Emir (06.03.2015): Iniciativa aprobada viola el derecho humano al agua, acusan académicos. *La Jornada*.

Orderud, Geir I./Polickova-Dobiasova, Berenika (2010): Agriculture and the Environment. *Journal of Environmental Policy & Planning* 12, 2: 201–221.

Organisation for Economic Co-Operation and Development (OECD) (2015): OECD Income Distribution database: Social Protection and Well-being: Income Distribution and Poverty by country. http://stats.oecd.org (accessed 8 October 2015).
Orlove, Ben/Caton, Steven C. (2010): Water Sustainability: Anthropological Approaches and Prospects. *Annual Review of Anthropology* 39, 1: 401–415.
Oßenbrügge, Jürgen (2003): Introduction: Globalisation, Urban Development and Inner Structures of Cape Town. In: Haferburg, C./Oßenbrügge, J. (Eds.): *Ambigous Restructurings of Post-Apartheid Cape Town,* Münster: Lit, 1–11.
Ostrom, Elinor (1990): *Governing the commons. The evolution of institutions for collective action.* Cambridge: Cambridge University Press.
Oswald Spring, Úrsula (2011a): Aquatic systems and water security in the Metropolitan Valley of Mexico City. *Current Opinion in Environmental Sustainability* 3, 6: 497–505.
Oswald Spring, Úrsula (Ed.) (2011b): *Water resources in Mexico. Scarcity, degradation, stress, conflicts, management, and policy.* Heidelberg, London, New York: Springer.
Painter, Joe (2000): Pierre Bourdieu. In: Crang, M./Thrift, N. (Eds.): *Thinking space.* London: Routledge, 239–259.
Parkin, Frank (2009): Strategien sozialer Schließung und Klassenbildung. In: Solga, H./Powell, J./ Berger, P. A. (Eds.): *Soziale Ungleichheit. Klassische Texte zur Sozialstrukturanalyse.* Frankfurt am Main: Campus, 155–166.
Parnreiter, Christof (2007): *Historische Geographien, verräumlichte Geschichte. Mexico City und das mexikanische Städtenetz von der Industrialisierung bis zur Globalisierung.* Stuttgart: Steiner.
Peña Ramírez, Jaime (Ed.) (2004): *El agua, espejo de los pueblos. Ensayos de ecología política sobre la crisis del agua en México en el umbral del milenio.* México, D.F.: Facultad de Estudios Superiores Acatlán; Plaza y Valdés.
Peña Ramírez, Jaime (2012): *Crisis del agua en Monterrey, Guadalajara, San Luis Potosí, León y la Ciudad de México (1950-2010).* México, D.F.: Universidad Nacional Autónoma de México.
Peña Ramírez, Jaime/López López, Adalberto (2004): Investigando sobre el agua: un análisis multidisciplinario. In: Peña Ramírez, J. (Ed.): *El agua, espejo de los pueblos. Ensayos de ecología política sobre la crisis del agua en México en el umbral del milenio.* México, D.F.: Facultad de Estudios Superiores Acatlán; Plaza y Valdés, 159–166.
Perec, Georges (2014): *Träume von Räumen.* Zürich: Diaphanes.
Perevochtchikova, María (2013): Retos de la información del agua en México para una mejor gestión. *Revista Internacional de Estadística y Geografía* 4, 1: 43–57.
Perló Cohen, Manuel/González Reynoso, Arsenio (2005): *Guerra por el agua en el Valle de México? Estudio sobre las relaciones hidráulicas entre el Distrito Federal y el Estado de México.* México, D.F.: Universidad Nacional Autónoma de México; Fundación Friedrich Ebert.
Pflieger, Géraldine/Matthieussent, Sarah (2008): Water and power in Santiago de Chile: Socio-spatial segregation through network integration. *Geoforum* 39, 6: 1907–1921.
Pichardo Hernández, Hugo/Hurtado Martín, Santiago (Eds.) (2010): *(In)justicia social, identidad e (in)equidad. Retos de la modernidad.* México, D.F.: Universidad Autónoma Metropolitana.
Pickvance, Christopher G. (1986): Comparative urban analysis and assumptions about causality. *International Journal of Urban and Regional Research* 10, 2: 162–184.
Pickvance, Christopher G. (2001): Four varieties of comparative analysis. *Journal of Housing and the Built Environment* 16, 1: 7–28.
Porto Gonçalves, Carlos W. (2009): De Saberes y de Territorios. Diversidad y emancipación a partir de la experiencia latino-americana. *Polis. Revista de la Universidad Bolivariana* 8, 22: 121–136.
Pradilla Cobos, Emilio (1994): Privatización de la infraestructura y los servicios públicos: Sus contradicciones. *Argumentos* 21: 57–79.
Procuraduría Federal del Consumidor (PROFECO) (April 2012): Calidad de productos para desinfectar agua y alimentos. *Revista del Consumidor:* 66–69.
Programa de las Naciones Unidas para el Medio Ambiente (PNUMA) (2003): *GEO Ciudad de México. Una visión teritorial del sistema urbano ambiental,* GeoCiudades. México, D.F.

Raffestin, Claude (2010): *Claude Raffestin – zu einer Geographie der Territorialität*. Stuttgart: Steiner.
Ramírez Cuevas, Jesús (11.09.2005): Cuando los ciudadanos tomaron la ciudad en sus manos. *La Jornada*.
Ranganathan, Malini (2014): Paying for Pipes, Claiming Citizenship: Political Agency and Water Reforms at the Urban Periphery. *International Journal of Urban and Regional Research* 38, 2: 590–608.
Reay, Diane (2005): Beyond Consciousness? The Psychic Landscape of Social Class. *Sociology* 39, 5: 911–928.
Reckwitz, Andreas (2002): Toward a Theory of Social Practices: A Development in Culturalist Theorizing. *European Journal of Social Theory* 5, 2: 243–263.
Rehbein, Boike (2006): *Die Soziologie Pierre Bourdieus*. Konstanz: UVK; UTB.
Reis, Nadine (2014): Coyotes, Concessions and Construction Companies: Illegal Water Markets and Legally Constructed Water Scarcity in Central Mexico. *Water Alternatives* 7, 3: 542–560.
Robinson, Jennifer (2011): Cities in a World of Cities: The Comparative Gesture. *International Journal of Urban and Regional Research* 35, 1: 1–23.
Robinson, Jennifer (2015): Thinking cities through elsewhere: Comparative tactics for a more global urban studies. *Progress in Human Geography*, online first.
Robles, Johana (09.04.2009): Sistema Cutzamala entra a niveles críticos. *El Universal*.
Robles, Johana (14.11.2013): Analizan Conagua y GDF nueva 'fuente' para la ciudad. *El Universal*.
Rodríguez Cerda, Oscar/Melo Carrasco, Myriam/Sánchez Bárcenas, Alma/García Mendoza, Karla/Lázaro Martínez, Dulce (2002): El agua: representaciones y creencias de ahorro y dispendio. *Polis. Investigación y Análisis Sociopolítico y Psicosocial (UNAM)*, 1: 29–44.
Rodriguez, Alfredo/Sugranyes, Ana (2005): *Los con techo. Un desafio para la politica de la vivienda social*. Santiago de Chile: Ediciones Sur.
Rodwan, John G. (2009): *Confronting Challenges. U.S. and International Bottled Water Developments and Statistics for 2008*. Alexandria: International Bottled Water Association.
Rodwan, John G. (2010): *Bottled Water 2009: Challenging Circumstances Persist: Future Growth Anticipated. U.S. and International Developments and Statistics*. Alexandria: International Bottled Water Association.
Rodwan, John G. (2011): *Bottled Water 2010: The Recovery Begins. U.S. and International Developments and Statistics*. Alexandria: International Bottled Water Association.
Rodwan, John G. (2012): *Bottled Water 2011: The Recovery Continues. U.S. and international developments and statistics*. Alexandria: International Bottled Water Association.
Rodwan, John G. (2013): *Bottled Water Industry Gathering Strength. 2012 Statistics Reveal Bottled Water's U.S. and International Growth*. Alexandria: International Bottled Water Association.
Rodwan, John G. (2014): *Bottled Water 2013: Sustaining Vitality. U.S. and International Developments and Statistics*. Alexandria: International Bottled Water Association.
Roebke, Thomas (27.02.2003): Wasser ohne Quelle. *Zeit Online*.
Rolshoven, Johanna (2008): The Temptations of the Provisional. *Ethnologia Europaea – Journal of European Ethnology* 37, 1-2: 17–25.
Romero Lankao, Patricia (2011): Missing the multiple dimensions of water? Neoliberal modernization in Mexico City and Buenos Aires. *Policy and Society* 30, 4: 267–283.
Romero Sánchez, Gabriela (07.04.2012): Los mayores efectos por la falta de agua se verán durante estos tres días. *La Jornada*.
Romero Sánchez, Gabriela/Camacho Servín, Fernando (17.03.2012): Conectarán la sierra de Santa Catarina al Cutzamala. *La Jornada*.
Royacelli, Geovana (12.01.2013): Satisfacen demanda de agua a bajo costo. *El Universal*.
Royacelli, Geovana (29.03.2013): Por corte, 441 pipas distribuyen agua en 13 delegaciones. *El Universal*.
Royacelli, Geovana/Cruz, Noé (28.03.2013): Enfrenta el DF megacorte de agua en 13 delegaciones. *El Universal*.
Royacelli, Geovana/Pantoja, Sara (31.03.2013): Hacen negocio con pipas del GDF. *El Universal*.

Roy, Ananya/AlSayyad, Nezar (Eds.) (2004): *Urban informality. Transnational perspectives from the Middle East, Latin America, and South Asia*. Lanham: Lexington Books.

Rubalcava, Rosa M./Schteingart, Martha (2012): *Ciudades divididas. Desigualdad y segregación social en México*. México, D.F. : El Colegio de México.

Saar, Martin (2008): Klasse/Ungleichheit. Von den Schichten der Einheit zu den Achsen der Differenz. In: Moebius, S./Reckwitz, A. (Eds.): *Poststrukturalistische Sozialwissenschaften*. Frankfurt am Main: Suhrkamp, 194–207.

Salazar Cruz, Clara E. (1999): *Espacio y vida cotidiana en la ciudad de México*. México, D.F.: El Colegio de México.

Salazar Cruz, Clara E. (2010): Movilidades cotidianas en ámbitos espaciales diferenciados de la ciudad de México. In: Garza, G./Schteingart, M. (Eds.): *Desarrollo urbano y regional*. México, D.F.: El Colegio de México.

Salazar Cruz, Clara E. (2014a): “El puño invisible” de la privatización. *Territorios* 16, 30: 69–90.

Salazar Cruz, Clara E. (2014b): Suelo y política de vivienda en el contexto neoliberal mexicano. In: Giorguli Saucedo, S. E./Ugalde, V. (Eds.): *Gobierno, territorio y población. Las políticas públicas en la mira*. México, DF: El Colegio de México.

Salgado, Agustín (08.06.2013): Intervendrá Conagua en conflicto por agua de manantiales en Cuajimalpa. *La Jornada*.

Sánchez Vega, José/Tay Zavala, Jorge/Aguilar Chiu, Artemisa/Ruiz Sánchez, Dora/Malagón, Filiberto/Rodríguez Covarrubias, José/Ordóñez Martínez, Javier/Calderon Romero, Leiticia (2006): Cryptosporidiosis and other intestinal protozoan infections in children less than one year of age in Mexico City. *The American Journal of Tropical Medicine and Hygiene* 75, 6: 1095–1098.

Santos, Milton (1996): *La Naturaleza del Espacio. Tecnica y Tiempo, Razón y Emoción*. Barcelona: Ariel.

Savage, Mike (2000): *Class analysis and social transformation*. Philadelphia: Open University Press.

Savage, Mike/Cunningham, Niall/Devine, Fiona/and others (Eds.) (2015): *Social Class in the 21st Century*. London: Penguin Publishing Group.

Savage, Mike/Devine, Fiona/Cunningham, Niall/Taylor, Mark/Li, Yaojun/Hjellbrekke, Johs/Le Roux, Brigitte/Friedman, Sam/Miles, Andrew (2013): A New Model of Social Class: Findings from the BBC’s Great British Class Survey Experiment. *Sociology* 0,0: 1–32.

Savage, Mike/Warde, Alan/Devine, Fiona (2005): Capitals, assets, and resources: some critical issues. *The British Journal of Sociology* 56, 1: 31–47.

Schatzki, Theodore (1996): *Social practices. A Wittgensteinian approach to human activity and the social*. New York: Cambridge University Press.

Schroer, Markus (2006): *Räume, Orte, Grenzen. Auf dem Weg zu einer Soziologie des Raums*. Frankfurt am Main: Suhrkamp.

Schteingart, Martha (1989): *Los productores del espacio habitable. Estado, empresa y sociedad en la ciudad de México*. México, D.F.: El Colegio de México.

Schteingart, Martha/Boltvinik, Julio/Duhau López, Emilio/Castillejas, Margarita (Eds.) (1997): *Pobreza, condiciones de vida y salud en la ciudad de México*. México, D.F.: El Colegio de México.

Schütze, Fritz (1983): Biographieforschung und narratives Interview. *Neue Praxis* 13, 3: 283–293.

Schwarz, Anke (2009): *Mi barrio más popular. Räumliche Reproduktion von Armut in Mexiko-Stadt*. Diplomarbeit. Hamburg: HafenCity Universität.

Schwarz, Anke (2013): Out All Day or Stay at Home: Mobility Patterns of the Urban Poor in Mexico City. *Trialog*, 110: 4–7.

Schwarz, Anke/Streule, Monika (2016): A Transposition of Territory. Decolonized Perspectives in Current Urban Research. *International Journal of Urban and Regional Research* 40 (5): 1000–1016.

Schwingel, Markus (2005): *Pierre Bourdieu zur Einführung*. Hamburg: Junius.

Sennett, Richard (1997): *Fleisch und Stein. Der Körper und die Stadt in der westlichen Zivilisation*. Frankfurt am Main: Suhrkamp.

Shapiro, E. Gary (1980): Is Seeking Help from a Friend Like Seeking Help from a Stranger? *Social Psychology Quarterly* 43, 2: 259.

Sheard, Sally (1994): Water and health. *Transactions,* 143: 141–163.

Shirley, Mary M. (Ed.) (2002): *Thirsting for Efficiency. The Economics and Politics of Urban Water System Reform*. Oxford: Pergamon Press; The World Bank.

Shiva, Vandana (2002): *Water wars. Privatization, pollution and profit*. London: Pluto Press.

Shove, Elizabeth/Pantzar, Mika (2005): *Consumers, Producers and Practices. Understanding the invention and reinvention of Nordic walking*. Journal of Consumer Culture 5, 1: 43–64.

Shove, Elizabeth/Pantzar, Mika/Watson, Matt (2012): *The dynamics of social practice. Everyday life and how it changes*. Thousand Oaks, London: Sage.

Silva Téllez, Armando (1992): *Imaginarios urbanos, Bogotá y São Paulo. Cultura y comunicación urbana en América Latina*. Bogotá, Colombia: Tercer Mundo Editores.

Silva Téllez, Armando (2003): *Bogotá imaginada*. Bogotá: Convenio Andrés Bello; Taurus.

Silva, Elizabeth B. (2005): Gender, home and family in cultural capital theory. *The British Journal of Sociology* 56, 1: 83–103.

Simon, Joel (1997): *Endangered Mexico. An environment on the edge*. San Francisco: Sierra Club Books.

Singh, Gitanjali M./Micha, Renata/Khatibzadeh, Shahab/Lim, Stephen/Ezzati, Majid/Mozaffarian, Dariush (2015): Estimated Global, Regional, and National Disease Burdens Related to Sugar-Sweetened Beverage Consumption in 2010. *Circulation* 132, 8: 639–666.

Sistema de Aguas de la Ciudad de México (SACM) (2005): *Plan Hidráulico Delegacional 2006–2012. Delegación Iztapalapa*. México, D.F.

Sistema de Aguas de la Ciudad de México (SACM) (2009): *Plan Hidráulico Delegacional 2010–2015. Delegación Cuauhtémoc*. México, D.F.: Sistema de Aguas de la Ciudad de México.

Sistema de Aguas de la Ciudad de México (SACM) (2012): *El Gran Reto Del Agua en la Ciudad de México. Pasado, Presente y Prospectivas de Solución para una de las Ciudades más complejas del Mundo*. México, D.F.: Sistema de Aguas de la Ciudad de México.

Sistema de Aguas de la Ciudad de México (SACM) (October 2012): *Analisis de la Calidad del Agua Octubre del 2012*. México, D.F.: Sistema de Aguas de la Ciudad de México.

Sistema de Aguas de la Ciudad de México (SACM) (2013): *Condiciones del Servicio para el Programa de Tandeo 2013*. México, D.F.: Sistema de Aguas de la Ciudad de México.

Sistema de Aguas de la Ciudad de México (SACM) (16.10.2013): *Comparecencia del Director General del SACMEX Ante la Comisión de Gestión Integral del Agua de la Asamblea Legislativa del Distrito Federal*. México, D.F.: Sistema de Aguas de la Ciudad de México.

Sistema de Aguas de la Ciudad de México (SACM) (2014): *Consumo medido de agua facturado en el año 2010 de usuarios que actualmente registran uso doméstico. Cifras por delegación y colonia*. México, D.F.: Sistema de Aguas de la Ciudad de México.

Sistema de Aguas de la Ciudad de México (SACM) (14.04.2014): *Calidad del Agua Suministrada en delegaciones y colonias de la Ciudad de México. Informe*. México, D.F.: Sistema de Aguas de la Ciudad de México.

Smith, Mick/Davidson, Joyce/Cameron, Laura/Bondi, Liz (Eds.) (2009): *Emotion, place and culture*. Burlington: Ashgate.

Smith, Neil (1984 (2010)): *Uneven development. Nature, capital, and the production of space*. London: Verso.

Soanes, Catherine/Stevenson, Angus (Eds.) (2008): *Concise Oxford English Dictionary*. Oxford: Oxford University Press.

Soja, Edward W. (1989): *Postmodern Geographies. The Reassertion of Space in Critical Social Theory*. London: Verso.

Soja, Edward W. (2010): *Seeking spatial justice*. Minneapolis: University of Minnesota Press.

Sorenson, Susan B./Morssink, Christiaan/Campos, Paola A. (2011): Safe access to safe water in low income countries: Water fetching in current times. *Social Science & Medicine* 72, 9: 1522–1526.

Sosa Rodríguez, Fabiola S. (2010): Exploring the risks of ineffective water supply and sewage disposal: A case study of Mexico City. *Environmental Hazards* 9, 2: 135–146.

Southerton, Dale/Chappells, Heather/van Vliet, Bastiaan J. M. (Eds.) (2004): *Sustainable consumption. The implications of changing infrastructures of provision*. Cheltenham: Elgar.

Steinfort, Lavinia (03.06.2014): Thessaloniki, Greece: Struggling against water privatisation in times of crisis. Transnational Institute. https://www.tni.org/en/article/thessaloniki-greece-struggling-against-water-privatisation-times-crisis (accessed 7 November 2015).

Steinmetz, George (2011): Bourdieu, Historicity, and Historical Sociology. *Cultural Sociology* 5, 1: 45–66.

Steinrücke, Margareta (2006): Habitus und soziale Reproduktion in der Theorie Pierre Bourdieus. In: Hillebrand, M./Krüger, P./Lilge, A./Struve, K. (Eds.): *Willkürliche Grenzen. Das Werk Pierre Bourdieus in interdisziplinärer Anwendung*. Bielefeld: Transcript.

Stoffer, Hellmut (1966): *Die Magie des Wassers. Eine Tiefenpsychologie und Anthropologie des Waschens, Badens und Schwimmens*. Meisenheim: Hain.

Strang, Veronica (2004): *The meaning of water*. Oxford, New York: Berg.

Streule, Monika (2015): *Patrones y trayectorias de una metropolis. Eine transdisziplinäre Analyse aktueller Urbanisierungsprozesse von Mexiko-Stadt*. PhD thesis. Zürich: Eidgenössische Technische Hochschule.

Sultana, Farhana (2011): Suffering for water, suffering from water: Emotional geographies of resource access, control and conflict. *Geoforum* 42, 2: 163–172.

Swyngedouw, Erik (2004): *Social power and the urbanization of water. Flows of power*. Oxford: Oxford University Press.

Swyngedouw, Erik (2006): *Power, Water and Money: Exploring the Nexus*. Human Development Report Office, Occasional Paper 14. New York: United Nations Development Programme.

Swyngedouw, Erik/Kaika, Maria/Castro, José E. (2002): Urban Water: A Political-Ecology Perspective. *Built Environment* 28, 2: 124–137.

Taylor, Vanessa/Trentmann, Frank (2011): Liquid Politics: Water and the Politics of Everyday Life in the Modern City. *Past & Present* 211, 1: 199–241.

Tejera Gaona, Héctor (2003): *"No se olvide de nosotros cuando esté allá arriba". Cultura, ciudadanos y campañas políticas en la ciudad de México*. México, D.F.: Universidad Autónoma Metropolitana.

Thornton, Nicole/Riedy, Chris (2015): A Participatory Mixed Methods Approach to Researching Household Water Use In Gosford, Australia. *ACME: An International E-Journal for Critical Geographies* 14, 3: 673–687.

Thrift, Nigel (1996): *Spatial formations*. London, Thousand Oaks: Sage.

Thrift, Nigel (2004): Intensities of Feeling: Towards a Spatial Politics of Affect. *Geografiska Annaler, Series B: Human Geography* 86, 1: 57–78.

Tortajada, Cecilia (2006): *Who has access to water? Case study of the Mexico City Metroplitan Area*, Human Development Report 2006, Occasional Paper. New York: United Nations Development Programme.

Tuckman, Jo (2012): *Mexico. Democracy interrupted*. New Haven: Yale University Press.

Turner, Stephen P. (1994): *The social theory of practices. Tradition, tacit knowledge, and presuppositions*. Chicago: University of Chicago Press.

Valdespino Gómez, José L./Parrilla Cerillo, María C./Sepúlveda Amor, Jaime/Díaz Ortega, José L./Camacho Amor, María d. L./Luna Jiménez, José L. (1987): Calidad del agua en la ciudad de México en relación a los sismos de septiembre de 1985. *Salud Publica de Mexico* 29, 5: 412–420.

Wallerstein, Immanuel (2004): *World-systems analysis. An introduction*. Durham: Duke University Press.

Warde, Alan (2005): Consumption and Theories of Practice. *Journal of Consumer Culture* 5, 2: 131–153.

Ward, Peter M. (1998): *Mexico City*. Chichester, New York, Weinheim: Wiley.

Waterson, Roxana (2005): Enduring Landscape, Changing Habitus: The Sa'dan Toraja of Sulawesi, Indonesia. In: Hillier, J./Rooksby, E. (Eds.): *Habitus: A Sense of Place*. Burlington: Ashgate, 334–354.

Watts, Jonathan (12.11.2015): Mexico City's water crisis – from source to sewer. *The Guardian*.

Weiss, Gail (2008): *Intertwinings. Interdisciplinary encounters with Merleau-Ponty*. Albany: State University of New York Press.

Wester, Philippus (2009): Capturing the waters: the hydraulic mission in the Lerma–Chapala Basin, Mexico (1876–1976). *Water History* 1, 1: 9–29.

White, Gilbert F./Bradley, David J./White, Anne U. (1972): *Drawers of water: Domestic water use in East Africa*. Chicago: The University of Chicago Press.

White, Robert (2000): The life of class: a case study in a sociological concept. *Journal of Sociology* 36, 2: 223–238.

Wilder, Margaret/Romero Lankao, Patricia (2006): Paradoxes of Decentralization: Water Reform and Social Implications in Mexico. *World Development* 34, 11: 1977–1995.

Wildner, Kathrin (2003): *Zócalo - die Mitte der Stadt Mexiko. Ethnographie eines Platzes*. Berlin: Reimer.

Wilk, Richard (2006): Bottled Water: The pure commodity in the age of branding. *Journal of Consumer Culture* 6, 3: 303–325.

Wills, Wendy/Backett-Milburn, Kathryn/Roberts, Mei-Li/Lawton, Julia (2011): The framing of social class distinctions through family food and eating practices. *The Sociological Review* 59, 4: 725–740.

Wilshusen, Peter R. (2010): The Receiving End of Reform: Everyday Responses to Neoliberalisation in Southeastern Mexico. *Antipode* 42, 3: 767–799.

Wiltse, Jeff (2007): *Contested waters. A social history of swimming pools in America*. Chapel Hill: University of North Carolina Press.

Wisner, Ben/Blaikie, Piers/Cannon, Terry/Davis, Ian (2004): *At risk. Natural hazards, people's vulnerability and disasters*. London: Routledge.

Witzel, Andreas (1985): Das problemzentrierte Interview. In: Jüttemann, G. (Ed.): *Qualitative Forschung in der Psychologie. Grundfragen, Verfahrensweisen, Anwendungsfelder.* Weinheim: Beltz, 227–255.

World Health Organization (WHO) (April 2006): *Country Cooperation Strategy at a glance. Mexico*. Geneva: World Health Organzation.

World Health Organization (WHO) (2011): *Guidelines for drinking-water quality*. Geneva: World Health Organzation.

World Health Organization (WHO)/Food and Agriculture Organization (FAO) (2007): *Waters*. Rome: Food and Agriculture Organization.

World Health Organization (WHO)/Water Engineering Development Centre (2011): *How much water is needed in emergencies. WHO technical notes on drinking-water, sanitation and hygiene in emergencies,* 9. Geneva: World Health Organzation.

Worthington, Andrew C./Hoffman, Mark (2008): An empirical survey of residential water demand modelling. *Journal of Economic Surveys* 22, 5: 842–871.

Zárate, Lorena (2010): Mexico City Charter: The Right to Build the City We Dream About. In: Sugranyes, A./Mathivet, C. (Eds.): *Cities for all. Proposals and Experiences towards the Right to the City*. Santiago de Chile: Habitat International Coalition, 259–266.

Zibechi, Raúl (2008): Espacio, territorios y regiones. *Revista Chuchará' y paso atrá'*, 18: 95–119.

Ziccardi, Alicia (2000): Delegación Iztapalapa. In: Garza, G. (Ed.): *La Ciudad de México en el fin del segundo milenio*. México, D.F.: El Colegio de México; Gobierno del Distrito Federal, 590–595.

Zihan, Zhang (22.07.2012): Untapped resource: Safe drinking tap water still elusive in Beijing despite new standards. *Global Times*.

Zimmerer, Jürgen (2004): Im Dienste des Imperiums. Die Geographen der Berliner Universität zwischen Kolonialwissenschaften und Ostforschung. *Jahrbuch für Universitätsgeschichte* 7: 73–100.

MEGACITIES AND GLOBAL CHANGE / MEGASTÄDTE UND GLOBALER WANDEL

herausgegeben von Frauke Kraas, Martin Coy, Peter Herrle und Volker Kreibich

Franz Steiner Verlag ISSN 2191-7728

1. Susanne Meyer
Informal Modes of Governance in Customer Producer Relations
The Electronic Industry in the Greater Pearl River Delta (China)
2011. 222 S. mit 15 Abb., 45 Tab., kt.
ISBN 978-3-515-09849-6
2. Carsten Butsch
Zugang zu Gesundheitsdienstleistungen
Barrieren und Anreize in Pune, Indien
2011. 324 S. mit 24 Abb., 11 Tab, 3 Ktn., 49 Diagr., kt.
ISBN 978-3-515-09942-4
3. Annemarie Müller
Areas at Risk – Concept and Methods for Urban Flood Risk Assessment
A Case Study of Santiago de Chile
2012. 265 S. mit 75 z.T. farb. Abb., kt.
ISBN 978-3-515-10092-2
4. Tabea Bork-Hüffer
Migrants' Health Seeking Actions in Guangzhou, China
Individual Action, Structure and Agency: Linkages and Change
2012. 292 S. mit 29 Abb., 37 Tab., kt.
ISBN 978-3-515-10177-6
5. Carolin Höhnke
Verkehrsgovernance in Megastädten – Die ÖPNV-Reformen in Santiago de Chile und Bogotá
2012. 252 S. mit 17 Abb., 10 Tab., kt.
ISBN 978-3-515-10251-3
6. Mareike Kroll
Gesundheitliche Disparitäten im urbanen Indien
Auswirkungen des sozioökonomischen Status auf die Gesundheit in Pune
2013. 295 S. mit 53 Abb., 13 Tab. und 45 Fotos, kt.
ISBN 978-3-515-10282-7
7. Kirsten Hackenbroch
The Spatiality of Livelihoods – Negotiations of Access to Public Space in Dhaka, Bangladesh
2013. 396 S. mit 31 z.T. farb. Abb., 10 Tab. und 57 z.T. farb. Fotos, kt.
ISBN 978-3-515-10321-3
8. Anna Lena Bercht
Stresserleben, Emotionen und Coping in Guangzhou, China
Mensch-Umwelt-Transaktionen aus geographischer und psychologischer Perspektive
2013. 445 S. mit 92 Abb., 6 Tab. und 9 Textboxen, kt.
ISBN 978-3-515-10403-6
9. Shahadat Hossain
Contested Water Supply: Claim Making and the Politics of Regulation in Dhaka, Bangladesh
2013. 336 S. mit 25 Abb., 1 Tab., 4 Textboxen und 51 z.T. farb. Fotos, kt.
ISBN 978-3-515-10404-3
10. Pamela Hartmann
Flexible Arbeitskräfte
Eine Situationsanalyse am Beispiel der Elektroindustrie im Perlflussdelta, China
2013. 201 S. mit 18 Abb., 29 Tab., 15 farbigen Fotos und 6 farbigen Karten, kt.
ISBN 978-3-515-10400-5
11. Juliane Welz
Segregation und Integration in Santiago de Chile zwischen Tradition und Umbruch
2014. 258 S. mit 19 Abb. und 41 Tabellen, kt.
ISBN 978-3-515-10467-8
12. Daniel Schiller
An Institutional Perspective on Production and Upgrading
The Electronics Industry in Hong Kong and the Pearl River Delta
2013. 253 S. mit 7 Abb. und 49 Tabellen, kt.
ISBN 978-3-515-10479-1
13. Benjamin Etzold
The Politics of Street Food
Contested Governance and Vulnerabilities in Dhaka's Field of Street Vending
2013. XVIII, 386 S. mit 103 z.T. farb. Abb. und 7 Karten, kt.
ISBN 978-3-515-10619-1

14. Tibor Aßheuer
Klimawandel und Resilienz in Bangladesch
Die Bewältigung von Überschwemmungen in den Slums von Dhaka
2014. 285 S. mit 21 Abb., 9 farb. Abb. und 37 Tab., kt.
ISBN 978-3-515-10786-0
15. Matthias Garschagen
Risky change?
Vulnerability and adaptation between climate change and transformation dynamics in Can Tho City, Vietnam
2014. 432 S. mit 38 Abb., 25 farb. Abb. und 18 Tab., kt.
ISBN 978-3-515-10876-8
16. Sebastian Homm
Global Players – Local Struggles
Spatial Dynamics of Industrialisation and Social Change in Peri-urban Chennai, India
2014. 219 S. mit 15 Abb., 26 Tab. und 10 Farbkarten, kt.
ISBN 978-3-515-10877-5
17. Sophie Schramm
Stadt im Fluss
Die Abwasserentsorgung Hanois im Lichte sozialer und räumlicher Transformationen
2014. 278 S. mit 38 Abb., 29 farb. Abb. und 5 Tab., kt.
ISBN 978-3-515-10905-5
18. Malte Lech
Institutions, Innovation and Regional Economic Change
2015. 220 S. mit 34 Abb. und 23 Tab., kt.
ISBN 978-3-515-11120-1
19. Markus Keck
Navigating Real Markets
The Economic Resilience of Food Wholesale Traders in Dhaka, Bangladesh
2016. XV, 240 S. mit 93 Abb. und 33 Tab., kt.
ISBN 978-3-515-11379-3
20. Alexander Follmann
Governing Riverscapes
Urban Environmental Change along the River Yamuna in Delhi, India
2016. 396 S. mit 72 Abb., 30 farb. Abb., 13 Tab. und 6 Farbkarten, kt.
ISBN 978-3-515-11430-1
21. Tobias Töpfer
Wem gehört der öffentliche Raum?
Aufwertungs- und Verdrängungsprozesse im Zuge von Innenstadterneuerung in São Paulo
2017. 350 S. mit 17 Abb., 31. farb. Abb. und 8 Tab., kt.
ISBN 978-3-515-11773-9